Food Protein Deterioration

Mechanisms and Functionality

Food Protein Deterioration
Mechanisms and Functionality

John P. Cherry, EDITOR

United States Department of Agriculture

Based on a symposium sponsored
by the ACS Division
of Agricultural and Food Chemistry
at the 182nd Meeting
of the American Chemical Society,
New York, New York,
August 23–28, 1981.

ACS SYMPOSIUM SERIES **206**

AMERICAN CHEMICAL SOCIETY

WASHINGTON, D. C.   1982

6936 7735

SEPIAE

CHEM

Library of Congress Cataloging in Publication Data

Food protein deterioration.
  (ACS symposium series, ISSN 0097–6156; 206)

  Includes bibliographies and index.

  1. Proteins—Deterioration—Congresses.
  I. Cherry, John P., 1941–   . II. American
Chemical Society. Division of Agricultural and Food
Chemistry. III. American Chemical Society. National
Meeting (182nd: 1981: New York, N.Y.) IV. Series.

TP453.P7F66   664   82–20739
ISBN 0–8412–0751–8   ACSMC8   206   1–444
                                        1982

ACS Symposium Series

M. Joan Comstock, *Series Editor*

FOREWORD

The ACS Symposium Series was founded in 1974 to provide a medium for publishing symposia quickly in book form. The format of the Series parallels that of the continuing Advances in Chemistry Series except that in order to save time the papers are not typeset but are reproduced as they are submitted by the authors in camera-ready form. Papers are reviewed under the supervision of the Editors with the assistance of the Series Advisory Board and are selected to maintain the integrity of the symposia; however, verbatim reproductions of previously published papers are not accepted. Both reviews and reports of research are acceptable since symposia may embrace both types of presentation.

CONTENTS

PREFACE

THE THEORIES AND CONCEPTS OF FOOD PROTEIN DETERIORATION are based on the composition, structure, and physicochemical deteriorative changes that occur in proteins during genetic and physiological processes. Endogenous and exogenous enzymatic activities, chemical reactions and modifications, pH changes, salt effects, storage-fungi contamination, and temperature dependency all contribute to making the study of food protein deterioration a vital science. The utilization of proteins as the functionally active constituents in food systems, the processes that control the degree of protein degradation, and the nutritional quality and bioavailability of deteriorated and modified proteins are topics that are included in this volume.

Proteins undergo denaturation by reaction with hydroperoxides which are formed by lipoxygenase and nonenzymatic oxidation of polyunsaturated fatty acids. Discoloration occurs as a result of the disruption of food systems by heating and/or blending during processing. Efforts to develop methods to characterize and control these reactions, and the objectionable odors and off-flavors developed during storage, handling, and processing have been fruitful.

Temperature also affects food protein deterioration. Below ambient temperature proteins can be affected adversely with notable changes in cellular systems, solubility, and enzymatic activity. These alterations can affect storage time, salt concentration, pH dehydration effects, and the effect of ice on membranes. Evidence has shown that during heat exposure, enzyme-resistant linkages form between polypeptides. Many physicochemical changes that are caused by heat-induced protein denaturation can result in texturation products having meat-like food properties.

In addition, brown-colored beneficial and deleterious products can be formed from carbonylamines which are products of the reaction between the aldehydes of sugars and the amino groups of amino acids. (The latter reaction is the first step in the series of reactions that are collectively known as the Maillard reaction.) These chemical reactions can be prevented by eliminating the aldehyde-containing compounds, or temporarily masking the ϵ-amino group of lysine during processing.

The functional properties of foods are influenced by the inherent molecular properties of proteins, the manner in which they interact with other protein and nonprotein constituents, and their behavior in prevailing environmental conditions. Most of the reactions that affect food functionality involve disulfide bonds, and the electrostatic and hydrophobic interactions of structure and surface properties of proteins. Chemical modification of these disulfide bonds results in the loss of both structural order and activity.

This volume confronts the issue that the nutritional quality of proteins can change as a result of structural modification or deterioration. There may be a small increase or decrease in protein digestibility as the proteins unfold or form aggregates with other protein or nonprotein constituents. Extremes in heat or chemical modification can result in a major loss of bioavailability when essential amino acids form nonreversible molecular complexes. Understanding the mechanisms by which proteins deteriorate during handling, storage, and processing will allow modification and/or control of these processes to protect the inherent functional and nutritional qualities of proteins in raw products. This will then allow for maximum production of quality foods.

JOHN P. CHERRY
Eastern Regional Research Center
United States Department of Agriculture
Philadelphia, PA 19118

June 1982

Genetic Mechanisms and Protein Properties with Special Reference to Plant Proteins

ASIM ESEN

Virginia Polytechnic Institute and State University, Department of Biology, Blacksburg, VA 24061

Continuous intake of high quality dietary protein in adequate amounts is prerequisite for normal growth and development in the young and for maintenance of nitrogen equilibrium in adult humans. The reason for this is that dietary protein provides certain specific (essential) amino acids that cannot be synthesized by man and other monogastric animals as well as indispensable nitrogen required for various physiological functions. Animal cells also cannot store amino acids as they do carbohydrates and lipids, which necessitates continuous protein intake. Malnutrition or undernutrition is a serious problem affecting nearly one half of the world's population. There is no consensus among nutritionists and sociologists on the causes of malnutrition. Some experts (1) attribute it to the lack of good quality protein for human consumption while others (2, 3, 4, 5, 6) to either protein and calorie deficiencies or calorie deficiency alone. Bressani and Elias (7) have recently emphasized the difficulty in identifying a simple nutrient as the causal factor in malnutrition. They concluded that the malnutrition problem is the end result of the interaction of many factors of which low food availability and intake are perhaps the most important.

On a worldwide basis, about 90% of the calories and 70% of the protein available for human consumption come directly from plants with the remainder from foods of animal origin (Table I). There is a high correlation between income level and the proportion of animal protein in the diet with the prosperous nations and also prosperous social and economic classes in a country receiving their protein primarily from foods of animal origin (8), namely milk, meat, egg, fish and their products. In developed countries large amounts of plant protein are fed to animals for meat production. This is a rather wasteful process from the standpoint of energetics since it takes about

0097-6156/82/0206-0001$08.75/0

eight kg of plant protein to produce one kg of animal protein.
Presently the world's population is about 4.5 billion and it is
estimated to reach 6 to 7 billion by the turn of the century.
Therefore, it is probable that the direct consumption of plant
protein by humans is likely to increase (9) and animal protein
is to become somewhat a luxury that only upper economic classes
and populations of certain developed countries can afford on a
regular basis.

Table I. Protein sources and intake (1975-77)

Region or Country	Daily Intake (gm)	Source % Plant	% Animal
World	69	65	35
Asia	58	79	21
Africa	59	80	20
S. America	66	56	44
N. & C. America	93	39	61
Europe	96	45	55
Oceania	96	35	65
USA	106	32	68
India	48	89	11
Zaire	36	80	20

Source: (96).

Regardless of its origin, plant or animal, a protein
source is expected to meet certain requirements in order to be
suitable for food or feed use.   First, it should have a high
protein content to provide the quantity of protein needed at
normal intake levels. Secondly, its proteins should be of high
nutritional quality to provide all essential amino acids at or
above recommended levels. Thirdly, the protein source must
have acceptable organoleptic properties (e.g., taste, color and
texture) in its edible state so that it can be ingested in
sufficient quantity alone or in mixtures with other dietary
components.   Fourthly,   its   protein   should   have   high
digestibility and biological value.   In other words,   the
protein must be resistant to or free of deteriorative changes
and biologically active protein and nonprotein substances which
unfavorably affect the digestibility and biological value.
Lastly, but not least, the protein source and its proteins
should have desirable functional properties to lend themselves

to processing, handling and storage with little or no deteriorative changes. If a protein source cannot satisfy one or more of these criteria, its deficiencies may be remedied by genetic or nongenetic (e.g. supplementation, improved processing and storage techniques) means. Animal proteins meet all or most of the requirements as ingredients with high nutritional quality and organoleptic properties. The only genetic improvement that can be made in these proteins is that of quantity by selecting and developing breeds with high production and feed utilization capability. Futhermore, animal proteins are exclusively structural and enzymatic in nature with well-defined metabolic activities and functions. Thus any attempt to change genetically their nutritional properties and functionality would be deleterious or lethal. However such constraints may not apply to the storage proteins of seeds where there is both need and potential for improvement of protein properties by genetic means. Therefore this review's focus will be on plant proteins with special reference to their deficiencies and possible genetic approaches to improve them.

There are about 80 thousand plant species with an organ or part suitable for use as food and many of them may be a potential protein source (10). However, only 12 species are the main stay of the world's food supply accounting for about 90% of the harvested products. Of these, cereal grains provide some 70% of the world's supply of plant protein and over 90% of the calories of plant origin. For example, rice alone is the main source of protein in Asia and provides 80% of the calories in the Asian diet. It is projected that the world demand for cereals in the year 2000 will be 2,422 billion metric tons assuming a net population increase rate of 0.7% in developed countries and 2.2% in developing countries (11). This will represent a 100% increase over 1970 production. The next major source of plant protein for humans is legume seeds, which account only for 18-20% of the total plant protein produced. Efforts directed towards improvement of quantity and properties of plant proteins by genetic means would have to be centered around the conventional crop plants such as cereals and legumes in order to achieve tangible results in the near future. There are basically two reasons for this observation. First, almost our entire repertoire of basic scientific information relative to plant genetics, breeding and biochemistry is derived from studies on traditional crop plants (e.g., corn, wheat, soybean, and others). This basic information constitutes the most valuable asset for further progress through research. Secondly, the traditional crop plants have been a part of social, economic and cultural life of mankind since the days of prehistoric agriculture; man knows how to produce, process, market, and consume them.

Theoretically speaking, genetic improvement of proteins would be a straightforward task. Protein is an immediate gene product whose amino acid sequence (primary structure) is colinear to the nucleotide sequence of the gene encoding it. Thus both the structure and function of a protein is under direct genetic control. From the standpoint of a geneticist, each of the different kinds of proteins is a heritable characteristic, one which is closest to the gene and one that can be described without any ambiguity. This is in contrast to many biological characteristics that geneticists have to work with that are far removed from the gene and cannot be often described and measured in precise quantitative or qualitative terms.

One may identify three major problems relative to plant proteins that are potentially amenable to a genetic solution. These are: low protein content in cereals and some legumes, poor protein quality in cereals and legumes, and the presence of protein and nonprotein components that are either toxic or affect unfavorably nutritional value and functionality of plant proteins and other essential nutrients. Within each problem area there are sub-problems some of which are amenable to solution, and some may defy a solution in view of biological and technical constraints.

Experimental Procedures

This chapter will attempt to present a catalogue of genetic strategies and approaches that may be used for the solution of the above mentioned problems relative to plant protein, and evaluate the potential and limitations of each approach based on representative experimental data from previous investigations. Finally the potential uses of recombinant DNA technology and genetic engineeering for plant protein improvement will be briefly mentioned. For additional information on topics discussed in this chapter and specific methodologies the reader is referred to numerous previous reviews (9, 10, 12, 13, 14, 15, 16, 17, 1:, 19, 20) and research and review papers that appeared in the publications of many proceedings and symposia held on this subject.

The genetic solution to these problems will rely upon the presence of sufficient genetic variation in the existing germ plasm sources. Equally important is the availability of simple, rapid, and inexpensive screening procedures to detect such variation. Extensive surveys of major cereals and legumes conducted within the past two decades for spontaneous high quantity and/or quality mutants have revealed a great deal of variation in the germ plasm (Table II). Moreover, additional variation has been generated successfully by employing the

various methods of induced mutagenesis (21, 22, 23, 24). In
order to make rapid progress and justify the expenditure of
financial and technical resources, high heritability of the
desirable characteristic chosen for improvement is a
prerequisite. Protein content seems to be a highly heritable
character and shows relatively a simple mode of inheritance.
For example, Johnson et al. (25) reported heritability
estimates as high as 83% for protein content in wheat.

Table II. Variation of protein content and limiting
amino acid in some cereals and legumes

Crop	Protein (%)	Limiting Amino Acid
		Lysine (g/16 g N)
Wheat	7-22	2.2-4.2
Rice	5-17	3.1-4.5
Corn	7-17	2.2-4.0
Barley	10-18	3.2-4.3
Sorghum	7-26	1.8-3.8
Oats	8-22	3.7-4.8
		Methionine (g/16 g N)
Soybean	35-52	1.0-1.6
Bean	19-31	0.5-2.1
Pea	17-31	-
Lentil	20-31	-
Cowpea	21-34	0.8-2.3
Chickpea	12-28	0.5-1.7

Source: (97, 98, 99).

Improvement of Protein Quantity

The protein content of cereal grains and other starch-
storing crops is relatively low. For example on the average
only 8% of the dry matter in rice, 12% in wheat, and 10% in
corn is protein. These levels may not necessarily be too low
in view of the fact that cereals are the primary source of
dietary energy. In fact if cereal proteins contained all the
essential amino acids in required proportions they would
actually meet the protein requirement even for infants and
growing children, the groups most vulnerable to protein
malnutrition. Of course, this would be true only if the
quantity of cereal consumed was adequate to meet the entire

calorie requirement (26). Some nutritionists (e.g., 6) argue
that most of the common plant staples, which of course includes
cereals,   "contain   sufficient utilizable   protein to   meet the
safe level requirements without any additions or supplements".
Munck (19)   states that there is no point for increasing the
protein quantity in cereals if such an increase is to be
achieved at the expense of yield and carbohydrates. Proponents
of this view believe that cereals should remain as energy foods
and legumes must be the primary source of plant protein.

Negative correlation between protein content and yield.
Higher protein content has almost always been found to be
negatively correlated with grain yield (27, 28, 29, 30, 31).
If this is indeed a universal phenomenon, the advocates for
increasing protein content in cereals cannot convincingly argue
their case before they resolve this problem.   The reason for
the inverse relationship between protein content and yield has
been explained by the energy requirements of various seed
biomass components (32, 33).   The amount of the primary
photosynthate, glucose, expended to produce a unit mass of
protein is twice as much as that needed to produce the same
unit mass of carbohydrate. To be more specific, 1 g of glucose
yields 0.83 g of carbohydrate and only 0.4 g of protein.   If
these theoretical calculations truly reflect what is happening
in the plant it would indeed be pointless to increase protein
content unless there is a premium paid for high protein to
offset yield loss.   However, the situation does not seem as
hopeless as the skeptics say it is.   The negative correlation
between yield and protein does not seem to be the rule.
Johnson et al. (25, 34) showed that the simultaneous progress
in grain yield and protein content was possible in wheat.   They
found that 4 out of 5 high protein selections derived from the
high protein cultivar Atlass 66 exceeded a popular Nebraska
cultivar, Scout 66, both in yield and protein content.
Likewise, new oats, rice, and corn cultivars and experimental
lines have been reported to consistently and significantly
exceed standard cultivars in protein content without loss in
yield (38, 39, 40).   These increases in protein content   do
not seem to be negatively correlated with yield (38, 39, 40).
However, some reports indicate considerably high negative
correlations especially in beans (41, 42). In soybean, protein
and oil contents tend to be negatively correlated but this
relationship may be of little consequence now since the oil is
no longer the more valuable product from soybean because the
value of protein exceeds that of oil.
Poey et al.   (43) and Robbelen (10)   postulate that the
inverse relationship between yield and protein is an artifact
of past selection practices which selected for only one

character at a time, either yield or protein. Implicit in this hypothesis is that it may be possible to develop high protein-high yield cultivars by selecting for both traits simultaneously.

Strategies for the improvement of protein quantity. 1) Increase yield per unit area. Increasing the grain yield per unit area would be the most straightforward way to increase protein production. This would require the development and use of high yield cultivars, namely, those with high photosynthetic efficiency, pest resistance, and harvest index. These high yield cultivars coupled with better cultural practices, e.g., better seed bed preparation, fertilization, irrigation, weed and pest control, and growth regulator application would further increase the yield per unit area. One potential drawback of this approach is that it may depress the protein percentage below acceptable levels especially in cereals, assuming that the inverse relationships between yield and protein is to operate. This might lead to a protein-deficient diet, although adequate in calories, in regions where cereals are the common staple. However, the development of high protein-high yield cultivars may alleviate this possibility. Increased yield per unit area is expected to be more successful in terms of protein production with legumes than with cereals. This is because protein content is less affected by yield and is inherently high in legumes.

2) Apply additional nitrogen-fertilizer at or after anthesis. Nitrogen fertilization, especially late applications at the time of anthesis, has been shown to be rather effective in increasing the protein content in cereals. Hucklesby et al. (44) observed 30-118% increase in protein production per hectare by late spring application of nitrogen to wheat. The increase was due to both yield and protein content. Somewhat similar but less dramatic results were obtained with corn by Hageman et al. (45). Zink (46) showed that late nitrogen application increased the kernel number, weight, and percent protein in corn by 15%, 12%, and 30%, respectively. These results suggested that soil nitrogen is depleted considerably by the vegetative growth prior to anthesis and late N application both provided additional reduced nitrogen from roots to the seed and prolonged the duration of nitrogen accumulation into seeds.

The development of hybrids and cultivars that will respond to nitrogen fertilization can further increase the effectiveness of nitrogen used in increasing protein content and production.

The application of additional nitrogen as a method to increase protein production may not be feasible at all in the

future in view of high cost of nitrogen fertilizers and their
application. In fact, the author personally believes that
future crop plant cultivars will be ones that perform best
under adverse environmental conditions with minimum care unless
new and cheap sources of energy are available for agricultural
use. Therefore plant geneticists and breeders should reassess
their goals of developing cultivars whose performance is
contingent upon cultural practices requiring high inputs of
energy.

3)   Abolish rate-limiting steps in protein biosynthesis
pathway. Conventional wisdom suggests that protein synthesis,
like any other biochemical process, is under genetic control.
This control may be exerted by regulating the quantity and/or
activity of any one or more of the components of the protein
synthesizing machinery, e.g. amino acids, ribosomes, messenger
ribonucleic acid (RNA), initiation, elongation and termination
factors, amino acyl transfer RNA synthetases, peptidyl
transferase, adenosine triphosphate, guanosine triphosphate,
and other unidentified factors. Figure 1 shows the pathway of
nitrogen assimilation from nitrate to amino acids and proteins
in plants (47). Abolishing the rate-limiting steps of protein
synthesis pathway by genetically altering the regulatory
components would be expected to increase the protein content.
The availability of substrates, amino acids, is probably the
most important rate-limiting factor in protein synthesis since
it affects not only the rate of synthesis of storage proteins
but also that of enzymes and other protein factors that
function directly in nitrogen assimilation. There is
considerable evidence for the regulation of amino acid
biosynthesis by allosteric feedback inhibition by the end-
product amino acids in plants (47, 48). Another logical step
for the regulation of the synthesis of amino acids is thought
to be the nitrate reductase reaction (45, 47). This stems from
the fact that it is the first enzyme in the nitrogen
assimilation pathway, inducible by the substrate and relatively
unstable under adverse environmental conditions. The
experimental data in support of this suggestion came from
studies of Hageman and his associates (46, 47, 50) who found
consistently positive correlations between nitrate reductase
activity, grain yield and protein content. In view of these
results, they suggested the use of nitrate reductase activity
as a screening tool for selecting high protein-high yield
genotypes.

        Most, or all, of the nitrogen reaching the seeds is in the
form of amino acids coming from leaves and roots (51). Amino
acids originating in leaves result primarily from proteolysis
of leaf proteins during senescence and also some from new amino
acid synthesis coupled to photosynthesis. But those coming

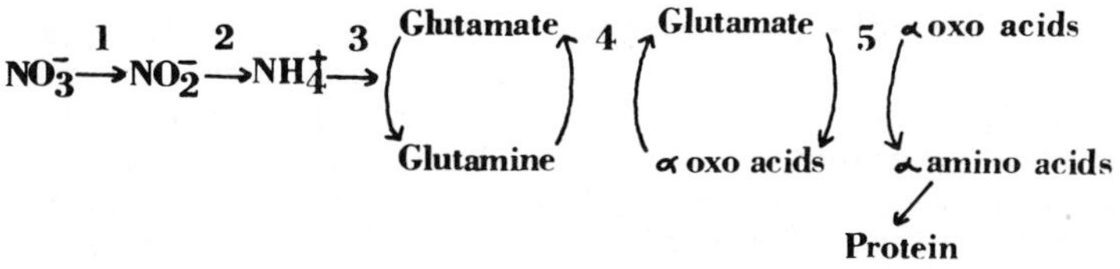

$$NO_3^- \xrightarrow{1} NO_2^- \xrightarrow{2} NH_4^+ \xrightarrow{3} \begin{array}{c} \text{Glutamate} \\ \\ \text{Glutamine} \end{array} \quad 4 \quad \begin{array}{c} \text{Glutamate} \\ \\ \alpha \text{ oxo acids} \end{array} \quad 5 \quad \begin{array}{c} \alpha \text{ oxo acids} \\ \\ \alpha \text{ amino acids} \end{array}$$

Protein

Figure 1.   The pathway of nitrate assimilation in plants. Key: 1, nitrate reductase; 2, nitrite reductase; 3, glutamine synthetase; 4, glutamic: 2-oxoglutarate amino transferase; and 5, transaminase (47).

from roots are from nitrate uptake and reduction.   Apparently,
there is no feedback control on amino acids from proteolysis in
leaves (48).   Oaks et al. (52) could detect very little nitrate
reductase activity in the corn endosperm during development
while high levels of other enzymes of nitrogen assimilation
were found.   The latter group was thought to be concerned with
providing amide nitrogen   in the form of glutamine and
asparagine for   conversions to other amino acids.   In wheat,
Kolderup (53) reported varying degrees of recovery in proteins
of labelled amino acids directly supplied to the developing
spikes.   The lowest recoveries were for aspartic acid, glutamic
acid, alanine, proline, and cysteine indicating considerable
conversions to other amino acids as well as respiratory loss.
The   highest   recoveries   were   for histidine,   valine,
phenylalanine, leucine and tryptophan indicating little or no
conversion.
        Obviously, the availability of a specific amino acid is
not the rate limiting factor for protein synthesis in the seed
but the amount of free amino acids and amide nitrogen that can
be transported from leaves and roots is.   These results suggest
that genetic manipulations to affect protein quantity should be
directed to selection of genotypes that are efficient in the
absorption,   reduction,   accumulation,   and translocation of
nitrogen.   Indeed this seems to be the case with some high
protein cereal variants.   Perez et al.   (54) reported that the
increase in protein content of the high protein rice was due to
the more efficient translocation of nitrogen from the leaf
blades to the developing grain.   Furthermore, the leaf blades
of high protein lines contained more protease activity than
those of the normal rice which in part, if not wholly, account
for increased amino acid translocation from leaves to seeds.
High protein wheat lines were also shown to be more capable of
nitrogen uptake and translocation than the standard lines (34).
Perhaps the best example of changing the protein content
dramatically is the well-known selection experiment in corn
which has been going on at the Illinois Agricultural Experiment
Station since 1896.   This legendary experiment selected corn
for high and low protein starting with a base population of
10.9% protein and raised the protein content to 26.6% in the
High Protein line and reduced it to 4.4% in the Low Protein
line after 70 generations of selection (31).   A thorough
investigation of alterations that occurred in the protein
synthesis machinery of these two lines during the course of
selection would be very informative.
        Although the abolition of rate-limiting steps can increase
protein content in the seed, unregulated protein synthesis may
be deleterious to other plant parts.   This is because it is
expected that genes regulating protein synthesis will be shared
by other plant parts.

4) Change seed morphology or relative proportions of seed
components. Selection for larger germ size in cereals is
likely to increase protein quantity and quality not only
because the germ contains twice as much protein as endosperm
but also has a better or complementary essential amino acid
pattern (19, 43). Furthermore there does not seem to be a
relationship between kernel weight and germ/endosperm ratio,
and the germ is rich in oil (30%); no reduction in yield and
calorie value is likely to occur (43). However benefits from
increased germ/endosperm ratio would be lost if the germ and
endosperm were separated and only the endosperm was utilized as
food. It was also suggested that the thickness of the protein
and lysine-rich aleurone layer in cereal seeds be increased
(19).

Improvement of Protein Quality

The major limitation of plant proteins of seed origin is
their low nutritional quality due to inadequate levels of
certain essential amino acids which cannot be synthesized by
monogastric animals including man. It is well known that the
major limiting essential amino acid in cereals is lysine and in
legumes it is methionine. The low nutritional quality of
cereal and legume protein resides in one or more predominant
storage protein fractions which are poor or devoid of one or
more essential amino acids. In the case of cereals, alcohol-
soluble proteins or prolamins are the predominant protein
fraction which make up 50-60% of the protein in the kernel.
Since prolamins are essentially devoid of lysine, a cereal
based diet will be lysine deficient and of poor quality. Table
III shows the prolamin content and biological value of various
cereals. These data clearly show a negative correlation
between prolamin content and biological value; low prolamin
cereals, rice and oats, have proteins with much higher
biological value than high prolamin cereals such as sorghum,
maize, and wheat.
In the case of legumes, the two major storage protein
fractions, 7S (vicilin) and 11S (legumin) globulins, contribute
60-90% of the protein in the cotyledon, depending on the
species. Of these the 7S globulin is extremely and 11S
globulin moderately low in sulfur amino acids, methionine plus
cysteine. Thus these two globulin fractions are responsible
for the low nutritional quality of legume proteins.
The improvement of protein quality of cereals and legumes
has long ranked as a high priority area in the research agenda
of nutritionists and plant breeders. This was primarily due to
emphasis given to this area by various international and
national agencies (Food and Agriculture Organization [FAO],

World Health Organization [WHO], International Atomic Energy
Agency [IAEA]), United States through the Agency for

Table III. Prolamin content of various cereals and
            biological value of their proteins

Cereal	Prolamin % of Total Protein	% Biological Value
Rice	5	67-89
Oats	12-15	75
Barley	40-52	74
Wheat	45-69	58-67
Corn	50-55	53-60
Sorghum	52-60	51

Source: (30, 75, 100).

International Development [US-AID], Federal Republic of Germany
through the Gesellschaft fur Strahlen-und Umweltforschung [GSF]
and various International crop improvement centers such as
International Maize and Wheat Improvement Center [CIMMYT] and
International Rice Research Institute [IRRI]) for solving
nutrition-related problems in developing countries.
     The intensive and well-coordinated research effort of the
last two decades has led to major discoveries in this field and
produced much useful data, including the development and use of
a whole array of screening methods to identify high quality
protein variants. Information and experience gained so far
have also resulted in a re-evaluation and modification of some
earlier objectives and premises. One major finding of the past
two decades has been that increased yield and protein content
almost invariably reduce protein quality. This is due to the
fact that genetic and environmental factors that increase the
yield and protein content also favor increased synthesis and
accumulation of poor quality, predominant protein fractions.
In fact this relationship has been known in cereals for some
time. Bishop (56) showed that as grain protein percentage
increased in barley, so did the relative proportion of the
lysine-poor prolamin fraction (Figure 2).
     The seed is essentially a dispersal organ which contains
the young sporophyte, embryo, and reserve nutrients needed for
its germination and early seedling growth. Storage proteins of
seeds then serve as nitrogen reserves. They are "definable
sets of organ-specific and tissue-specific gene products

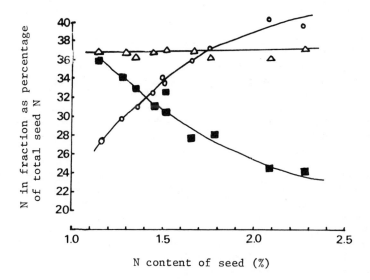

Figure 2.   The effect of protein content on the relative proportions of different protein fractions in barley (56, 101). Key: ○, hordein; ■, salt-soluble; and △, glutelin.

sequestered and stored in membrane-bound protein bodies for utilization at a subsequent stage in development" (57). These are a complex mixture of different proteins which differ in structure, size, charge, amino acid composition, solubility and other physiochemical properties. Osborne (58), in his pioneering research separated seed proteins into 4 fractions based on solubility in various solvents: albumins (water-soluble), globulins (salt-soluble), prolamins (alcohol-soluble) and glutelins (alkali-soluble) (Table IV). These data show that relative proportions of different Osborne fractions vary among cereal species. Also, it should be mentioned that different fractions contain varying amounts of the limiting amino acid lysine, the prolamin fraction being the poorest as mentioned before. The comparable solubility fractions of legume proteins are given in Table V.

The improvement of nutritional quality of seed proteins primarily hinges upon genetic manipulations changing the relative proportions of different solubility fractions, or of individual polypeptides in these fractions. The objective is to increase the relative proportion of proteins with the most desirable essential amino acid pattern or to decrease or eliminate those with the least desirable pattern. This will have to be done in a way not to unfavorably affect grain yield and protein functionality.

Strategies for the improvement of protein quality. 1) Change the relative proportions of different solubility fractions. Reducing the relative proportion of prolamins and increasing that of albumins, globulins, or glutelins is almost certain to improve the quality of cereal proteins since these latter fractions are relatively lysine-rich. The feasibility of this approach has been confirmed with the discovery of the high-lysine mutant opaque-2 in corn (59), 'Hiproly' in barley (60), and IS-11167 and 1S-11758 in sorghum (61). Additional high-lysine mutants of barley (21, 62) and sorghum (24) were obtained through the use of chemical mutagens. Neither the extensive screening of over 17,000 wheat accessions from the World Collection nor the work with mutagens produced a high lysine wheat mutant comparable to those in corn, barley and sorghum (63). Interestingly, the reduction of prolamin content has already occurred and become established in nature with two cereal genera, oats (Avena) and rice (Oryza), accounting for the greater biological value of their protein. Both in oats and rice, reduction in prolamin is accompanied by an increase in the glutelin fraction. As a result of this, oats and rice have the highest glutelin content among cereals. There is a striking similarity among high lysine mutants of maize, barley, and sorghum. They are single gene mutations where the mutant allele is recessive to its normal counterpart.

Table IV.   Proportions of protein solubility fractions
            of cereals

| Cereal | Solubility Fractions (g/100 g Protein) | | | |
	Albumins	Globulins	Prolamins	Glutelins and Residue
Barley	5	15	40	40
Wheat	5	10	45	40
Corn	5	5	50	40
Rice	5	10	5	80
Oats	1	13	18	68
Sorghum	8	8	52	32

Source: (97).

Table V.   Proportions of protein solubility fractions
           of legumes

| Legume | Solubility Fractions (g/100 g Protein) | | |
	Albumins	Globulins	Glutelins
Mung Bean	4	67	29
Broad Bean	20	60	15
Pea	21	66	12
Peanut	15	70	10
Soybean	10	90	0

Source: (97).

Also, they either have floury (opaque) endosperms of normal
size (corn and sorghum) or shrunken kernels (barley).   All but
'Hiproly' barley have shifts in relative proportions of the
endosperm protein solubility fractions in comparison to their
normal counterparts.   This reduction, in relative and absolute
terms, of the lysine deficient prolamin fraction, is
accompanied by an increase in albumin plus globulin and
glutelin fractions which are relatively lysine-rich.   Nelson
(64) postulated that the high lysine mutants in cereals are
regulatory mutants in which the lesion causes reduced prolamin

synthesis and indirectly enhances the synthesis of certain albumin and globulin polypeptides. The 'Hiproly' barley attained its high lysine content without any marked decrease in prolamin but by increasing the quantity of four lysine-rich components in the albumin fraction (15). Another similarity shared by these mutants is their undesirable agronomic characteristics most important of which is low yield. For example, the opaque-2 corn lines had lower yield, kernel density, higher moisture content (at maturity), grain cracking and greater susceptibility to certain pests than normal corn.

Consequently, none of the high-lysine mutants have gained commercial acceptability to any significant degree in spite of their well documented nutritional superiority. In fact, the inferior agronomic performance of high-lysine cereals has had an unfavorable effect on support for high quality protein research as disillusioned sponsors have begun to discontinue or reduce funding. A partial, if not complete, explanation for the inverse relationship between low yield and high lysine trait in cereals has recently been offered. Mehta et al. (65) proposed that the major constraints on grain yield are premature termination of grain development and the slow rate of protein synthesis which encompasses the enzymes catalyzing the starch biosynthesis pathway. Tsai et al. (66, 67) found positive correlation between zein (prolamin) content, kernel weight, and grain yield in corn. They proposed that an important function of zein and glutelin is to serve as a nitrogen sink during seed development by increasing the translocation of sucrose and other solutes from leaves to seeds. Implicit in this suggestion is that prolamin and glutelin are indispensable for high yield. If this is universally the case, then low prolamin-high lysine cereal types may never achieve commercial status.

There are other ways of changing the relative proportions of endosperm proteins to improve protein quality. One way this can be accomplished is by breeding genotypes with low prolamin and high glutelin. If reduction in prolamin were to be compensated only by a relative and net increase in glutelin, one nitrogen sink would be replaced by another one and thus no reduction in yield might occur. The fact that rice and oats have already adopted this scheme makes it a promising one to pursue. In practice this could be accomplished in two steps: a regulatory gene mutation to reduce the prolamin content drastically (step 1) followed by a different regulatory mutation to increase the glutelin content (step 2). The first step has already been accomplished in the high-lysine mutants mentioned above. In fact, these mutants should be screened for vitreous (corn and sorghum) or plump (barley) kernels some of which might be high glutelin mutants and some unwanted revertants.

2) Increase the relative proportions of individual polypeptides with the most desirable amino acid profile. The basis for this approach is the assumption that there exist individual polypeptides, within each solubility fraction except the prolamine fraction of cereals, which either have a rather desirable essential amino acid profile or one that is rich in the limiting amino acid(s). If the quantity of such individual polypeptides were to be increased without changing the relative proportions of the solubility fractions, a significant improvement in protein quality would occur. Alternatively, decreasing the quantity of individual polypeptides with the least desirable amino acid profile or their elimination would yield the same result. This approach would be applicable not only to cereals but also to legumes and other grains. For example, selecting for relatively sulfur amino acid-rich polypeptides of the 11S globulin, or against sulfur amino acid-poor polypeptides of the 7S globulin would improve protein quality in legumes. The major drawback of this approach is that one or a few mutational events are not expected to produce dramatic changes in protein quality unless the polypeptides in question are coded by homologous genes regulated coordinately by one regulatory gene. Apparently, "Hiproly" barley is such a mutant whose high lysine content is due to the increased amounts of four lysine-rich polypeptides in the albumin fraction (15).

Progress under this approach would be slow and require the use of an assay procedure for the limiting amino acid(s) being selected for. This is because one is selecting for or against many structural gene products simultaneously, each with small desirable or undesirable additive effects on the overall amino acid composition. Rapid progress would be made if there were a simple and specific assay (e.g. immunochemical assay) for the desirable or undesirable polypeptides in question. However, this would require that each of the different major polypeptides in different solubility fractions be purified and characterized.

This approach, in spite of its limitations, may prove to be useful in improving protein quality in polyploid species such as wheat in which mutations causing drastic changes in amino acid composition are difficult to detect and induce as well as in legumes whose different solubility fractions do not differ considerably in their content of the limiting amino acids.

3) Eliminate major polypeptides with the least desirable amino acid profile or change their amino acid composition through induced mutations. This strategy is to eliminate a predominant polypeptide that is deficient in the limiting amino acid(s) by deleting the gene encoding it or inducing a nonsense

mutation towards the 3' terminus of the gene (the terminus corresponding to the N-terminus of the polypeptide). To the same end, point mutations can be induced such that a nonessential amino acid will be replaced with the limiting amino acid(s) or they result in a reading frame shift at a point beyond which some of the new sets of triplet codons may now specify the limiting amino acid(s). The success of this type of approach depends on how many structural genes encode a given predominant polypeptide class. The fewer the number of different structural genes the easier it would be to affect desirable changes in the amino acid composition by point mutations.

There is apparently both size and charge heterogeneity among the major storage protein polypeptides such as prolamin of cereals and globulins of legumes (57). Although the genetic basis for this heterogeneity has not been rigorously defined, it appears to come from homologous structural loci which arose by duplication from an ancestral gene and then evolved independently. Another possibility is that structural genes for each major polypeptide within a class may exist in many tandem copies like ribosomal RNA genes. If many reiterated homologous structural genes code for the predominant storage polypeptides with less favorable amino acid profile, there is very little hope of correcting their deficiencies by induced or spontaneous point mutations. It has been widely believed that storage proteins tolerate and accumulate random mutations rather readily since their sole function is to serve as the source of free amino acids and amide nitrogen during seed germination. Righetti et al. (68) calculated that, if zein genes accepted missense mutations randomly, a zein molecule would have sixteen lysine residues (as opposed to none as it now seems to have) and eight arginine residues (as opposed to 2 to 3). This implies that there are certain strong selective constraints on the amino acid composition and sequence of storage proteins and that these constraints may involve mostly their transport, assembly and packaging rather than their ultimate function.

4) Induce, select and use variants that are overproducers of key amino acids. The regulation of amino acid biosynthesis in prokaryotic and unicellular eukaryotic organisms by feedback inhibition is well documented. The mutants that have lost the ability to regulate the biosynthesis of a particular amino acid due to either constitutive enzyme synthesis or the synthesis of an altered enzyme insensitive to feedback inhibition are known to be the overproducers of that amino acid. It is now established that higher plant cells regulate amino acid biosynthesis in a manner similar to that in bacteria (49). Therefore, it should be possible to produce mutant plant cells

that are overproducers of lysine in cereals and sulfur amino
acids in legumes, select them by using analogues in the culture
media and regenerate plants from them. This possibility, first
suggested by Nelson (12) and further elaborated later (18, 64),
may   be   a   real   one   in   view   of   the   isolation   of
5-methyltryptophan resistant cell cultures of tobacco and
carrot with an altered anthranilate synthetase that is
insensitive to feedback inhibition by tryptophan (69, 70).
Similarly, valine-resistant tobacco has been obtained from
selected cell cultures (71). There are several observations
and   reasons   that suggest that this approach may not be
successful. First, Kolderup (53) showed that all ^{14}C-labeled
amino acids but histidine supplied to the wheat spikes were
converted to other amino acids; the site of interconversions
was very likely to be the endosperm. Furthermore, the
exogenously supplied amino acids did not seem to have an effect
on the relative rate of synthesis of the different protein
fractions. Earlier Sodek and Wilson (72) reported similar
results by exogenously supplying labeled lysine and leucine to
developing   corn   ears.   These   results   suggest   that
overproduction of an amino acid will neither result in its
storage in the seed nor in the synthesis of proteins that are
rich in it. Second, production of an amino acid in excess of
the normal amount may be toxic or detrimental to other plant
tissues or organs even though it is desirable in the seed.
Third, the overproducer phenotype observed in suspension
cultures or calli may not be expressed in specialized tissues
or organs such as the seed. Fourth, except in the case of a
mineral nutrient deficiency such as sulfur, the availability of
a particular amino acid may not be rate-limiting because of
possible interconversions. For example, detached wheat spikes
could develop normally when supplied with glutamine or
asparagine as the only nitrogen source. And fifth, presently
it is not possible to regenerate plants from cell cultures of
any of the agronomically important crop plants, although whole
plants by culturing the excised anthers have been obtained
(73).

Improvement of Digestibility, Availability and Functionality

The poor nutritional quality of plant proteins is not
always explained by their inadequate quantity and amino acid
composition. There are a variety of protein and nonprotein
factors that unfavorably affect the digestibility of proteins
and availability of essential amino acids in seeds of cereals,
legumes and oil crops. Some of these antinutritional factors
affect also the digestion and utilization of nonprotein
nutrients. Others may be toxic or cause pathological

conditions. The effect of these factors on protein digestion
and utilization is either through direct deteriorative changes
and antiproteolytic activity or indirect ones, such as the
undesirable deteriorative changes caused by heating in an
effort to inactivate or detoxify the antinutritional and toxic
substances.

Because of the high priority given to the improvement of
protein quantity, quality and grain yield, antinutritional and
toxic factors have often been ignored, though not
intentionally. Therefore, if plant breeders find that their
new varieties with improved protein quantity and quality do not
perform up to expectations in feeding trials, they should not
be surprised.

The most important antinutritional and toxic factors
encountered in seeds are protease inhibitors, amylase
inhibitors, phytohemagglutinins, phenolic compounds (tannins,
gossypol, resorcinol, chlorogenic acid, etc.), glycosides
(goitrogens, cyanogens), metal ion-binding substances (phytic
acid), antivitamins, lathyrogens, favism factors, allergens,
flatulance factors and other unidentified protein and
nonprotein substances. Their structure, mode of action,
distribution, and other effects have been reviewed (74, 75,
76). The importance of these factors is expected to increase
in the years ahead along with increased use of proteins of
plant origin. Needless to say there are many other
antinutritional factors yet to be identified because in many
cases undesirable effects could not be fully accounted for by
the known factors.

There are only a few limited surveys of different plant
taxa conducted to investigate the distribution of
antinutritional and toxic substances (77, 78, 79, 80, 81).
These surveys indicated considerable variation between and
within species with respect to occurrence and the level of
activity. However, very little information is available on
their mode of inheritance. Jaffe and Brucher (82) found 4
types of phytohemagglutinins varying in toxicity in beans.
Crosses between varieties with toxic and nontoxic
phytohaemogglutinins revealed a simple (monogenic) mode of
inheritance where toxicity was dominant to nontoxicity. Pull
et al. (83) reported that several soybean lines lacked the seed
lectin. Genetic studies (84) showed the occurrence of this
lectin to be simply inherited where the presence was dominant
to the absence. Recently, Vodkin (85) has shown that the
lectinless cultivars lacked functional mRNA suggesting that the
genetic lesion affects the transcription. Similarly, it was
shown that the Kunitz trypsin inhibitor occurred in three
different forms, distinguishable by electrophoresis, in
soybeans and some accessions completely lacked this protein

(79). Again, genetic studies indicated monogenic inheritance where the presence of any of three electrophoretic variants was dominant to their absence (86). In some instances, the problem caused by antinutritional and toxic factors has been solved by genetic means. A case in the point is that of the gossypol problem in cottonseed. This phenolic substance is toxic to nonruminant animals and reduces the nutritional value and functionality of cottonseed protein (87). Gossypol is stored in pigment glands in the seed. The discovery of spontaneous and induced glandless mutants and breeding of the glandless varieties alleviated the problem (88). Another example of successful genetic intervention was in the rapeseed which contain goitrogenic factors that inhibit the binding of iodine in the thyroid gland. There is considerable variation in the rapeseed germplasm with respect to goitrogenicity and plant breeders have been able to develop varieties with low goitrogen content (89).

Some antinutritional and toxic substances can be reduced or inactivated by heat processing which explains the reason for their being a lesser potential problem. Therefore, most antinutritional and toxic effects occur in animals whose feed is not ordinarily cooked. However, heating also causes many undesirable deteriorative changes in proteins, e.g. loss of solubility, digestibility and covalent modification in essential amino acids, lysine in particular. Moreover, some factors cannot be inactivated and destroyed by heat processing. Thus, it would be best to alleviate the problem by genetic means. Strategies to be adapted to this end will depend on various factors such as the availability of sufficient genetic variation in the germ plasm and information on the identity of an unwanted substance and biosynthetic pathway producing it, as well as on the availability of screening procedures. Consequently, the first logical step would be to purify and characterize all antinutritional and toxic factors of importance. One can then develop screening procedures for them, study their distribution in the germ plasm, and define breeding goals and strategies. Another point worthy of determining in advance is the function of these substances in the plant and how essential they are so that one will not have to be in a situation where other problems have been created while trying to solve one. For example, the significance of the presence of protease inhibitors to the plant and of their heterogeneity is not well understood.

Often protein and nonprotein antinutritional and toxic substances have been implicated in the defense of plants against insects and disease agents. Alpha amylase inhibitors in wheat apparently provide some resistance to insects that attack the grain because most such insects seem to have high

-amylase activities inhibited by wheat amylase inhibitors.
Another example is that of gossypol in cotton.   When glandless
cotton was introduced to solve the gossypol problem,   it was
realized that glandless types were more susceptible to cotton
pests than the normal cultivars which store gossypol in their
glands (90).   Fortunately, it was possible to correct this
problem by breeding plants with fewer glands in seeds but with
the normal number in vegetative parts.   It is conceivable that
some unwanted protein and non-protein substances represent
rudimentary genetic functions from   the past evolutionary
history which are now meaningless and dispensable especially
under man's care (76).

The unwanted protein substances will be much easier to
deal with genetically than nonprotein ones because the former
are direct gene products whose reduction, elimination,   or
inactivation by genetic means can be monitored through a
variety of methods, e.g. specific serological tests.   In doing
so, one can exploit existing genetic variation or generate
variation by induction with mutagens.   For example,   some
protease inhibitors may be inactivated by the alteration of a
single amino acid at a critical site required for activity or
binding, assuming that their expression is specific to the seed
and they are not needed for a critical function in other
organs.   In legumes,   it would actually be beneficial to
maintain trypsin inhibitors in an inactive form because they
are rich in cysteine.

As for the nonprotein antinutritional and toxic
substances,   the most suitable genetic manipulation targets
would be the enzymes in the biosynthetic pathways that produce
them or those in the catabolic pathways that utilize them.   The
objective would be to find or induce mutations to alter the
regulatory or structural gene products such that either the
synthesis is blocked or the product is catabolized or converted
to an inactive form.   Alternatively,   one can also affect the
anatomical and morphological structures in which unwanted
products are sequestered.   As mentioned before, this approach
has been employed in cotton where the pigment glands storing
the gossypol have been eliminated through the discovery and
induction of glandless mutants.

Improvement of Protein Properties by Genetic Engineering

A paper by Cohen et al.   (91) in 1973 reported the
"construction of biologically functional" genomes in vitro by
cutting and splicing genomes from two different sources.   This
marked the advent of a new era in genetics, the era of "genetic
engineering" or "recombinant DNA" technology.   It has now
become a routine laboratory procedure to isolate genes from

different organisms, cleave them with specific restriction enzymes and splice the resulting fragments to obtain recombinant DNA molecules. Such recombinant DNA molecules could be inserted either the source organism from which they came or into totally unrelated organisms. In other words, genetic engineering is not restricted by any taxonomic and phylogenetic barriers in construction, modification, transmission, and expression of the genetic material. For example the genes from a corn plant can be transferred to the bacterium Escherichia coli as easily as the genes from a related bacterial species. The rapid development in the recombinant DNA technologies in the past 8-9 years caused much concern, controversy and excitement encompassing all segments of the society. The initial fears that it could result in the creation of genetic monstrosities, intentionally or unintentionally, which might threaten the present biological and social order on the earth have now subsided. Intensive research efforts directed at the application of the recombinant DNA technologies to the solution of various problems facing the mankind are currently in progress in many industrial and academic laboratories.

The potential applications of the recombinant DNA technologies to the improvement of quality, quantity, and utilization of plant proteins are numerous. It is not unrealistic to perceive the transfer of the genes coding for lysine-rich storage proteins from legumes to cereals and those coding for methionine-rich proteins from cereals to legumes. Likewise it is conceivable that regulatory genetic elements could be transferred or inserted into the cereal or legume genomes to affect favorably the quality and quantity of seed proteins. In fact, in the distant future it may even be possible to synthesize the desirable genes in the laboratory and insert them into the genome of cereals and legumes. However, the recombinant DNA technology, although a most useful and exciting research tool, is still in its infancy in terms of application. At this writing , there is not a single genetic engineering product or procedure that can compete with conventional products or procedures in the marketplace. However, there is no reason to conclude that there will not be one soon. The application of genetic engineering techniques in plant genetics may be hampered for some time because of the lack of suitable vectors, screening, and culture procedures to regenerate plants from individual cells or calli.

For additional information on genetic engineering and recombinant DNA technology the reader is referred to recent texts and review articles (92, 93, 94, 95).

Conclusions

Proteins are primary gene products whose structure and function are direct reflections of the information encoded in the genetic material. Therefore, most problems associated with quality, quantity, utilization, deterioration and processing of plant proteins should lend themselves to solution by genetic manipulation. In fact, a genetic solution to biological problems would be most lasting and least expensive in the long run. There are numerous examples in the literature illustrating the feasibility and possibility of improving the quality and quantity of plant proteins. Further success in this area will depend upon (1) the availability of simple, rapid and inexpensive screening procedures, (2) overcoming the inverse relationship between protein quality, quantity, and yield, and (3) a thorough understanding of key biochemical-genetic and physiological events that affect protein content, composition and utilization.

The question of whether deteriorative changes that occur in proteins under in vivo and in vitro conditions can be influenced by genetic manipulations is worthy of consideration. The molecular biology logic dictates that all post-translational side chain modifications and proteolytic changes that occur in vivo are under genetic control. On the other hand the deteriorative changes that occur in proteins during the storage and processing of foods and feedstuffs are not specifically coded by the genetic material although the potential to undergo such unintended changes is dependent upon the primary structure of the protein. Thus, in theory it is possible to prevent unwanted deteriorative changes, to promote desirable deteriorative changes, and to affect both nutritional value and functionality of food proteins through genetic manipulation. The question is: is it worth the time, money and efforts that need to be expended? It has already been discussed that elimination or inactivation of specific antinutritional and toxic protein and nonprotein substances can be realistically done and is worth the effort. Technically and financially it makes sense to eliminate tannins from sorghum grain or reduce them to a nutritionally insignificant level by breeding tanninless or low tannin cultivars. This would improve not only the functionality and organoleptic properties of sorghum but also the many undesirable deteriorative changes resulting from protein-tannin interactions reducing digestibility and utilization of the protein. It also makes sense to use the genetic approach to breed soybean cultivars with inactive trypsin inhibitors, wheats with improved nutritional and baking quality, or rapeseed lacking goitrogens. But it does not make sense to try to genetically tailor every different polypeptide in the soybean seed so as to withstand denaturation during heat processing. Also it is not feasible to try to breed cereal and legume types to meet short-lived functionality needs of the food systems. In other words many undesirable changes that occur in proteins during storage and processing, desirable deteriorative changes and certain functional properties may be controlled and produced more efficiently through improvement in storage and processing technologies than through plant breeding.

There are reasons to believe that the recombinant DNA technology may be successfully applied to the solution of many problems in agriculture, including the improvement of the quality, quantity and utilization of plant proteins in the years ahead. However, there are also reasons to be skeptical of sensational claims and projections one hears and reads about this new technology and its applications. Therefore, it would be unwise to scale down or abandon conventional research efforts and approaches because some wise men or women project breakthroughs in genetic engineering that are imminent to supplant conventional approaches.

Literature Cited

1. Pearson, P. B. In "Improving Plant Protein by Nuclear Techniques"; International Atomic Energy Agency: Vienna, 1970; p. 433.
2. Sukhatme, P. V. Nutr. Rev., 1970, 28, 223.
3. McLaren, D. S. Lancent, 1974, 2, 93.
4. Harper, A. E.; Hegsted, D. M. In "Improvement of Protein Nutriture"; National Academy of Sciences: Washington, D.C. 1974; p. 184.
5. Hegsted, D. M. Amer. Sci., 1978, 66, 61.
6. Payne, P. R. In "Plant Proteins"; Norton G., Ed.; Butterworths, London, 1978; p. 247.
7. Bressani, R.; Elias, L. G. In "Seed Protein Improvement in Cereals and Grain Legumes"; International Atomic Energy Agency: Vienna, 1979; p. 3.
8. Altschul, A. M. In "Genetic Improvement of Seed Proteins"; National Academy of Sciences: Washington, D.C. 1976; p. 5.
9. Rhoades, A. P.; Jenkins, G. In "Plant Proteins"; Norton G., Eds.; Butterworths, London, 1978; p. 207.
10. Robbelen, G. In "Seed Protein Improvement in Cereals and Grain Legumes", Vol. 1; International Atomic Energy Agency: Vienna, 1979; p. 27.
11. Aziz, S. In "Proceedings of the World Food Conference of the 1976"; Iowa State University Press: Ames, Iowa, 1976; p. 15.
12. Nelson, O. E. Adv. Agron., 1969, 21, 171.
13. Wareing, P. F. Adv. Sci., 1970, 27, 38.

14. Kamra, O. P. Z. Pflanzenzucht., 1971, 65, 293.
15. Munck, L. Hereditas, 1972, 72, 1.
16. Mertz, E. T. In "Protein Nutritional Quality of Foods and Feeds" Vol. 1.; Friedman, M., Ed.; Marcel Dekkers, Inc. New York, 1975; p. 1.
17. Pomeranz, Y. In "Protein Nutritional Quality of Foods and Feeds" Vol. 1; Friedman, M., Ed.; Marcel Dekker Inc: New York, 1975; p. 13.
18. Nelson, O. E. In "Genetic Improvement of Seed Proteins"; National Academy of Sciences: Washington, D. C.   1976; p. 383.
19. Munck, L. In "Evaluation of Seed Protein Alterations by Mutation Breeding".   International Atomic Energy Agency: Vienna, 1976; p. 3.
20. Sigurbjornsson, B.; Brock, R. D.; Hermelin, T. In "Seed Protein Improvement in Cereals and Grain Legumes" Vol. 1; International Atomic Energy Agency: Vienna, 1979; p. 387.
21. Ingverson, J.: Koie, B.; Doll, H. Experientia, 1973, 29, 1151.
22. Harn, C.; Won, J. L.; Choi, K. T.   In "Breeding for Seed Protein   Improvement   Using   Nuclear   Techniques"; International Atomic Energy Agency: Vienna, 1975; p. 17.
23. Tanaka, S.   In "Breeding for Seed Protein Improvement Using Nuclear Techniques"; International Atomic Energy Agency: Vienna, 1975; p. 176.
24. Axtell, J. D. In "Evaluation of Seed Protein Alterations by Mutation Breeding";   International Atomic Energy Agency: Vienna, 1976; p. 45.
25. Johnson, V. A.; Mattern, P. J.; Schmidt, J. W.   In "Seed Proteins", Inglett, G. E., Ed.; Avi Publishing Company: Westport, Conn. 1972; p. 126.
26. Jansen, G. R.   In "Seed Protiens"; Inglett; G. E., Ed.; Avi Publishing Company: Westport, Conn. 1972; p. 19.
27. East, E. M.; Jones, D. F. Genetics. 1920, 5, 543.
28. Frey, K. J.; Brimhall, B.; Sprague, G. F. Agron. J., 1949, 41, 399.
29. Frey, K. J. In "Alternate Sources of Protein for Animal Production"; National Academy of Sciences: Washington, D. C. 1973; p. 9.
30. Whitehouse, R. N. H. "The Biological Efficiency of Protein Production"; Jones, J. G. W., Ed.; Cambridge University Press, London, 1973, p. 83.
31. Dudley, J. W.; Lambert, R. J.; Alexander, D. E. In "Seventy Generations of Selection for Oil and Protein in Maize"; Dudley, J. W., Ed.; Crop Science Society of America: Madison, Wisc., 1974; p. 181.

32.   Penning de Vries, F. W. T.; Brunsting, A. H. M.; Van
       Loar, H. H. J. Theor. Biol., 1974, 45, 339.
33.   Bhatia, C. R.; Rabson, R. Science, 1976, 194, 1418.
34.   Johnson, V. A.; Mattern, P. J.; Wilhelmi, K. D.; Kuhr, S.
       L.   In "Seed Protein Improvement by Nuclear Techniques";
       International Atomic Energy Agency: Vienna, 1978; p. 23.
35.   Juliana, B. O. In "Seed Proteins"; Inglett, G. E., Ed.;
       Avi Publishing Company: Westport, Conn. 1972; p. 114.
36.   Pollmer, W. G.; Klein, D.; Bjarnason, M.; Eberhard, D.
       Z. Acker Pflbau., 174, 140, 241.
37.   Schrickel, D. J.; Clark, W. L.   In "High Quality Protein
       Maize"; Dowden, Hutchinson & Ross, Inc.:   Stroudsburg,
       Pa., 1975; p. 389.
38.   Bond, D. A. Proc. Nutri. Soc., 1970, 9, 74.
39.   Shannon, J. G.; Wilcox, J. R.; Probst, A. H.   Crop Sci.,
       1972, 12, 824.
40.   Ali-Khan, S. T.; and Youngs, C. G.   Can. J. Pl. Sci.,
       1973, 53, 37.
41.   Tandon, O. B.; Bressani, R.; Schrimshaw, N. S.; LeBeau,
       F. J. Agric. Food Chem., 1957, 5, 137.
42.   Leleji, O. I.; Dickson, M. H.; Huckler, L. R.   Crop Sci.,
       1972, 12, 168.
43.   Poey, F. R.; Bressani, R.; Garcia, A. A.; Garcia, M. A.;
       Elias, L. G. In "Seed Protein Improvement in Cereals and
       Grain Legumes" Vol. 1;   International Atomic Energy
       Agency: Vienna, 1979; p. 369.
44.   Hucklesby, D. P.; Brown, C. M.; Howell, S. E.; Hageman,
       R. H. Agron. J., 1971, 63, 274.
45.   Hageman, R. H.; Lambert, R. J.; Loussaert, D.; Dalling,
       M.; Klepper, L.   A.   In "Genetic Improvement of Seed
       Proteins"; National Academy of Sciences: Washington, D.
       C., 1976; p. 103.
46.   Zink, F.   In "Seed Protein Improvement in Cereals and
       Grain Legumes" Vol. 1;   International Atomic Energy
       Agency: Vienna, 1979; p. 273.
47.   Hageman, R. H.   "Nitrogen Assimilation of Plants";
       Hewitt, E. J.   and Cutting, C. W., Eds; Academic Press:
       London, 1979; p. 591.
48.   Miflin, B. J. In "Genetic Improvement of Seed Proteins";
       National Academy of Sciences: Washington, D. C.   1976;
       p. 135.
49.   Miflin, B. J.; Bright, S. W. I.; Davies, H. M.; Shewry,
       P. R.; Lea, P. J.   In "Nitrogen Assimilation of Plants";
       Hewitt, E. J. and Cutting, C. V., Eds.; Academic Press,
       London, 1979; p. 335.
50.   Beevers, L.; and Hageman, R. H.   "The Biochemistry of
       Plants"; Miflin, B. J., Ed.;   Academic Press, New York,
       1980; p. 116.

51. Pate, J. S. "Recent Aspects of Nitrogen Metabolism"; Hewitt, E. J. and Cutting, C. V., Eds.; Academic Presss: London, 1968; p. 219.

52. Oaks, A.; Jones, K. E., Ross, D. W.; Boesel, I.; Lenz, D.; Misra, S. In "Seed Protein Improvement in Cereals and Grain Legumes", Vol. 1; International Atomic Energy Agency: Vienna, 1979; p. 179.

53. Kolderup, F. In "Seed Protein Improvement in Cereals and Grain Legumes"; International Atomic Energy Agency: Vienna, 1979; p. 89.

54. Perez, C. M.; Cagampang, G. B.; Esmanna, B. V.; Monseratta, R. U.; Juliana, B. O. Plant Physiol. Lancaster, 1973, 51, 537.

55. Tanaka, S. In "Seed Protein Improvement by Nuclear Techniques"; International Atomic Energy Agency: Vienna, 1978; p. 199.

56. Bishop, L. R. J. Ins. Brewk., 1928, 34, 101.

57. Thomson, J. A.; Doll, H. In "Seed Protein Improvement in Cereals and Grain Legumes" Vol. 1; International Atomic Energy Agency: Vienna, 1979; p. 109.

58. Osborne, T. B. "Vegetable Proteins"; Longmans, Green and Co: London, 1924.

59. Mertz, E. T.; Bates, L. S.; Nelson, O. E.; Science, 1964, 145, 279.

60. Munck, L.; Karlsson, K. E.; Hagberg, A.; Eggum, B. O. Science, 1970, 168, 985.

61. Singh, R.; Axtel, J. D. Crop Sci., 1973, 13, 535.

62. Doll, H. In "Proceedings of 3rd International Barley Genetics Symposium", Verlag Karl Thiemig: Munich, 1976; p. 542.

63. Johnson, V. A.; Lay, C. L. J. Agric. Food Chem., 1974, 22, 558.

64. Nelson, O. E. In "Seed Protein Improvement in Cereals and Seed Legumes" Vol. 1; International Atomic Energy Agency: Vienna, 1979; p. 79.

65. Mehta, S. L.; Dongre, A. B.; Johari, R. P.; Lodha, M. L.; Naik, M. S. In "Seed Protein Improvement in Cereals and Grain Legumes"; Vol. 1; International Atomic Energy Agency; Vienna, 1979; p. 241.

66. Tsai, C. Y.; Huber, D. M.; Warren, H. L. Crop Sci., 1978, 17, 399.

67. Tsai, C. Y.; Huber, D. M.; Warren, H. L. Plant Physiol., 1980, 66, 330.

68. Righetti, P. G.; Gianazza, E.; Viotti,A.; Soave, C. Planta, 1977, 136, 115.

69. Widholm, J. M. Biochim. Biophys. Acta, 1972, 279, 48.

70. Widholm, J. M. Can. J. Bot., 1976, 54, 1523.

71. Bourgin, J. P. Molec. Gen. Genet., 1978, 161, 225.

72. Sodek, L.; Wilson, C. M. Arch. Biochem. Biophys., 1970, 140, 29.
73. Vasil, I. K.; Ahuja, M. R.; Vasil, V. Adv. Gen., 1979, 20, 127.
74. Liener, I. E. "Plant Proteins"; Norton, G., Ed.; Butterworths; London, 1978; p. 117.
75. Silano, V. In "Nutritional Evaluation of Cereal Mutants"; International Atomic Energy Agency: Vienna, 1977; p. 13.
76. Rackis, J. J. In "Soybeans: Chemistry and Technology Vol. 1. Proteins"; Smith, A. K. and Circle, S. J., Eds.; AVI Publishing Co Inc.: Westport, Conn. 1978; p. 159.
77. Montgomery, R. D. "Toxic Constituents of Plant Foodstuffs"; Liener, I. E., Ed.; Academic Press: New York, 1969; p. 143.
78. Kakade, M. L.; Hoffa, D. E.; Liener, I. E. J. Nutr., 1972, 103, 1772.
79. Hymowitz, T. Crop Sci., 1973, 13, 420.
80. Liener, I. E. J. Agric. Food Chem., 1974, 22, 17.
81. Jaffe, W. G. In "Nutritional Improvement of Food Legumes by Breeding"; Milner, M., Ed.; John Wiley & Sons: New York, 1975; p. 43.
82. Jaffe, W. G.; Brucher, O. Arch. Latinoamer. nutr., 1972, 22, 267.
83. Pull, S. P.; Pueppke, S. G.; Hymowitz, T.; Orf, H. J. Science, 1978, 200, 1277.
84. Orf, J. H.; Hymowitz, T.; Pull, S. P.; Pueppke, S. G. Crop Sci., 1978, 18, 899.
85. Vodkin, L. Plant Physiol., 1981, 68, 766.
86. Orf, J. H.; Hymowitz, T. Crop Sci., 1979, 19, 107.
87. Blouin, F. A.; Zarrins, Z. M.; Cherry, J. P. "Protein Functionality in Foods". Cherry, J. P., Ed.; ACS Symposium Series, Washington, D.C., 1981, p. 21.
88. King, E.E.; Leffler, H. R. In "Seed Protein Improvement in Cereals and Grain Legumes"; International Atomic Energy Agency: Vienna, 1979; p. 385.
89. Stefansson, B. R.; Kondra, Z-P. Can. J. Pl. Sci., 1975, 55, 343.
90. Ridgway, R. L.; Bailey, J. C. In "Glandless Cotton: Its Significance, Status, and Prospects"; USDA, SEA, NPS: Washington, D.C. 1978; p. 118.
91. Cohen, S. N.; Chang, A. C. Y.; Boyer, H. W.; Helling, R. B. Proc. Natl. Acad. Sci., 1973, 70, 3240.
92. Boyer, H. W.; Nicosia, S. "Genetic Engineering"; Elsevier/North-Holland Biomedical Press: New York, 1978.
93. Setlow, J. K.; Hollaender, A. "Genetic Engineering", Vol. 1; Plenum Press: New York, 1979.

94. Setlow, J. K.; Hollaender, A. "Genetic Engineering",
    Vol. 2; Plenum Press: New York, 1980.
95. Dolly, E. L.; Eveleigh, D. E.; Montenecourt, B. S.;
    Stokes, H. W.; Williams, R. L. Food Technol., 1981, 35,
    26.
96. Food and Agriculture Organization of the United Nations
    (FAO). "Production Yearbook, Vol. 33; FAO: Rome, 1979;
    p. 251.
97. Boulter, D.; Derbyshire, E. In "Seed Proteins"; Norton
    G., Ed.; Butterworths: London, 1978; p. 1.
98. Kaul, A. K. In "Food Proteins"; Pirie, N. W., Ed.;
    Butterworths: London, 1975; p. 1.
99. Sinha, S. K. "Food Legumes"; Food Agriculture
    Organization of the United Nations: Rome, 177.
100. Mosse, J. Fed. Proc., 1966, 25, 1663.
101. Byers, M.; Kirkman, M. A.; Miflin, B. J. In "Seed
    Proteins"; Norton, G.; Ed.; Butterworths: London, 1978;
    p. 227.

RECEIVED September 14, 1982.

Controlling Enzymic Degradation of Proteins

T. RICHARDSON

University of Wisconsin—Madison, Department of Food Science and the
Walter V. Price Cheese Research Institute, Madison, WI 53706

Traditionally, the food industry has used a limited number
of processes for controlling endogenous (in the raw material) and
exogenous (added or adventitious) enzymic activities in foods
including proteolysis. Control of enzymic activities in foods
usually depends upon controlling the enzymic environment so as to
maximize or prevent the action of enzymes on their substrates.
All food scientists recognize that the activities of exogenous
enzymes added to foods can be controlled by manipulating such
factors as pH, temperature, presence of co-factors or inhibitors,
etc. On the other hand exogenous, adventitious enzymes arising
from spoilage microorganisms can often be controlled by the
judicious use of sanitizers and pH controls which may be a
function of product formation, water activity, osmolarity, etc.
The control of endogenous food enzymes, however, is often very
complex and requires a thorough knowledge of enzymic activities
and how they are integrated into the post-harvest and post-mortem
physiologies of foods. Control of these metabolic (usually
catabolic) activities by the food scientist can have a profound
influence on the quality and storage stability of various foods.
Controlled atmosphere storage and packaging of various foods is
an obvious example of environmental control of endogenous enzymic
activities (1). Oxidative and fermentative metabolic reactions
are of extreme importance in the conversion of muscle to meat (2)
and in the maintenance of quality in post-harvest fruits and
vegetables (3). Endogenous enzymic reactions are of great
importance in the maturation of dates and tea as well as other
foods.

Endogenous and exogenous proteolytic enzymes are well-docu-
mented as factors controlling the tenderization of meat (2).
Blanching of vegetables and fruits is the time-honored method for
inactivating endogenous enzymes to enhance stability of frozen
foods (1).

Within the intact tissues of foods, numerous enzymes are
compartmentalized or constrained by intracellular membranes to
physically prevent their interaction with substrates. This

latency is destroyed after harvest or slaughter as the membranes
deteriorate, thereby allowing admixture of enzymes with cellular
substrates (2, 3, 4). This results in the degradation of
cellular structural elements which may be desirable for meat
tenderization or undesirable in the textural deterioration of
fruits. Thus, the control of enzymic latency in food processing
can be extremely important. Recent research suggests some
possibilities for controlling the latency of exogenous and endogenous enzymes in food systems.

Ideally, the food scientist would like to control the
various enzymic reactions affecting food quality thereby leading
to products with better texture, color, flavor and nutritional
value. At the present time, this ideal is unobtainable; however,
further research will undoubtedly lead to additional ways for
controlling enzymic activities in foods. The subsequent
discussion will explore some possibilities for manipulating
enzymic activities which may serve as a basis for additional
research in this area. In many cases, it is necessarily speculative but the speculation, I believe, is based on acceptable
chemical, physical and biological principles. Since this volume
is dedicated to the degradation of proteins in foods, the
following text will be restricted primarily to a discussion of
the control of proteolytic activities in foods.

Need to Control Proteolysis

The control of proteolytic activity in food materials should
be considered in the broadest possible terms and not only in its
prevention. The regulation of the extent and timing of
proteolysis by endogenous and exogenous enzymes could be of prime
importance in processing and storage of foods. If one could
selectively initiate, enhance, retard or inhibit various
proteases, a greater flexibility in food processing, in product
development and in storage variables might be possible. The
requisite control might be achieved by manipulating enzymic
latency physically or chemically, by judicious use of natural or
synthetic, digestible inhibitors, by engineering exogenous
proteases with recombinant DNA methods or by various other
techniques discussed in the succeeding sections.

A few examples will serve to illustrate why the control of
proteolytic enzymes, or the lack of it, is important in food
systems. The substitution of fungal rennets for calf-rennet in
cheesemaking created an unexpected problem in the control of
proteolytic activity in the cheese whey. Fungal rennets appearing
in the whey after curd formation are substantially more stable
than calf rennet to thermal inactivation (5). Consequently,
residual fungal proteolytic activity after conventional thermal
processing of the whey led to obvious problems when the dried
whey was subsequently used in products containing casein.
Ideally, these proteases should be controlled with a minimum heat

treatment to save energy and minimize damage to whey proteins. Recently, fungal rennets more labile to heat treatments have become commercially available.

Activities of heat-stable proteases and lipases from psychrotrophic microorganisms (6, 7) proliferating in milk stored at low temperatures may adversely affect milk quality and cheese yields. Inexpensive, practical control of psychrotrophs and/or their heat-stable proteolytic and lipolytic enzymes would be highly desirable.

Bovine plasmin is secreted from the blood into milk where it acts primarily on β-casein to generate a series of more hydrophobic γ-caseins and polar peptides (8, 9). Apparently, the plasmin can attack the milk proteins while they are still in the lumen of the mammary gland. In addition, proteolysis can continue, after milking, in the raw, stored product. The amount of plasmin in milk seems to increase in late lactation milk and sufficient plasminolysis of the milk proteins may lead to defects in texture and loss of yield in the resultant cheese (10, 11). Since the plasmin has a specificity similar to trypsin, its activity has been inhibited with soybean trypsin inhibitor. The control of endogenous plasmin activity in milk would definitely benefit the dairy industry.

The foregoing three examples relating to control of added, microbial and endogenous enzymes serve to illustrate that the control of enzymic activity is important in all three general areas of food enzymology.

In Table I are listed some major possibilities for controlling proteolytic degradation of food proteins by enhancing, retarding, inhibiting or delaying enzymic activities. These are based in part on observations from the literature and will be discussed in the following sections.

Enzymic Latency

Latent activity of enzymes will be defined as potential activity which becomes expressed upon removal of physical and/or chemical constraints. Control of enzymic latency should allow release of enzyme at definite points in a food process or during storage to effect desired changes. Conversely, latent enzyme inhibitors may also be controlled to inhibit enzymic activity after a desired interval. Various types of potential enzymic latency are listed in Table II.

Lysosomal latency. A classical example of enzymic latency is the various hydrolytic enzymes that are sequestered within intracellular lysosomal particles (2, 4, 12). Included among the hydrolytic enzymes retained within the lysosomal membranes are the proteolytic cathepsins. Since these lysosomal proteases have pH optima in the acidic region around 5, there has been much discussion concerning their relevance in the economy or turnover of cells with a presumed, nominal pH near neutrality. This has been especially true in research on meat tenderization whereby the post-mortem liberation of latent catheptic activity has been implicated in degradation of muscle proteins to enhance the tenderization of meat (2). However, the low pH optima of these enzymes coupled with the sparse numbers of lysosomal particles and relatively low catheptic activity in muscle tissue have led some scientists in meat research to question their importance in the tenderization of meat (2, 4, 13, 14). Nevertheless, recent research by Dutson et al. (15) on the increase in meat tenderness resulting from high voltage electric stimulation of ovine carcasses suggests that this treatment releases latent catheptic activity by rupturing lysosomal particles. An attendant decrease in the pH of the muscle tissue apparently favors the rupture of lysosomes and catheptic-mediated hydrolysis of muscle proteins leading to increased tenderization of the meat. However, it must be emphasized that the release of lysosomal enzymes result-ing from electrical stimulation is only one of several possible mechanisms suggested for this tenderizing effect. In addition, other endogenous proteases have also been involved in meat tenderization (2).

Table I.   Possibilities for controlling exogenous and endogenous
           proteases.

1) Manipulation of enzymic latency in situ, ex vivo and in vivo.

2) Chemical and biological engineering of enzymes.

3) Use of natural and synthetic inhibitors.

4) Control of electrostatic interactions that may affect enzymic-substrate and post-enzymic product interactions.

Thus, the meat scientist may be able to control latency of lysosomal enzymes in muscle tissue by electrical stimulation. In addition, other factors of a biochemical nature may contribute to the stability or lack of it in cellular membranes that delimit intracellular enzymic reactions. For example, Lawrie (2) briefly discussed the liberation of lysosomal enzymes as a result of vitamin E deficiency or of excess vitamin A in animal tissues.

Polyunsaturated fatty acids of membrane lipids are known to enhance fluidity of cellular membranes implying a decrease in their physical stability (16). This has obvious implications in the post-mortem latency of lysosomal enzymes in fish muscle. Mitochondria from fish liver swell much more rapidly than those from rat liver (17). Also, one might ask whether the latency of lysosomal enzymes in the muscles of ruminants fed encapsulated polyunsaturated oils to yield more unsaturated meat (and presumably membranes) (18, 19) would be different than the normal, more saturated counterpart.

Some meat scientists and muscle biologists have suggested that lysosomal enzymes released from phagocytic cells within muscle tissue may play a role in the tenderization of meat. Catheptic activity in phagocytic cells is very high (4). If the migration of the phagocytic cells into muscle tissue could be controlled, it may be possible to regulate, in part, the tenderization process. In this regard, chemotactic peptides (e.g., N-formyl-methionyl-leucyl-phenylalanine) that attract phagocytic cells have been isolated and characterized (20). If these peptides could be selectively delivered to skeletal muscles, they might prove useful in manipulating the migration of phagocytic cells into muscle tissue thereby partially controlling levels of lysosomal cathepsins. Thus, possibilities exist, albeit remote at this point, for controlling post-mortem lysosomal enzymic latency in muscle tissue. Latent enzymic activities are also very important in the post-harvest physiology of plant tissue (3). It is less clear, however, how these activities might be better controlled.

Table II. Enzymic latency.

1) Control of lysosomal proteases in vivo and ex vivo.

2) Microencapsulation of proteases (controlled release).

3) Control of zymogen activation.

4) Latency of chemical derivatives of proteases (chemical latency).

5) Latency of polycationic proteins - protease complexes.

6) Genetic engineering of food microorganisms. Proteolytic activity after autolysis or secretion of enzymes.

Chemical latency. An excellent example of chemical latency arises out of the pre-slaughter, intravenous injection of the sulfhydryl protease papain into animals, again to enhance tenderization of meat via post-mortem proteolytic activity. In this case, the enzyme injected before slaughter of the animal is disseminated by the circulation of the animal throughout its musculature. The uniform distribution of enzyme within the tissues results in post-mortem tenderization of the resultant meat. Injection of the free, active enzyme results in stress to the animal accompanied by internal hemorrhaging leading to rejection of the carcass. However, if the essential thiol group in the active site of papain is reversibly blocked by thiol-disulfide interchange, this inactive enzyme can be administered to the animal without ill effects. The latent papain activity is subsequently regenerated in situ by reducing agents in the meat (perhaps by glutathione while cooking) (21). If the

essential functional groups for enzymic activity are known, it
should be possible to introduce chemical latency into virtually
any exogenous enzyme used in food processing.

The acid proteases used in the manufacture of cheese may
offer opportunities for latent proteolysis in the ripening of
cheese. Conversion of essential carboxylate residues in the
acid proteases to labile anhydrides or esters may allow slow but
sustained release of proteolytic activity within the cheese. In
this regard, it should be useful to discuss the inhibition of
chymosin activity upon treating the enzyme with dansyl chloride
(22, 23, 24). Since lysine residues in proteins are readily
dansylated, an essential lysine was proposed for chymosin
activity (22, 23). However, the slow regeneration of latent
proteolytic activity upon storage of the dansylated chymosin (23)
and rapid reactivation upon treatment with NH_4OH (24) were not
consistent with proposed dansylation of an essential lysine
residue, which yields a very stable covalent derivative.
Subsequently, formation of a labile, dansylated histidine residue
in chymosin was suggested (24) to explain the regeneration of
activity. However, in view of the essentiality of the γ-carboxyl
groups of two aspartate residues in acid proteases (25), it is
plausible that the dansyl chloride reacted with an essential
carboxylate anion as shown in Figure 1 to yield a mixed anhydride
thereby inactivating the chymosin. This could subsequently react
with ammonium hydroxide or a lysine residue to regenerate
proteolytic activity and yield, in the latter case a dansylated-
lysine. This suggestion is consistent with the mechanism for
hydrolysis of sulfite esters by pepsin proposed by Kaiser and
Nakagawa (26) who were able to infer anhydride intermediates by
using strong nucleophiles as trapping agents (Figure 2).

Microencapsulation. Enzymes have been entrapped within
microcapsules using a variety of methods (27, 28) to yield so-
called microcells and immobilized enzymes for biomedical or food
processing purposes. The feasibility of encapsulating enzymes
or cell-free extracts containing mixtures of enzymes into edible
microcapsules of milk fat has been established by Olson and coworkers (29, 30, 31). These workers are developing a cheese
ripening system whereby cell-free extracts of microorganisms can
be encapsulated with substrates and co-factors for generation of
flavors such as diacetyl and acetoin. The milk-fat coated
microcapsules containing the appropriate enzyme systems are added
to milk and retained in the resulting cheese. The juxtaposition
of enzyme, substrates and co-factors allows the enhancement and
regulation of flavor development in the cheese. Proteolytic
enzymes could be encapsulated and the latent activity released by
thermal or lipolytic treatment of the product after an appropriate
period.

Physical latency. Although enzymic compartmentalization in

$$\text{Chym-}\overset{\overset{\text{O}}{\|}}{\text{C}}\text{-O}^- \ + \ \text{DanSO}_2\text{Cl} \ \longrightarrow \ \text{Chym-}\overset{\overset{\text{O}}{\|}}{\text{C}}\text{-O-SO}_2\text{-Dan} \ + \ \text{HCl}$$

(mixed anhydride)

$$\text{Chym-}\overset{\overset{\text{O}}{\|}}{\text{C}}\text{-O-SO}_2\text{-Dan} \ \xrightarrow{\text{NH}_4\text{OH}} \ \text{Chym-}\overset{\overset{\text{O}}{\|}}{\text{C}}\text{-O}^- \ + \ \text{Dan-SO}_3^-$$

Figure 1. Suggested mechanism for inhibition of chymosin by dansyl chloride (top) and subsequent reactivation (bottom). Although aromatic sulfonates are good leaving groups (56), it is possible that an incoming amine could attack the sulfur moiety in such an asymmetric anhydride, yielding the appropriate sulfonamide (67).

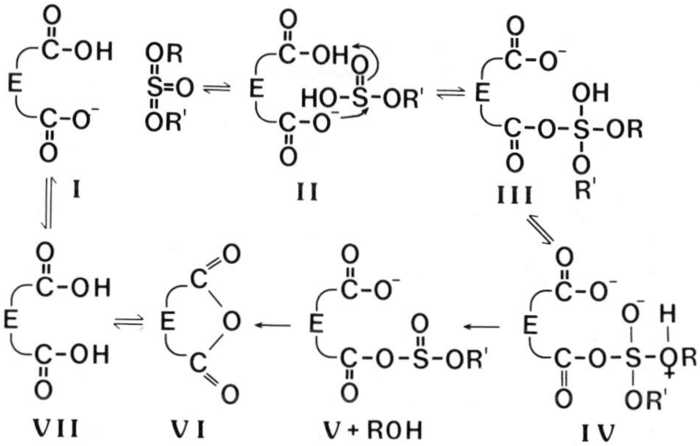

Figure 2. Proposed mechanism for the pepsin-catalyzed hydrolysis of sulfite esters. (Reproduced, with permission, from Ref. 26. Copyright 1977, Plenum Publishing Corp.)

lysosomes and microencapsulation of enzymes are forms of physical
latency, this section will be restricted to latency via formation
of digestible, macromolecular aggregates. Generally, most
proteins and enzymes carry a net negative charge at neutral pH
values (32). It follows that there are few proteins and enzymes
with net positive charges in food systems. However, positively
charged proteins (pI = 7 to 10) can be prepared by simply
esterifying or amidating free carboxyl groups (33, 34, 35).
These positively charged proteins avidly interact with negatively
charged caseins, for example, to yield a precipitated complex
(33, 34, 35). Thus, it seems possible to form insoluble,
inactive complexes between positively charged proteins and
negatively charged proteases such as porcine pepsin (pI = 2.2)
and bovine chymosin (pI = 4.6). Proteolysis of the complexed
polycation by the entrapped protease should yield the latent
activity. The putative latency might be useful in controlling
ripening of cheese. Digestible, macromolecular complexes with
other food-related enzymes such as the amylases should also be
possible to prepare.

     Zymogen activation. As is well known, many proteolytic
enzymes are secreted as inactive zymogens. Subsequent cleavage
of an activation peptide from the N-terminus, for example, of
pepsinogen (36) or prochymosin (autocatalytically or by other
proteases) results in realization of proteolytic activity
(Figure 3). The rates of activation of the foregoing zymogens
are affected by a number of environmental factors such as pH,
ionic strength, temperature, etc. (25, 37). Apparently,
electrostatic interactions between the covalently bound activation peptide sequence and the active site of the protease
sterically prevents enzyme-substrate interactions (25). With
pepsinogen, at least, the released activation peptide can, at
relatively high molar ratios, still electrostatically bind to
the enzyme as an inhibitor (Figure 4). Guanidation of the
activation peptide enhanced the inhibition (38) suggesting that
chemical synthesis and modification of peptides may yield useful
protease inhibitors.
     The control of zymogen activation by other than environment-
al factors might be accomplished by varying the sequence of the
activation peptide with recombinant DNA techniques. For example,
Nishimori et al. (39) recently cloned the gene for calf prochymosin into Escherichia coli. It is well known that many of
the acid proteases are genetically related with long amino acid
sequences conserved (40) (Table III). However, there are also a
number of non-homologous regions that apparently are not
essential for proteolytic activity. Therefore, these non-
homologous regions within the various acid proteases or their
activation peptides might be altered using point mutations and
recombinant DNA techniques (41, 42, 43) to engineer the rate of
zymogen activation and perhaps the activity of the proteolytic

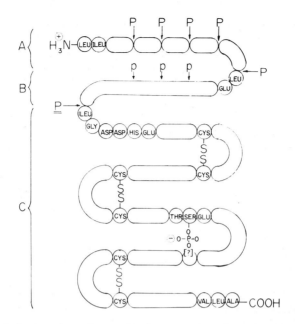

Figure 3. Activation of pepsinogen. In the conversion of pepsinogen to pepsin, hydrolysis occurs at several points (P) releasing peptides (A), a pepsin inhibitor (B), and pepsin (C). (Reproduced, with permission, from Ref. 36. Copyright 1960, Academic Press.)

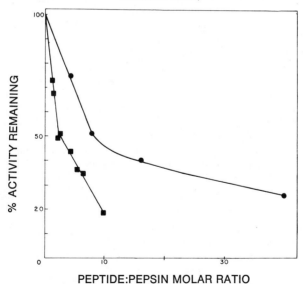

PEPTIDE:PEPSIN MOLAR RATIO

Figure 4. Inhibition of pepsin by varying amounts of its activation. Key: ■, peptide 1-16; and ●, its guanidinated derivative. (Reproduced, with permission, from Ref. 38. Copyright 1977, Plenum Publishing Corp.)

enzyme itself. Other possibilities for genetic engineering of
proteases will be discussed below.

The foregoing examples serve to illustrate how the regulation of the latency of enzymes, especially proteases, may be
attainable to facilitate food processing and to control storage.

Table III. Homology in some acid proteases (40).

Homology in Bovine and Porcine Pepsins		Homology in Bovine Chymosin and Pepsin	
% Homology	Sequences	% Homology	Sequences
67	Activation Peptide	40	1-47
83	48-111	67	48-111
100	259-265	86	259-265
79	355-373	57	355-373

Chemical Engineering

The chemical modification of enzymes is probably of limited
practical value. However, it may help to understand such factors
as how protease-protein interactions may be affected by their
relative pI values. In addition, chemical modifications of
proteases may provide worthwhile information for engineering
proteases using recombinant DNA methods.

It is likely that any attempts to substantially modify the
surface charge of enzymes by esterification of carboxyl groups
or by acylation of amino groups, for example, may lead to
disruption of their three dimensional structures and thereby
inactivate them. For example, extensive phosphorylation, amida-
tion or esterification of β-lactoglobulin tends to disrupt its
globular structure (35, 44). Nevertheless, Johansen et al. (45)
found that succinylated, nitrated and iodinated preparations of
subtilisin Carlsberg were more active towards clupein and
gelatin but not casein or N-benzoyltyrosine ethyl esters as
compared to the native enzyme. By the same token, Mitz and
Summaria (46) reported increased activity of soluble chymotrypsin-
cellulose derivatives in relation to the native enzyme.

There are a number of examples of chemical modifications of
enzymes that suggest possibilities for engineering enzymes by
recombinant DNA techniques. For example, polypeptidyl enzymes
have been studied in some detail (47). In general, the enzyme
is reacted with an appropriate amino acid N-carboxyanhydride to
yield a product whereby polyamino acid chains are covalently
bound to the enzyme to render it more cationic, anionic or
hydrophobic, etc.

Soluble, polypeptidyl proteases are generally more resistant to inactivation by autolysis or by proteolysis resulting from other enzymes. Also, polypeptidyl proteases may have much different activities on the same substrates compared to the native enzymes. Poly-DL-t-leucyl chymotrypsin had 64% of the esterase activity of chymotrypsin acting on L-phenylalanyl-ethyl ester. On the other hand, it was 90% more active than chymotrypsin as an amidase acting on benzoyl-L-phenylalanylhydroxamide (48).

Poly-L-lysyl-ribonuclease (P-L-RNase) has been prepared as indicated in Figure 5 to yield a positively charged enzyme (49) which, of course, interacts with polyanionic ribonucleic acid. As shown in Table IV, the P-L-RNase acts differently on various substrates as the pH is changed compared to the unmodified enzyme indicating a marked alteration in interactions with substrate. Although various P-L-RNase preparations possessed only 4 to 25% of the native activity on RNA, P-L-RNase displays maximal activity at ionic strengths 2 to 6 times greater than for the native enzyme at pH 5 or 8 (49). Optimum pH at low ionic strengths was shifted in the same direction as that of the isoelectric point of the modified enzyme for RNA. Removal of lysyl peptides by tryptic treatment of P-L-RNase tended to reverse the changes in enzymic properties.

Interestingly, disulfide bridges in the P-L-RNase (as with the control RNase) could be reduced and subsequently reformed by oxidation to completely recover activity.

Table IV.   Relative rates of hydrolysis of nucleoside 2'-3' cyclo-phosphates by polypeptidyl RNase. Native RNase = 100% (49).

Preparation (Number of Lysines/RNase)	pH 5.0		pH 7.0		pH 8.0	
	$C-c-P^1$	$U-c-P^2$	$C-c-P^1$	$U-c-P^2$	$C-c-P^1$	$U-c-P^2$
p-Lys-RNase (21 lysines/RNase)	53	125	60	65	107	100
p-Lys-RNase (26 lysines/RNase)	88	250	112	95	120	108
p-DL-Ala-RNase	84	42	177	47	210	25

[1] Cytidine 2'-3' cyclic phosphate

[2] Uridine 2'-3' cyclic phosphate

The foregoing data suggest modification of proteases using genetic engineering techniques to alter their pI, interactions with substrates, kinetic parameters, hydrophobicity, pH optimum,

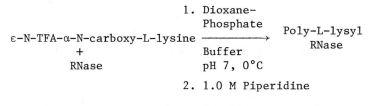

Figure 5.   Synthesis of polylysyl ribonuclease (49).

thermal stability, etc.   This offers numerous opportunities for
designing novel enzymes to be used in food processing.   It may
eventually be possible to manipulate secondary and tertiary
structures of enzymes using genetic and computer techniques which
could define the effects of changes in primary amino acid
sequences on three dimensional structure (50).

Biological Engineering

In addition to modifying putative non-homologous or other
regions of the primary sequence of zymogens and proteases with
point mutation-recombinant DNA techniques, other possibilities
are suggested by the previous discussion of polypeptidyl enzymes.
For example, insertion into a cloning vector of the appropriate
oligonucleotide sequence coding for polylysine, polycysteine,
polyleucine, polyglutamic acid, etc. covalently bound to the
C-terminus of the protease might allow the production of food-
related enzymes with vastly altered properties which could be
exploited.   The regeneration of activity after oxidation of the
inactive, reduced P-L-RNase discussed previously suggests that
polyamino acids added to proteases by genetic manipulation
should not affect folding of the modified proteases and oxidation of thiol groups to yield an active three-dimensional
structure.   Recombinant DNA technology is currently at a stage
where the aforementioned modifications of enzymes are clearly
possible.

Natural and Synthetic Protease Inhibitors

There is a wide range of proteases that are inhibited by
naturally occurring protein inhibitors in plants (51).   Inhibitors are known to occur in certain animal tissues and fluids
as well (52).   Evidently there are numerous inhibitors specific
for proteases with different active sites such as the serine
proteases and the acid proteases.   There is a substantial amount
of information on the structures and active sites of these
natural inhibitors available with which to design potential
synthetic peptide inhibitors (51, 53).

Although the use of these natural inhibitors in foods may
not meet with the approval of the U.S. Food and Drug Administration, a detailed knowledge of their structures and mechanisms
of action may eventually allow the synthesis of reversible and
digestible peptide inhibitors that could prove useful.   For
example, pepstatin analogs, pepstatin-like peptides, or modified
activation peptides from zymogens may be used to control acid
proteases in the production of cheese.   Or peptides patterned
after the active site in trypsin inhibitors could inhibit unwanted plasmin activity in milk mentioned previously.

Pepstatin is a pentapeptide, isolated from various species
of actinomycetes, which strongly inhibits several acid

proteases; the K_i for inhibition of pepsin is 10^{-10} M (25, 54). Pepstatin, which contains a unique, central statyl residue involved in interactions with the active sites of acid proteases has the following structure (25, 54):

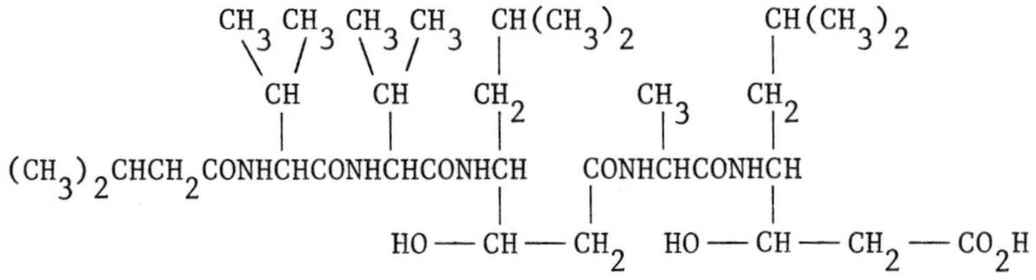

$$
\begin{array}{l}
\ \ \ \ \ \ \ \ CH_3\ CH_3\ CH_3\ CH_3\ \ \ CH(CH_3)_2 \ \ \ \ \ \ \ \ \ CH(CH_3)_2 \\
\ \ \ \ \ \ \ \ \ \ \backslash\ /\ \ \backslash\ /\ \ \ \ \ \ \ | \ \ \ \ \ \ \ \ \ \ \ \ \ \ \ \ \ \ | \\
\ \ \ \ \ \ \ \ \ CH\ \ \ CH\ \ \ CH_2 \ \ \ \ \ \ CH_3\ \ \ \ CH_2 \\
\ \ \ \ \ \ \ \ \ \ |\ \ \ \ \ |\ \ \ \ \ | \ \ \ \ \ \ \ \ \ \ \ | \ \ \ \ \ \ | \\
(CH_3)_2CHCH_2CONHCHCONHCHCONHCH \ \ \ CONHCHCONHCH \\
\ |\ \ \ \ \ |\ \ \ \ \ | \ \ \ \ \ \ \ \ \ \ \ | \ \ \ \ \ \ | \\
\ \ \ \ \ \ \ \ \ \ \ \ \ \ \ HO-CH-CH_2 \ \ HO-CH-CH_2-CO_2H
\end{array}
$$

There are a number of novel features in the structure of pepstatin that are important in the inhibition of pepsin (Figure 6). It is a very hydrophobic peptide that has poor solubility in water. The central statyl residue (Figure 6) is thought to combine with the active site of pepsin to mimic the transition state during normal peptic proteolysis. Pepstatin has thus been referred to as a "transition state" inhibitor (25). The central hydroxyl group is essential for assuming a pseudo-transition state and is, therefore, necessary for inhibition of pepsin (25, 54, 55). The terminal carboxyl group is not required for inhibitory activity. Under comparable conditions, pepstatin inhibits chymosin at 48% of the inhibition for pepsin, suggesting a certain structural specificity for inhibition that might be exploited (55). Although there are a number of unique features about the structure of pepstatin, the judicious use of model-building combined with appropriate peptide chemistry might allow the synthesis of reversible, digestible inhibitors of acid proteases for use in foods.

Chemical modifications of pepstatin such as those suggested in Figure 7 may help define structure-activity relationships as well as in designing some novel inhibitors of acid proteases. Removal of the essential hydroxyl group (dideoxypepstatin) increases the K_i for porcine pepsin 2,000 fold (54). In Figure 7, reactions are proposed whereby a thiol, amino or other nucleophilic group might be substituted for the essential hydroxyl function. These modifications rely on the good leaving properties of aromatic sulfonates (56). Such structural alterations should lead to a better understanding of molecular requirements for producing pepstatin analogs.

Various types of esters involving the pepstatin hydroxyl group should be easily obtainable to impart latent inhibitors or other features to these derivatives. For example, phosphoryla-tion of the hydroxyl group (Figure 7) potentially results in a more water-soluble, inactive but latent inhibitor. Slow hydrolysis of the phosphate ester in acidic foods (optimum pH for hydrolysis of phosphate esters ~4-6) (57) should release active inhibitor. The non-essential carboxyl group of pepstatin can be

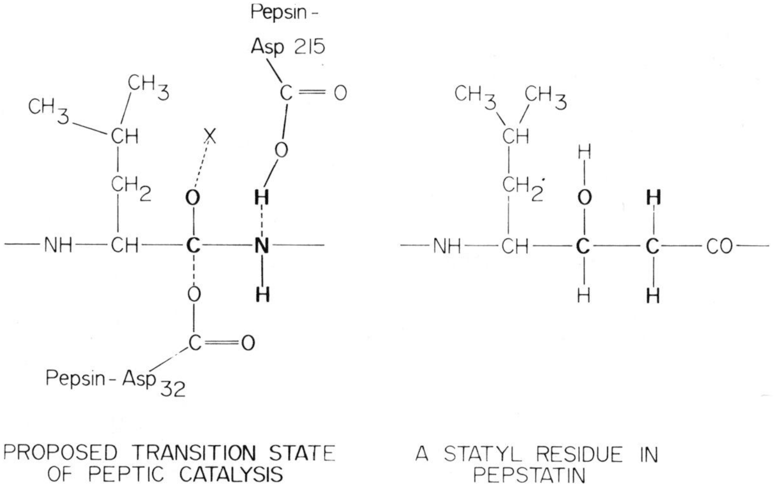
Pepsin - Asp 215
Pepsin - Asp 32
PROPOSED TRANSITION STATE OF PEPTIC CATALYSIS
A STATYL RESIDUE IN PEPSTATIN

Figure 6. Inhibition of pepsin by pepstatin. (Reproduced, with permission, from Ref. 55. Copyright 1977, Plenum Publishing Corp.)

1)   Modification of essential -OH on statyl residue:

$$\text{Pepstatin -OH} \xrightarrow[-HF]{+ \ C_6H_5CH_2SO_2F} \text{Pepstatin-O-SO}_2CH_2C_6H_5$$

$$\text{-C}_6H_5CH_2SO_3^- \qquad\qquad \Big\downarrow \qquad \begin{array}{c} + \ AcSH \\ or \\ + \ NH_3 \end{array}$$

$$\underline{\text{Pepstatin-SH}} \xleftarrow[-Ac]{H_2O} \text{Pepstatin-S-Ac}$$

$$\text{or}$$

$$\underline{\text{Pepstatin-NH}_2}$$

2)   Active esters of essential -OH group:

A. $\text{Pepstatin-OH} + POCl_3 \xrightarrow{+ \ OH^-} \text{Pepstatin-O-}\overset{\overset{\displaystyle O}{\|}}{\underset{\underset{\displaystyle O^-}{|}}{P}}\text{-O}^- + 3Cl^-$

B. $\text{Pepstatin-O-}\overset{\overset{\displaystyle O}{\|}}{\underset{\underset{\displaystyle O^-}{|}}{P}}\text{-O}^- \xrightarrow[+ \ H_2O]{+ \ H^+} \text{Pepstatin OH} + HO\text{-}\overset{\overset{\displaystyle O}{\|}}{\underset{\underset{\displaystyle O}{|}}{P}}\text{-O}^-$

slow?

3)   Derivatives of non-essential carboxyl group:

$\text{Pepstatin-COO}^- + X^- \longrightarrow \text{Pepstatin-COX}^*$
(activated carboxyl group)

$\text{Pepstatin-COX}^* + Z: \longrightarrow \text{Pepstatin-COZ}$

Figure 7.   Possible chemical modifications of pepstatin.

activated or otherwise modified to immobilize the inhibitor or to
couple it to undigestible macromolecules such as carboxymethyl
cellulose for use in foods. This latter compound would impart
water solubility to the hydrophobic inhibitor. Immobilized
pepstatin has been used to isolate acid proteases (58).

Pepsin is known to be inhibited by high molar ratios of its
activation peptide (38). When the activation peptide is
guanidated, its inhibitory activity can be more than doubled
(Figure 4). The sequence of this peptide and relevant chemical
modifications suggest routes for preparation of effective,
digestible synthetic inhibitors.

Trypsin is inhibited by oligomers of homoarginine (n = 10)
with a K_i approaching 10^{-5} M (53) (Figure 8). This oligomeric
substrate analog for trypsin apparently binds to the active site
of trypsin where the geometry for rapid hydrolysis is not
adequate.

Synthetic peptide inhibitors with high affinities for the
active site of proteases are also possible based on affinity
labeling techniques (59).

There are thus numerous possibilities for designing
synthetic peptide (60) protease inhibitors that could prove
useful in food systems. Again based on an understanding of the
peptide chemistry of inhibitors, recombinant DNA technology could
probably be employed to eventually produce desirable peptide
inhibitors using conventional fermentation technology. Solid-
phase oligonucleotide synthetic procedures are progressing
quickly and it is rapidly becoming possible to synthesize
oligonucleotides, coding for a particular amino acid sequence,
for insertion into a cloning vector (61, 62).

Control of Interactions

A few, brief examples from the literature should be
sufficient to illustrate the potential of directly managing
enzyme-substrate interactions and indirectly controlling
subsequent product interactions. Since virtually all food
proteins have a net negative charge at the pH of most foods,
enzymes engineered to be more positively or negatively charged
or more hydrophobic may be retained more or less in a food as
desired. For example, the amount of milk-clotting or lipolytic
enzymes retained in cheese curd could be governed by charge or
functional groups on a modified enzyme. Interactions of this
type might be important in subsequent ripening or aging of the
cheese. Holmes et al. (63) have demonstrated the differential
retention of various, proteolytic milk-clotting enzymes in
cheese curd.

Marshall and Green (64) have studied in some detail the
effects of cationic substances including proteins on the rennet
clotting time of milk. Normal renneting of milk is thought to
result in a decreased charge repulsion between caseinate

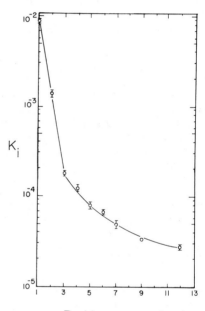

Figure 8.   Inhibition of trypsin by poly-homoarginine. (Reproduced, with permission, from Ref. 53. Copyright 1974, Springer-Verlag New York, Inc.)

particles from the enzymic-mediated release of a hydrophilic,
negatively charged peptide. The decreased charge repulsion
favors coagulation of the modified caseinate particles to
yield curd. Marshall and Green (64) observed that as increasing
amounts of cationic materials were added to the milk to progressively neutralize the net negative charge on the casein
micelles, the rate of enzymic-mediated coagulation of the
casein increased with the reduced net negative charge in the
system. Many of these observations have been verified and
extended by DiGregorio and Sisto (33, 34) and by Mattarella (35),
who studied the interactions of positively charged β-lactoglobulin derivatives with caseinate systems. This has resulted in
the use of positively charged proteins as electrostatic
coagulants of milk which dramatically increased the yield of
proteins in cheesemaking (33, 34). However, it is not known
what the effects would be on the quality of resultant cheese.

Kang and Kepplinger (65) showed that addition of a cationic
peptide, palmitoyl-L-lysyl-L-lysine ethyl ester, to a pepsin
solution before addition to milk inhibited the clotting activity
of pepsin. On the other hand, when the peptide was added first
to the milk, an enhanced rate of clotting was observed. This is
an extreme example of how proteolysis and subsequent product
interactions might be regulated electrostatically.

Controlling the pattern of proteolysis in protein
hydrolysates by relying on steric or charge effects between
protein and an immobilized protease is a distinct possibility.
By analogy, the patterns of starch hydrolysis are substantially
different depending upon whether a soluble or immobilized α-
amylase is used (66).

Conclusions

Control of endogenous and exogenous enzymic activity in
foods must be extended beyond current, conventional methods to
realize advances in food processing and storage. Suggestions in
this review for manipulating food enzymes are based on accepted
physical, chemical and biological principles. Among
possibilities discussed are: (1) management of enzymic latency
in situ, ex vivo and in vivo; (2) engineering of enzymes using
chemical and biological techniques; (3) control of enzymes using
natural and synthetic inhibitors; and (4) alteration of electrostatic and hydrophobic forces on exogenous enzymes and within
foods that may regulate enzymic-substrate and post-enzymic
product interactions. Although proteolytic enzymes are specifically discussed, many of the concepts can be generalized to
other food-related enzymes.

Future food scientists may thus be able to exert better
control over exogenous and adventitious enzymes in foods by
designing and engineering the enzymes as well as by using
acceptable enzymic inhibitors. Powerful techniques such as

those used for chemical modification of enzymes and inhibitors, and solid phase synthesis of polypeptides and oligonucleotides coupled with emerging recombinant DNA technology can be brought to bear to eventually realize better control of enzymic reactions in food processing and in food deterioration.

Regulation of endogenous enzymes in foods is obviously more difficult and must await a better understanding of the physiology of post-harvest and post-mortem processes.

Acknowledgments

This contribution was made possible by support from the Walter V. Price Cheese Research Institute and from the College of Agricultural and Life Sciences, University of Wisconsin-Madison, Madison, WI 53706.

Literature Cited

1. Potter, N.N. "Food Science"; AVI Publishing Co., Inc.: Westport, CT, 1968; p 149, 191.
2. Lawrie, R.A. "Meat Science", 3rd Ed.; Pergamon Press: New York, NY, 1979; p 348.
3. Schwimmer, S. "Post-Harvest Biology and Biotechnology"; Hultin, H.O.; Milner, M., Eds.; Food and Nutrition Press, Inc.: Westport, CT, 1978; p 317.
4. Tappel, A.L. "The Physiology and Biochemistry of Muscle as a Food"; Briskey, E.J.; Cassens, R.G.; Trantman, J.C., Eds.; Univ. Wisconsin Press: Madison, WI, 1966; p 237.
5. Thunell, R.K.; Duersch, J.W.; Ernstrom, C.A. J. Dairy Sci. 1979, 62, 373.
6. Adams, D.M.; Bramley, T.G. J. Dairy Sci. 1981, 64, 1951.
7. Marshall, R.T.; Marstiller, J.K. J. Dairy Sci. 1981, 64, 1545.
8. Eigel, W.N.; Hofmann, C.J.; Chibber, B.A.K.; Tomich, J.M.; Keenan, T.W.; Mertz, E.T. Proc. Natl. Acad. Sci. U.S. 1979, 76, 2244.
9. Humbert, G.; Alais, C. J. Dairy Res. 1979, 46, 559.
10. Barry, J.G.; Donnelly, W.J. J. Dairy Res. 1980, 47, 71.
11. Donnelly, W.J.; Barry, J.G.; Richardson, T. Biochim. Biophys. Acta 1980, 626, 117.
12. Richardson, T. "Principles of Food Science. I. Food Chemistry"; Fennema, O.R., Ed.; Marcel Dekker, Inc.: New York, NY, 1976; p 285.
13. Canonico, P.G.; Bird, J.W.C. J. Cell Biol. 1970, 45, 321.
14. Pennington, R.J.T. "Proteinases in Mammalian Cells and Tissues"; North-Holland Publ. Co.: New York, NY, 1977; p 528.
15. Dutson, T.R.; Smith, G.C.; Carpenter, Z.L. J. Food Sci. 1980, 45, 1097.

16. Mead, J.F.; Fulco, A.J. "The Unsaturated and Polyunsaturated Fatty Acids in Health and Disease"; C.C. Thomas: Springfield, IL, 1976; p 122.

17. Richardson, T.; Tappel, A.L. J. Cell Biol. 1962, 13, 43.

18. Bitman, J.; Wrenn, T.R.; Dryden, L.P.; Edmondson, L.F.; Yoncoskie, R.A. "Microencapsulation"; Vandegaer, J.E., Ed.; Plenum Press: New York, NY, 1974; p 200.

19. Scott, T.W.; Cook, L.J.; Mills, S.C. J. Am. Oil Chem. Soc. 1971, 48, 358.

20. Snyderman, R.; Goetzl, E.J. Science 1981, 213, 830.

21. Kang, C.K.; Warner, W.D.; Rice, E.E. U.S. Patent 3,818,106, 1974.

22. Hill, R.D.; Laing, R.R. Nature 1966, 210, 1160.

23. Hill, R.D.; Laing, R.R. Biochim. Biophys. Acta 1967, 132, 188.

24. Rickert, W. Biochim. Biophys. Acta 1970, 220, 628.

25. Tang, J., Ed. "Acid Proteases: Structure, Function and Biology"; Adv. Exp. Med. Biol., Vol. 95; Plenum Press: New York, NY, 1977; p 355.

26. Kaiser, E.T.; Nakagawa, Y. "Acid Proteases: Structure, Function and Biology"; Tang, J., Ed.; Adv. Exp. Med. Biol., Vol. 95; Plenum Press: New York, NY, 1977; p 159.

27. Chang, T.M.S. "Enzyme Engineering", Vol. 2; Pye, E.K.; Wingard, L.B., Eds.; Plenum Press: New York, NY, 1973; p 419.

28. May, S.W.; Li, N.N. "Enzyme Engineering", Vol. 2; Pye, E.K.; Wingard, L.B., Eds.; Plenum Press: New York, NY, 1973; p 77.

29. Magee, E.L., Jr.; Olson, N.F. J. Dairy Sci. 1981, 64, 600.

30. Magee, E.L., Jr.; Olson, N.F. J. Dairy Sci. 1981, 64, 611.

31. Magee, E.L., Jr.; Olson, N.F. J. Dairy Sci. 1981, 64, 616.

32. Malamud, D.; Drysdale, J.W. Anal. Biochem. 1978, 86, 620.

33. DiGregorio, F.; Sisto, R. U.K. Pat. Appl. 2,052,515, 28 Jan., 1981.

34. DiGregorio, F.; Sisto, R. J. Dairy Res. 1981, 48, 267.

35. Mattarella, N. Ph.D. Thesis, 1981, University of Wisconsin, Madison, WI 53706.

36. Bovey, E.A.; Yanari, S.S. "The Enzymes", Vol. IV; Boyer, P.D.; Lardy, H.; Myrback, K., Eds.; Academic Press: New York, NY, 1960; p 63.

37. Ruenwongsa, P.; Chvlavatnatol, M. "Acid Proteases: Structure, Function and Biology"; Tang, J., Ed.; Adv. Exp. Med. Biol., Vol. 95; Plenum Press: New York, NY, 1977; p 329.

38. Kumar, P.M.H.; Ward, P.H.; Kassell, B. "Acid Proteases: Structure, Function and Biology"; Tang, J., Ed.; Adv. Exp. Med. Biol., Vol. 95; Plenum Press: New York, NY, 1977; p 211.

39. Nishimori, K.; Kawaguchi, Y.; Hidaka, M.; Uozumi, T.;
    Beppu, T. J. Biochem. 1981, 90, 901.
40. Foltmann, B.; Pedersen, V.B. "Acid Proteases: Structure,
    Function and Biology"; Tang, J., Ed.; Adv. Exp. Med. Biol.,
    Vol. 95; Plenum Press: New York, NY, 1977; p 3.
41. Green, C.; Tibbetts, C. Proc. Natl. Acad. Sci., USA 1980,
    77, 2455.
42. Shortle, D.; Koshland, D.; Weinstock, G.M.; Botstein, D.
    Proc. Natl. Acad. Sci., USA 1980, 77, 5375.
43. Kudo, I.; Leineweber, M.; Raj Bhandary, U.L. Proc. Natl.
    Acad. Sci., USA 1981, 78, 4753.
44. Woo, S.; Creamer, L.K.; Richardson, T. J. Agric. Food Chem.,
    In Press, 1982.
45. Johansen, J.T.; O'Hesen, M.; Svendsen, I. Biochim. Biophys.
    Acta 1967, 139, 211.
46. Mitz, M.A.; Summaria, L.J. Nature 1961, 189, 576.
47. Silman, H.I.; Sela, M. "Poly-α-Amino Acids"; Fasman, G.D.,
    Ed.; Marcel Dekker, Inc.: New York, NY, 1967; p 605.
48. Becker, R. "Polyamino Acids, Polypeptides and Proteins";
    Stahmann, M.A., Ed.; Univ. Wisconsin Press: Madison, WI,
    1962; p 301.
49. Frensdorf, A.; Sela, M. J. Biochem. 1967, 1, 267.
50. Wade, N. Science 1981, 213, 623.
51. Richardson, M. Food Chem. 1980-81, 6, 235.
52. Fritz, H.; Tschesche, H.; Green, L.J.; Truscheit, L., Eds.
    "Proteinase Inhibitors"; Proc. 2nd Int. Res. Conf.;
    Springer-Verlag: New York, NY, 1974; p 311.
53. Rigbi, M.; Elkana, Y.; Segal, N.; Kliger, D.; Schwartz, L.
    "Proteinase Inhibitors"; Proc. 2nd Int. Res. Conf.; Fritz,
    H.; Tschesche, H.; Green, L.J.; Truscheit, L.; Eds.;
    Springer-Verlag: New York, NY, 1974; p 541.
54. Rich, D.H.; Sun, E.; Sengh, J. Biochem. Biophys. Res.
    Commun. 1977, 74, 762.
55. Marciniszyn, J., Jr.; Hartsuck, J.A.; Tang, J. "Acid
    Proteases: Structure, Function and Biology"; Tang, J.,
    Ed.; Adv. Exp. Med. Biol., Vol. 95; Plenum Press: New
    York, NY, 1977; p 199.
56. Morrison, R.T.; Boyd, R.N. "Organic Chemistry"; Allyn and
    Bacon, Inc.: Boston, NY, 1973; p 458.
57. Bruice, T.C.; Benkovic, S.J. "Bioorganic Mechanisms, Vol.
    II"; W.A. Benjamin, Inc.: New York, NY, 1966; p 3.
58. Kobayashi, H.; Murakami, K. Agric. Biol. Chem. 1978, 42,
    2227.
59. Jakoby, W.B.; Wilchek, M. "Methods in Enzymology"; Colowick,
    S.P.; Kaplan, N.O., Eds.; Academic Press: New York, NY,
    1977; p 774.
60. Stewart, J.M.; Young, J.D. "Solid Phase Peptide Synthesis";
    W.H. Freeman and Co.: San Francisco, CA, 1969.
61. Duckworth, M.L.; Gait, M.J.; Goilet, P.; Hong, G.F.; Singh,
    M.; Richards, T.C. Nucleic Acid Res. 1981, 9, 1691.

62. Matteucci, M.D.; Caruthers, M.H. J. Am. Chem. Soc. 1981, 103, 3185.
63. Holmes, D.G.; Duersch, J.W.; Ernstrom, C.A. J. Dairy Sci. 1977, 60, 862.
64. Marshall, R.J.; Green, M.L. J. Dairy Res. 1980, 47, 359.
65. Kang, Y.; Kepplinger, J. Unpublished Observations, 1981.
66. Reilly, P.J. "Immobilized Enzymes for Food Processing"; Pitcher, W.H., Jr., Ed.; CRC Press, Inc.: Boca Raton, FL, 1980; p 113.
67. Schröder, E.; Lübke, K. "The Peptides", Vol. 1; Academic Press: New York, 1965; p 96.

RECEIVED August 30, 1982.

Effects of Lipid Oxidation on Proteins of Oilseeds

ROBERT L. ORY and ALLEN J. ST. ANGELO

United States Department of Agriculture, Southern Regional Research Center,
Agricultural Research Service, New Orleans, LA 70179

Peanuts (groundnuts) are one of the world's major oilseeds and are grown in tropical and subtropical countries. World production of peanuts is annually about 19 million tons (1), with most of them grown in India, China, Africa, and the United States (2). The United States produces about 10% of the world supply and is the only country that consumes most of the crop as whole nut products. Most countries crush peanuts for the edible oil. Peanuts contain 50-55% oil and 27-30% protein. The major proteins of peanuts are the storage globulins, arachin and conarachin, that make up almost 75% of the total proteins. After removal of the oil, the oil-free peanut meal contains 55-60% protein that can be used to raise the nutritional quality of protein-deficient diets or to provide alternate sources of low-cost functionally useful protein. Nutrition, however, is not simply a matter of supplying individual nutrients. Good nutrition involves groups of nutrients whose functions are closely related, each providing a necessary part of the balanced diet.

Storage oil and protein in oilseeds like soybeans and peanuts are in close proximity and, under normal conditions, no changes occur in either the protein or the oil, which contains polyunsaturated fatty acids. Peanut oil contains varying amounts of oleic and linoleic acids; soybean oil contains these two, plus linolenic acid. It is the latter two fatty acids that can be oxidized to hydroperoxides and their breakdown products (i.e.: ketones, aldehydes, alcohols, etc.) which, in turn, can react with terminal functional groups of amino acids in proteins and enzymes. Such interactions of lipid peroxides and the secondary products with proteins can alter the functional and nutritional properties, in addition to affecting their flavor.

Lipoxygenase is the principal enzyme that catalyzes oxidation of polyunsaturated fatty acids in oilseeds. Denatured metalloproteins are the primary nonenzymatic catalysts. Soybean lipoxygenase is possibly the most studied oilseed lipoxygenase. Peanuts also contain the enzyme, about a fifth of the activity in

soybeans ($\underline{3}$, $\underline{4}$). This enzyme acts on linoleic and linolenic acids
to produce hydroperoxides which, in turn, can react with terminal
functional groups of amino acids in proteins (i.e.: -SH, -OH,
-COOH, -NH$_2$) to alter their functional, flavor, and nutritional
properties. Hydroperoxides also break down into carbonyl compounds that react readily with these amino acids; especially when
catalyzed by metalloproteins. Some effects of protein interactions with lipid peroxides and their secondary products and
methods for studying them are described.

Experimental Procedures

Interactions of lipid peroxides with proteins can be measured
by several methods: thin layer chromotography of the amino acids,
lipid peroxides, and their reaction products ($\underline{24}$); dual staining
of gel electrophoretic protein patterns for both protein and associated lipid ($\underline{13}$); and flourescence spectroscopy of the extracted
proteins ($\underline{23}$).

Catalysts of Lipid Oxidation

Enzymatic Oxidation. Lipoxygenase (E.C. 1.13.1.13) is the
principal enzyme for catalyzing oxidation of polyunsaturated
fatty acids in vegetable oils. It is present in virtually all
oilseeds but has been studied more in soybeans than in other
oilseeds ($\underline{3}$). This enzyme is substrate-specific in that it
attacks only cis, cis,-1,4- pentadiene bonds, such as those in
linoleic and linolenic acids. The primary products are optically
active cis-trans conjugated hydroperoxides ($\underline{5}$). Once formed,
hydroperoxides can be degraded further into various aldehydes,
ketones, and alcohols by enzymatic and/or nonenzymatic catalysts.
It is these secondary products, plus the intact hydroperoxides
that can interact with proteins to lower both nutritional and
flavor quality of a food product.
Soybean lipoxygenase, because of its higher acitivty, has
been the most completely investigated of the seed lipoxygenases
($\underline{3,8,9}$). It is not particulate and can be extracted with water.
It has a molecular weight of 100,000, consists of a single polypeptide chain, exists in at least two isomeric forms, and has a
pH range from 6-9, depending upon the isomer being studied. With
linoleic acid as substrate, it forms predominantly the C-13 hydroperoxide, rather than the C-9 isomer ($\underline{3}$). Chan and coworkers
($\underline{10,11}$) examined methyl esters of the soybean lipoxygenase-formed
C-9 and C-13 linoleate hydroperoxides after thermal decomposition.
The volatile degradation products included many of the shorter
chain aldehydes that have been implicated in flavor problems of
plant food products.
Peanut lipoxygenase has not been as well characterized as
soybean lipoxygenase but various workers ($\underline{4,12-17}$) have identified
some similarities in the two enzymes. The partially purified

enzyme has a general pH optimum of 6.2 and is rather heat-labile, losing all activity at temperatures above 40°C. Sulfhydryl reducing agents (2,3-dimercaptoethanol and dithiothreitol) inhibit activity by 55-100%. Sanders et al. (18) isolated 3 isozymes from raw peanuts; two having a pH optimum of 6.2 and the third, at pH 8.3. Molecular weight of all isomers was 73,000. The alkaline pH enzyme was reported to be CN-tolerant but the acid isozymes were inhibited by NaCN in their tests. However, work by Siddiqui and Tappel (16) and St. Angelo and Kuck (19) showed no inhibition of peanut lipoxygenase by CN when the system was buffered at pH 6.2 to offset the NaCN-induced rise in pH. Increasing concentration of NaCN caused a rise in pH, which then decreased activity of the acid pH isozyme.

Nonenzymatic Oxidation. As was shown earlier by St. Angelo, et al. (6,7), catalysis of lipid peroxidation/ degradation by nonenzymatic agents, such as hemeproteins and metals, can induce faster deterioration in a high oil/protein product like peanut butter, than that caused by lipoxygenase. Relative rates of lipid peroxidation shown in Table I illustrate the faster rates catalyzed by metal ions, Fe and Cu, and the metalloproteins, peroxidase and tyrosinase. Depending upon the types of handling, processing and/or storage conditions, it is these latter

Table I. Effect of various catalysts on rates of peroxidation of fatty acids in stored peanut butter

Catalyst	Peroxide Value (meq/kg)		
	initial	28 days	meq. increase
None	30.0	31.1	1.1
NaCl (0.04 mmole)	14.3	24.9	10.6
Cupric acetate (0.02 mmole)	11.8	33.1	21.3
FeCl$_3$ (0.02 mmole)	10.3	36.8	26.5
Soybean lipoxygenase[1]/	7.0	21.0	14.0
Tyrosinase	6.3	24.9	18.6
Peroxidase	5.4	27.0	21.6
Boiled Peroxidase	7.3	26.6	19.3

1/ Concn. of enzymes was 0.1% (w:v).

catalysts that frequently cause greater deterioration of foods through lipid peroxide-protein interactions.

Efffects of Lipid Peroxide-Protein Interactions

Advanced stages of lipid peroxidation cause flavor problems that are well known. Rancidity in a vegetable oil or oil-containing food will preclude any attempts to consume the food

because of objectionable odors and off-flavors. These off-flavors
may or may not be lipid-protein interactions, depending upon the
extent of the rancidity. Effects on nutritional value of proteins, however, may not be as obvious without chemical analysis
of the products. In general, the protein quality of oilseed
proteins is, with very few exceptions, lower than animal proteins
(20). Oilseed proteins have inherent amino acid deficiencies
that differ between seeds. For example, soybeans are low in
methionine but have sufficient lysine. Sesame is low in lysine
but is high in sulfur-containing amino acids, whereas cottonseed
and peanuts are lower in both of these amino acids (20). These
deficiencies can be easily improved by blending with other
proteins having the complimentary amino acid but, if large
amounts of lipid peroxides and their secondary products are
present, they can bind to terminal reactive groups of cysteine
and methionine or other amino acids (Table II) to inactivate
them, if present in enzymes, or lower nutritional value, if
present in food proteins.

Table II. Terminal reactive groups of amino acids in proteins

Amino Acid	Reactive Group
ARG, LYS	$-NH_2$
CYS	$-SH$
GLU, ASP	$-COOH$
SER, THR, TYR	$-OH$
MET	$-SCH_3$
HIS	$=NH$

Proteins are the most important source of amino acids. Some
amino acids can be produced in the body but the essential amino
acids must be obtained from the diet as protein, peptides, or
free amino acids. Of these, it is lysine, methionine, and
cysteine that are most limiting in oilseeds and are the ones that
are inactivated by lipid peroxides. If peroxidized or autoxidized
lipids are ingested, they can also bind to enzymes and inhibit
their activity. This was shown by Matsushita (21), who examined
the specific interactions of linoleic acid hydroperoxide and its
secondary products with trypsin, pepsin, lipase, and RNase. He
found a correlation between the incorporation of autoxidized
lipids into the enzymes and their inactivation caused by resulting damage to the amino acids: aspartic and glutamic acids,
threonine, cysteine, methionine, leucine, tyrosine, lysine, and
histidine. Methionine, cysteine, histidine, lysine, and tyrosine
were the most labile to hydroperoxide damage, whereas lysine,

histidine, and methionine were the most susceptible to interactions with the peroxide secondary products (e.g.: aldehydes, ketones, etc.)

Karel, et al. (22) also showed that peroxidizing methyl linoleate could react with proteins and amino acids. They followed the reaction of peroxidizing linoleate with lysozyme by electron spin resonance (ESR). Peroxides did not break disulfide bonds but several reaction products were identified from histidine, methionine, and lysine interactions. In addition to the thorough characterization of ESR signals in peroxide-lysozyme interactions, they also examined ESR spectral characteristics of gelatin, bovine serum albumin, casein, lactalbumin, gliadin, trypsin, several other enzymes and free amino acids. Results were similar to those obtained with lysozyme. Lipid peroxides and/or free radicals can readily abstract hydrogen from -SH groups in proteins but they do not easily break -SS- bonds.

As noted earlier (13), lipoxygenase is the prime suspect for catalyzing lipid oxidation in raw peanuts but this enzyme is destroyed by roasting temperatures. In cooked or roasted peanut products, lipid oxidation is catalyzed by nonenzymic catalysts such as metalloproteins, free Fe and Cu.

One of the primary methods for identifying lipid-protein interactions in crude protein extracts is polyacrylamide gel electrophoresis, employing dual staining for both protein and lipid. Amido Black and Coomassie Blue are ideal stains for protein, but they do not identify those proteins with associated lipid. Sudan stains are the principal stains employed in histochemical analyses for lipids but they are not sensitive enough for oilseed proteins separated by gel electrophoresis. We, therefore, developed a procedure using Rhodamine 6G or Oil Red O stain, allowing the gels to soak in the lipid stain at $37^{\circ}C$ overnight. This produced satisfactory lipid-stained gels as shown in Figure 1. Total proteins are stained with amido black. Lipid-stained gels show only those bands that contain lipid material associated with protein. Only three proteins of peanuts appear to bind lipid peroxides. Of these, the principal protein that binds lipids appears to be the major storage protein, arachin.

This general type of lipid-protein staining pattern for raw peanuts also appears in roasted peanuts (Figure 2). Peanuts that are roasted for candy or peanut butter manufacture undergo some protein denaturation, which can lower protein solubility. Because of this, there are fewer protein bands in gel patterns of roasted peanut proteins, but lipid-staining bands still appear. Only two major protein bands appear in roasted peanut extracts (compared to three in raw peanuts) and these same two lipid-protein complexes are evident in Oil Red O-stained gels of both raw and roasted peanuts. Lipid bands are similar in both raw and roasted peanuts, in that arachin is still the principal protein that binds to lipid peroxides. The conarachin fraction of peanut proteins does not appear to bind to the lipid peroxides.

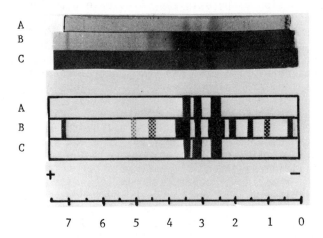

Figure 1. Dual staining of gel electrophoretic patterns of raw peanut proteins with oil red O (A), amido black (B), and rhodamine 6G (C). Migration is toward the anode.

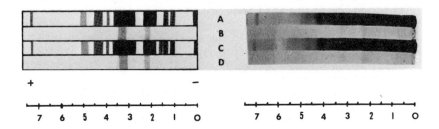

Figure 2. Dual staining of gel electrophoretic patterns of roasted peanut proteins with amido black (A and C) and oil red O (B and D). A and B are freshly roasted peanuts; C and D are rancid roasted peanuts, 1 year old. Migration is toward the anode.

Thin layer chromatography (TLC) is another method that can be used to identify reactions between peroxidized lipids and amino acids but, because of different solubilities of the amino acids and the lipids in aqueous and organic solvents, this technique has not been used much. Reactants had to be separated on two or more plates or in 2-phase systems to insure separation of both materials. To avoid the use of several separations and/or 2-phase systems that required an additional development, Kuck, et al. (24) developed a single phase system for TLC analysis of free amino acids, lipid peroxides, and the lipid peroxide-amino acid complex (Figure 3). The developing solvent of petroleum ether: diethyl ether: glacial acetic acid (60:40:1) separated the linoleate hydroperoxide-amino acid complex (A), linolete hydroperoxide (D), oxidized linoleic acid (E), and the free amino acids threonine and lysine (F). B and C are unknown minor products. Identities of the separated materials on the TLC plates were confirmed by infrared spectral analysis and mass spectrometer fragmentation patterns. Polyacrylamide gel electrophoresis and TLC therefore, can provide useful information on the effects of lipid peroxides binding to proteins and amino acids, but these methods do not identify changes in protein conformation by changes in size and charge of the proteins in the gels.

We also employed fluorescence spectroscopy to compare fresh and peroxidized peanut proteins (23). The NaCl-soluble proteins were extracted and scanned. Figure 4 illustrates fluorescence spectra of salt-soluble proteins from fresh raw peanuts (curve A) and rancid raw peanuts stored 4.5 months at 30°C (B), plus freshly roasted peanuts (C) and 12-month old roasted peanuts (D). These scans show that both raw and roasted peanuts have some natural fluorescence which can be quenched or decreased by apparent lipid peroxide interactions (curves B and D). Both curves of rancid peanut proteins (B and D) showed some decrease in intensity compared to the fresh samples (A and C). Malonaldehyde is one product of lipid peroxidation that fluoresces but this apparent quenching suggests that other nonfluorescing lipid peroxide secondary products may be formed in greater amounts than is malonaldehyde.

Increasing lipid peroxidation in both raw and roasted peanuts can also affect solubility of salt-soluble proteins (14, 23). Peanuts were ground in a food blender to disrupt the cells and promote lipoxygenase activity on the linoleic acid of the storage oil. Ground samples were stored under separate conditions for 4.5 months: (1) at 4°C in a sealed glass jar, (2) at 30°C in an open, gauze-covered jar, and (3) at 30°C in an open, gauze-covered jar with 10% rancid oil added to enhance lipid-protein interactions. Results in Table III illustrate the recoveries of soluble proteins from hexane-deoiled meals prepared from these samples. As absorbance (lipid peroxidation) increased, so did the amounts of salt-soluble proteins extracted from the ground peanuts. Conversely, the amounts of

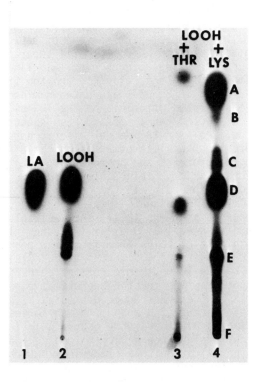

Figure 3. Thin layer chromatographic separation of linoleate hydroperoxide, amino acids, and their complexes. Key to spots: 1, pure linoleic acid; 2, oxidized linoleic acid; 3, threonine; and 4, lysine. A–F and the developing solvent are defined in the text.

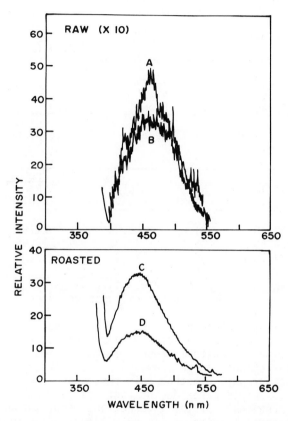

Figure 4. Fluorescence spectra of proteins extracted from fresh and rancid peanuts. Key: A, fresh raw peanuts; B, rancid raw peanuts; C, freshly roasted peanuts; and D, rancid roasted peanuts.

salt-extracted residue (oil-and protein-extracted) remaining
showed a decrease. The reason for the increased solubility of
proteins after association with lipid peroxides is unknown; the
amounts of bound lipid are not sufficient to account for the
increase.

Table III.   Effects of lipid peroxidation on protein solubility
in ground-up raw peanuts after 4.5 months storage.
(A) Initial recoveries.

Sample	% Oil removed	% Deoiled meal	CDHP Units [1]	%Protein in meal
1	46.10	53.90	0.53	46.64
2	45.09	54.91	1.21	44.47
3	48.45	51.55	1.89	44.16

(B) Recoveries after NaCl extraction of proteins from deoiled
meals.[2]

				Protein Content	
Sample	%NaCl-insoluble	%NaCl-soluble	%Total recovered	% NaCl-insoluble	% NaCl-soluble
1	37.48	31.60	69.08	22.99	88.48
2	35.80	36.55	72.35	18.96	87.86
3	34.30	39.17	73.47	17.29	88.01

(1) Conjugated diene hydroperoxide units of peroxidation.
(2) Values shown are those after dialysis and freeze-drying.

Conclusions. Lipid peroxides formed by both enzymatic and
nonenzymatic catalysts can react with proteins, very likely
through terminal functional groups of amino acids. These lipid-
protein interactions can be measured by gel electrophoresis, thin
layer chromatography, electron spin resonance, and fluorescence
spectroscopy. Lipid peroxide interactions can affect several properties of proteins. If the interactions occur through functional
groups of essential amino acids, nutritional value will be
lowered. If the binding of lipid peroxides with arachin involves
lysine, this would be very undesirable since lysine is already
low in this protein. If binding affects amino acids in active
sites of enzymes, these enzymes would have diminished activity.
Peroxidation also appears to decrease fluorescence of oilseed
proteins but it seems to increase their solubility.

LITERATURE CITED

1. Lusas, E. W. J. Amer. Oil Chem. Soc., 1979, 56, 425.
2. Anonymous. Southeast Peanut Farmer, 1978, 7(5), 15.
3. Veldink, G. A.; Vliegenthart, J. F. G; Boldingh, J. Prog. Chem. Fats Other Lipids, 1977, 15, 131.
4. St. Angelo, A. J.; Ory, R. L. J. Agr. Food Chem., 1979, 27, 229.
5. Privett, O. S.; Nickel, C.; Lundberg, W. O.; Boxer, P. D. J. Amer. Oil Chem. Soc., 1955, 32, 505.
6. St. Angelo, A. J.; Ory, R. L. J. Amer. Oil Chem. Soc., 1975, 52, 38.
7. St. Angelo, A. J.; Ory, R. L.; Brown, L. E. Oleagineux, 1973, 28, 351.
8. Axelrod, B. In "Food Related Enzymes"; Whitaker, J. R., Ed., American Chemical Society: Washington, D.C., 1974, p. 324.
9. Tappel, A. L. In "The Enzymes" 2nd ed., vol. 8; Boyer, P. D., Ed., Academic Press: New York, 1963; p. 275.
10. Chan, H. W. S.; Levett, G. Lipids, 1976, 12, 99.
11. Chan, H. W. S.; Prescott, F. A. A. Biochim. Biophys. Acta, 1975, 380, 141.
12. Dillard, M. G.; Hendrick, A. S.; Koch, R. B. J. Biol. Chem., 1960, 236, 37.
13. St. Angelo, A. J.; Kuck, J. C.; Ory, R. L. J. Agr. Food Chem., 1979, 27, 229.
14. St. Angelo, A. J.; Ory, R. L. Peanut Sci., 1975, 2, 41.
15. Pattee, H. E.; Singleton, J. A. J. Amer. Oil Chem. Soc., 1977, 54, 183.
16. Siddiqi, A. M.; Tappel, A. L. J. Amer. Oil Chem. Soc., 1957, 34, 529.
17. St. Angelo, A. J.; Ory, R. L. In "Symposium: Seed Proteins"; Inglett, G. E., Ed., Avi Publish. Co.: Conn., 1972; p. 284.
18. Sanders, T. H.; Pattee, H. E.; Singleton, J. A. Lipids, 1975, 10, 681.
19. St. Angelo, A. J.; Kuck, J. C. Lipids 1977, 12, 682.
20. Molina, M. R.; Bressani, R. Dev. Food Sci., 1979, 2, 39.
21. Matsushita, S. J. Agr. Food Chem., 1975, 23, 150.
22. Karel, M.; Shaich, K.; Roy, R. B. J. Agr. Food Chem. 1975, 23, 159.
23. St. Angelo, A. J.; Ory, R. L. J. Agr. Food Chem., 1975, 23, 141.
24. Kuck, J. C.; St. Angelo, A. J.; Ory, R. L. Oleagineux, 1978, 33, 507.

RECEIVED May 10, 1982.

Discoloration of Proteins by Binding with Phenolic Compounds

F. A. BLOUIN, Z. M. ZARINS, and J. P. CHERRY[1]

United States Department of Agriculture, Southern Regional Research Center, Agricultural Research Service, New Orleans, LA 70179

Phenolic compounds that occur in plant materials have very diverse chemical structures. Some are simple benzene derivatives, like the phenolic acids, p-hydroxybenzoic acid and caffeic acid (Figure 1). Others contain more complex ring systems such as the flavonoid quercetin and the terpenoid gossypol (Figure 1). Still others are extremely complex in structure. The vegetable tannin, procyanidin, is thought to be an oligomer of flavan-3-ol units (Figure 1). Lignin is a network type polymer composed of substituted cinnamyl alcohols (Figure 1).

Phenolic compounds are found in all plant tissues. Griffiths (1), for example, found gentisic acid in the wood, bark, roots, stems, leaves, flowers, pod wall, cotyledons and testa of the cacao tree (Table I). Leucoanthocyanins and epicatechin were also present in most of the parts of this particular plant.

The concentration of phenolics in plant materials can be quite high. Lignin represents 20-30% of the dry weight of the woody stems of trees. The dry matter of tea leaves is 40% polyphenols of the tannin type. Chlorogenic acid, one of the commonly occurring phenolic esters, is present in fruits at about a 0.1% level whereas in sunflower seed its concentration can be as high as 2-3%.

In the natural state, the low molecular weight phenolic compounds are generally bound to carbohydrate moieties. For example, one of the commonly occurring flavonoids is rutin (Figure 2). It is a quercetin 3-O-rutinoside; a disaccharide of glucose and rhamnose is bound to the 3-hydroxyl of the C-ring of the quercetin. Even the polymeric polyphenol, lignin, is covalently bound to the hemicellulose components of the cell walls. Phenols sometime occur in the free state but usually in storage tissues such as seeds, in dead tissues such as the heartwood of trees, or in diseased plant tissues.

[1] Current address: United States Department of Agriculture, Eastern Regional Research Center, Philadelphia, PA 19118.

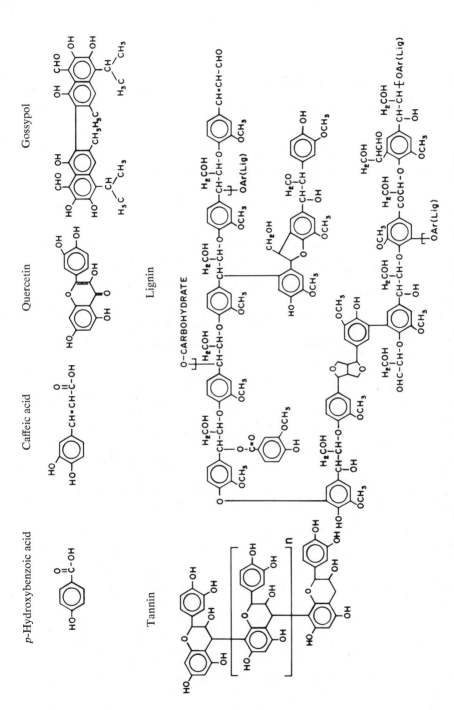

Figure 1. Structures of various types of phenolic compounds found in plant materials.

Table I. Distribution of Flavonoid Compounds in the Tissues of the Cacao Tree (Treobroma cacao) (1).

Plant part	Compound in						
	Hydrolysed extract					Unhydrolysed extract	
	Quercetin	Caffeic acid	p-Coumaric acid	Gentisic acid	Leuco-anthocyanin	(-)-Epi-catechin	Cyanidin glycoside
Sap wood				++	++	+	
Heart wood				++	+	+	
Bark				++	+	+	
Roots		+++		+	+	+	
Green stem		++	++	+	++	++	
Young leaf	++	+++	+	++	++	+++	++
Old leaf	+	+++	+++	+++	+	+	
Flower	+++	++	++	+	++	+	++
Pod wall	+	++	+	+	+	++++	
Cotyledon	+	++	+	+	++	+	++++
Testa				+			

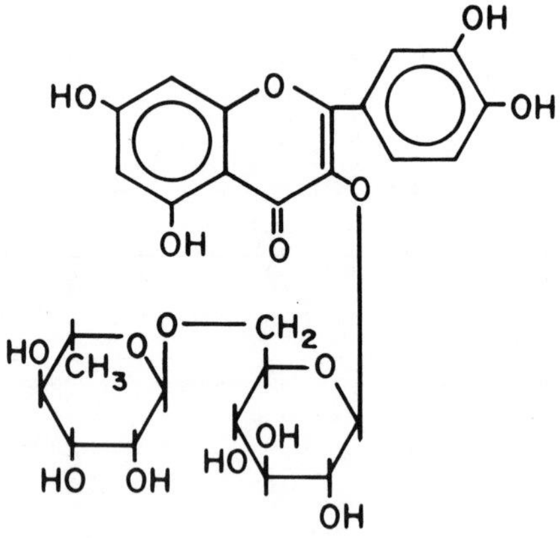

Figure 2.   Structure of rutin.

Discoloration

When plant tissues are disrupted by grinding, heating or blending with other ingredients such as occurs in food processing, the phenolic components usually undergo enzymatic and nonenzymatic reactions that can cause undesirable discoloration in the processed food. Figure 3 illustrates the effect of several types of plant protein products on the color of biscuits (2). Soybean and peanut flours used at the 20% replacement level do not cause a color problem. Sunflower, alfalfa leaf and cottonseed flours, on the other hand, cause a marked discoloration of the biscuits. In the case of the sunflower seed flour, this discoloration is believed to be due to chlorogenic acid. The coloration caused by the alfalfa leaf protein product has been attributed mainly to chlorophyll pigments (3) but polyphenols are also thought to contribute to the discoloration (4). Work at the Southern Regional Research Center (SRRC) has shown that flavonoids, gossypol and possibly other phenolic compounds contribute to the color observed when cottonseed flours are used in foods (2).

Interactions

There are two major types of interactions between proteins and phenolic compounds (5). First, phenols combine reversibly with proteins by hydrogen bonding. Second, they interact irreversibly by oxidation and condensation-type reactions.

Infrared spectral data indicate that hydrogen bonding between phenols and N-substituted amides is one of the strongest types of H-bonds (6). Research on collagen, vegetable tannins and synthetic polymers (5) has established the importance of the peptide linkage in the formation of H-bonded complexes between tannins and proteins. The interaction is probably between the peptide oxygen and the hydrogen of the phenol. Condensed tannins are bound to proteins almost independent of pH below pH 7.0 to 8.0. This bonding is thought to involve only un-ionized phenolic hydroxyl groups. Hydrolyzable tannins, on the other hand, are strongly bound to proteins at pH 3.0 to 4.0 but the binding decreases above pH 5.0. In this case, the stronger H-bonding has been attributed to the un-ionized carboxyl groups and the weaker H-bonding to the un-ionized hydroxyl groups of the tannins.

Oxidation of phenols is the first step in the irreversible type protein-phenol interaction. This oxidation can be enzymatic (as illustrated in Figure 4) or nonenzymatic. The primary reaction products are quinones which are themselves very reactive compounds. These quinones can then react directly with protein groups or can first undergo polymerization and/or oxidation-reduction reactions to produce secondary reaction products which can form covalent bonds with the proteins. The work of Mason and Peterson (7) has shown that sulfhydryl and N-terminal amino groups are the most reactive. The epsilon-amino group of lysine also

*Figure 3.   Biscuits containing 100% wheat and 20% plant-protein products.
(Reproduced from Ref. 2. Copyright 1981, American Chemical Society.)*

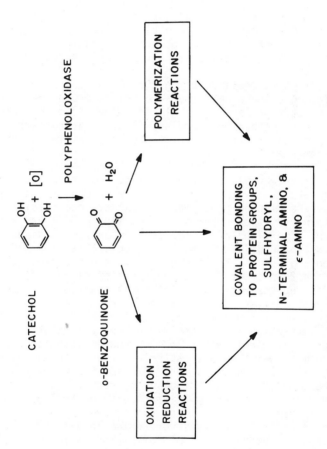

Figure 4.   Reaction sequences involved in covalent bonding of phenols to proteins.

reacts with quinones but more slowly. These quinone-type
reactions are most commonly cited as causing discoloration
problems in plant food products.

Cottonseed Flours

One of our interests at SRRC is the use of high protein
cottonseed flours as edible food products. Glanded cottonseeds
contain high concentrations of the phenolic compound gossypol,
enclosed in pigment glands. This compound is toxic to humans but
an edible flour can be made by removal of the glands in a process
developed at SRRC called the Liquid Cyclone Process (8).
Glandless cottonseeds do not contain gossypol and an edible flour
is prepared from this material by simple hexane extraction
methods.

Biscuits prepared using 20% cottonseed flours as a
replacement for wheat flour are yellow-brown in color (Figure 3).
The biscuit prepared with glanded flour is much darker in color
than that containing glandless flour. Our research work was to
isolate and identify the pigments responsible for this color
problem.

Experimental. Cottonseed flour was prepared from glanded
seeds at SRRC by the Liquid Cyclone Process of Gardner, et al.
(8). The glanded flour had a proximate composition of 57.3%
protein, 32.1% carbohydrate, 1.6% lipid, 7.7% ash and 2.3% crude
fiber. The glandless cottonseed flour was obtained from the
Plains Cooperative Oil Mill, Lubbock, Texas and had a proximate
compositon of 56.6% protein, 29.9% carbohydrate, 3.7% lipid, 7.5%
ash and 2.3% crude fiber. Analysis of glanded flour indicated
0.04% free gossypol and 0.15% total gossypol. Free and total
gossypol analyses of glandless flour indicated that its gossypol
content was below the detectable levels of the methods.

Extraction with petroleum ether, chloroform, 85% aqueous
isopropyl alcohol, water and 10% sodium chloride were conducted
at room temperature with stirring at a solid to solvent ratio of
1:10. Extraction times were 1 hr for the water and salt solution
systems and 24 hr for the other solvents. Solids were separated
from extracts either by filtration or centrifugation and residues
were washed with a volume of solvent equal to the original volume
used. Insoluble residues were dried in air (to remove petroleum
ether), in vacuum oven (chloroform) or by freeze drying after
removal of alcohol by rotary evaporation or salt by dialysis.
Solvents were removed from the extracts by rotary evaporation
and/or freeze drying and salt by dialysis.

Pepsin digestions were conducted on the salt insoluble
residues at 37°C for 24 hr at pH 2 using 1:100 ratio of enzyme to
substrate. Cellulase treatments were at 45°C for 24 hr at pH 4.5
with an enzyme ratio of 1:50. Digestions with amylase were run
at 75°C for 4 hr at pH 6.5 with an enzyme ratio of 1:25.

Insoluble residues were separated from digests by centrifugation, washed with water, neutralized and freeze dried. Digests were also neutralized and freeze dried.

Insoluble amylase residues were treated with dimethyl-sulfoxide (DMSO) with stirring for 1 hr at 160°C. The insoluble residue was collected on a filter, washed with 50% methanol, dispersed in water, neutralized and freeze dried. The DMSO extract was poured into a large volume of 50% methanol and stirred for about 1 hr. The precipitated solids were separated from the solution by centrifugation, washed with 50% methanol, dispersed in water, neutralized and freeze dried (fraction A). The non-precipitable fraction was isolated by rotary evaporation of the methanol-water-DMSO solution (fraction B).

Wheat flour biscuits were prepared with 20.0 g wheat or composite flour, 1.0 g baking powder, 0.5 g salt, 4.5 g shortening, and 15.0 g fluid milk. Biscuits were prepared from plant-protein flours based on 20% replacement for wheat flour. In biscuits prepared with fractions isolated from cottonseed flours, the quantities used were calculated using the percentages these fractions represented of the original flour, e.g. the salt solution soluble fraction of glanded flour was 35.7% of the original flour and 7.1 % (20 x .357) replacement for wheat was used to prepare the biscuit.

Other methods and procedures are fully described by Blouin et al. (9,10).

Solvent Extraction Steps. Glandless and glanded cottonseed flours were subjected to a series of five mild extraction steps. Biscuits were prepared with the extracted flours and in most cases with the extracts (Figure 5). The quantity of cottonseed material used in preparation of a biscuit was proportional to the amount that the fraction represented in the original flour. These percentages of the orginal flour are shown in Figure 5 in parentheses. In this way, the preparation of biscuits was used as a guide to determine if the pigments responsible for the color had been modified or lost and to establish in which fractions they occurred.

Figure 5 illustrates that extraction of glandless flour with petroleum ether and chloroform does not alter the color of bis-cuits. After extraction with aqueous alcohol, the yellow color is present in the biscuit containing the extract and the brown color is present in the biscuit containing the insoluble flour residue. We were able to establish that the yellow is caused by flavonoids present in the flours (9). Seven flavonol compounds were isolated and identified as glycosides of quercetin and kemp-ferol. No strong interactions between these phenolic components and the cottonseed proteins were evident since they were readily removed by simple room temperature extraction. The pigments responsible for the brown color, however, appear to be strongly bound to the more insoluble flour components. They remain in the

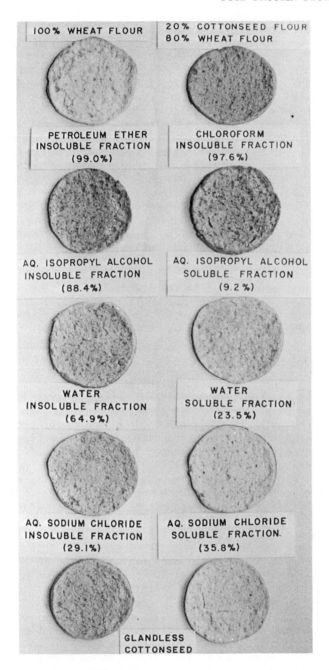

Figure 5. Biscuits containing glandless cottonseed flour fractions isolated in five solvent extraction steps. Values in parentheses represent the amount of cottonseed material used in preparation of the biscuit.

insoluble flour fractions after extractions with water and 10%
sodium chloride solution. At this point in the isolation procedure, the brown color causing pigments are contained in a
fraction which represents 29% of the orginal flour.

Figure 6 shows biscuits prepared from glanded flour fractions obtained by the same five extraction steps. The yellow
color due to the flavonoids is in the biscuit containing the
aqueous alcohol extract and the brown coloration remains in the
insoluble fraction through the salt solution extraction step.
Our initial assumption about the more intense brown color in
glanded flour was that it was caused by gossypol bound to the
cottonseed proteins. The original flour contained 0.1% bound
gossypol measured by an analytical method developed by W. Pons
and co-workers at SRRC (11).

This analytical method assumes gossypol is bound as a Schiff
base to the epsilon-amino groups of the lysine in proteins
(Figure 7). Although this bonding is not of the type previously
described for phenol-quinone-protein systems, it is a type quite
common in biological systems (12). In this analytical method,
the cottonseed flour is treated with 3-amino-1-propanol and
acetic acid in dimethylformamide (DMF) solution. Under these
conditions, it is assumed that the gossypol becomes bound to the
aminopropanol as the Schiff base of this low molecular weight compound. This gossypol-aminopropanol derivative can then be washed
from the flour and the gossypol content measured spectrophoto-
metrically. This reaction appeared to be a good means to test
the role of bound gossypol in the color problem.

The flour residues after extraction with salt solution were
treated with 2% aminopropanol and 10% acetic acid in DMF solution
for 15 min at 90°C. The treated fractions were washed, dried and
used in biscuits (Figure 8). Most of the brown color was removed
from the glanded flour fraction by this treatment. The color of
the biscuit prepared with the glandless flour fraction was not
altered by this aminopropanol treatment. These results suggest
that bound gossypol is responsible for the intense brown color
observed in the glanded cottonseed flour-containing biscuit.

Enzyme Digestion Steps. To further concentrate and isolate
the brown pigments in glandless and glanded flours, the salt
solution insoluble residues were subjected to a series of four
sequential enzyme treatments. Biscuits prepared from the soluble
and insoluble fractions of glandless flour are shown in Figure 9.
Pepsin digestion solubilized about 50% of the salt solution
insoluble residue. Treatment of the pepsin residue with a cellulase enzyme, which also contained hemicellulase activity, removed
an additional 50-60% from the insoluble fraction. This was
followed by a second pepsin digestion which removed 20-30% from
the cellulase residue. Treatment with amylase solubilized an
additional 20-25%. At each step of these treatments, biscuit
preparation showed that the brown color-causing pigments remained

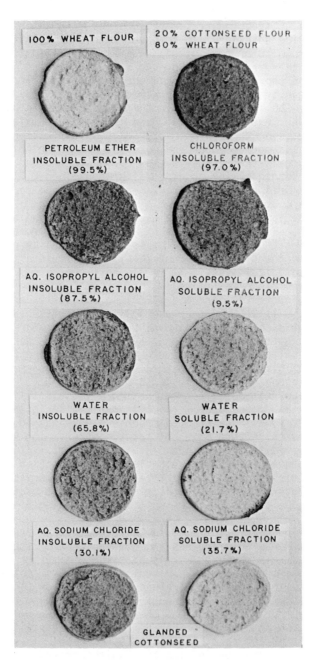

Figure 6. Biscuits containing glanded cottonseed flour fractions isolated in five solvent extraction steps.

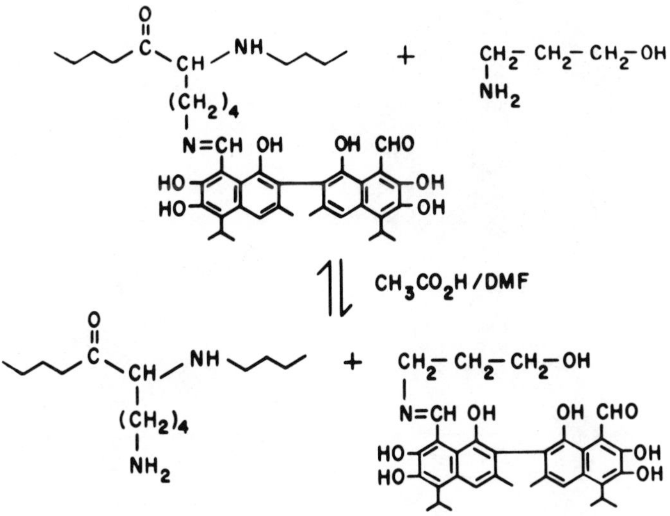

Figure 7. Reaction of protein-bound gossypol with 3-amino-1-propanol in the presence of acetic acid in DMF solution. (Reproduced, with permission, from Ref. 10. Copyright 1980, Institute of Food Technologists.)

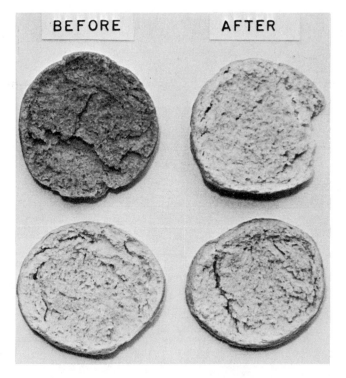

Figure 8.   Biscuits containing the salt solution insoluble fractions of glanded (top) and glandless (bottom) cottonseed flour before and after treatment with amino-propanol.

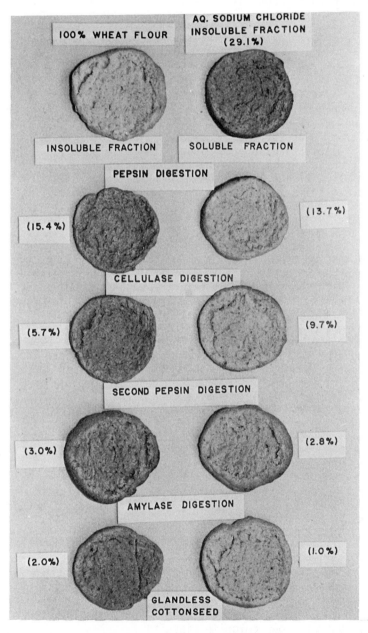

Figure 9. Biscuits containing glandless cottonseed flour fractions isolated in four enzyme digestion steps.

in the insoluble fractions. The amylase residue represented 2%
of the original flour. Glanded cottonseed flour responded in an
identical manner when subjected to the same four enzyme treatments. Biscuits prepared with these fractions are shown in
Figure 10. The last insoluble fraction in this case was 2.6% of
the original flour.

   Dimethylsulfoxide Treatment Step. The last step in the isolation procedure was to treat the amylase residues from glanded
and glandless flours with DMSO at 160°C for 1 hr. This treatment
solubilized 70 to 75% of the residues. For both flours, the DMSO
solution was filtered off to isolate the insoluble fraction. A
second fraction was isolated by precipitation on mixing of the
DMSO solution with a large volume of 50% methanol (soluble fraction A). A third fraction was obtained by removal of the
solvents on a rotary evaporator (soluble fraction B).
   Biscuits prepared from these DMSO treated fractions, illustrated in Figure 11, showed some coloration in all six biscuits.
However, a marked difference in color distribution is now evident
between the two types of flour. For the glanded flour, the
intense brown color is clearly present in the DMSO soluble fraction A. With the glandless flour, the major proportion of the
brown pigments is in the insoluble fraction. Also, a brown color
of similar hue and intensity is observed in the biscuit
containing the insoluble fraction of glanded cottonseed flour.
This indicated that the pigments responsible for the brown color
in glandless flour are also present in glanded flour. Chemical
analyses indicated that these insoluble fractions are about 40%
carbohydrate (presumably polysaccharide), 30% protein and 5% fat.
As yet, we have no experimental evidence that the brown color in
these insoluble fractions is caused by phenolic compounds.
Because of the extremely insoluble nature of these residues, however, we suspect that lignins or condensed tannins are present.
   The DMSO soluble fraction A of glanded flour is about 70%
protein. It also contains about 7% carbohydrates and 8% fat.
The fact that this fraction, which contains the intense brown
color characteristic of glanded flour, is mainly protein, is in
agreement with the assumption that gossypol bound to proteins is
responsible for this discoloration.
   The DMSO soluble fraction A of glandless flour is about 50%
protein. This lower proportion of protein is due to a higher fat
content (18%). This high lipid content and the amino acid profile data suggests that the DMSO soluble A fractions are mainly
membrane type lipoproteins (13).
   The DMSO soluble B fractions appear to be mixtures of proteins, 35 to 50%, and carbohydrates, 14 to 19%. They cause the
least amount of brown color in biscuits.
   The infrared spectra of these six fractions, shown in
Figure 12, mainly confirm the chemical data previously described.
The spectra of the DMSO insoluble residues are identical for both

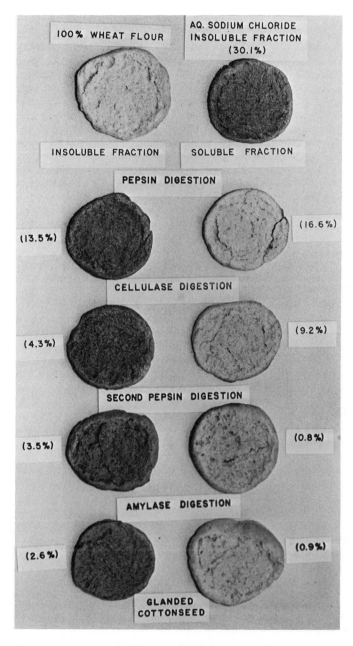

Figure 10. Biscuits containing glanded cottonseed flour fractions isolated in four enzyme digestion steps.

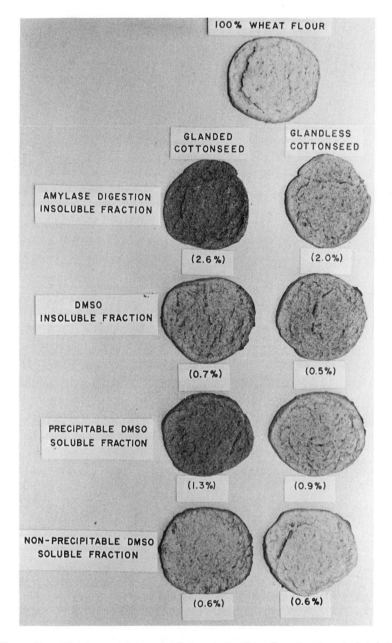

Figure 11.   Biscuits containing glanded and glandless flour fractions isolated in DMSO treatment step. Precipitable DMSO soluble fraction is Fraction A and non-precipitable DMSO soluble fraction is Fraction B in text.

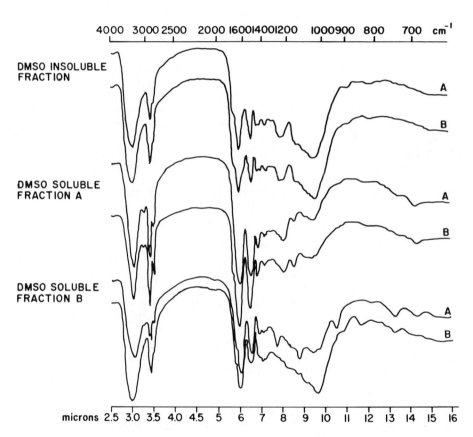

Figure 12.   IR spectra of DMSO treated fractions of glanded (A) and glandless (B) cottonseed flours.

flours. The -OH, -CH and -C-O- absorption regions indicate a predominance of carbohydrate components. For the DMSO soluble A
fractions, the -NH and amide I and II absorption regions indicate
a predominance of protein components. The -CH absorption region
confirms the presence of greater amounts of lipids in the glandless flour. There is, however, no indication of the presence of
the intense brown color causing components in the spectrum of the
DMSO soluble fraction A of the glanded flour. The spectra of the
DMSO soluble B fractions again confirm the presence of protein
and carbohydrate mixtures in these fractions. There are some differences between these two spectra which are unexplainable at
this time.

The ultraviolet-visible spectra of the DMSO soluble
fractions, shown in Figure 13, do indicate significant differences
between the two types of flours. The concentrations used were
0.2 mg/ml for the four flour fractions and 0.02 mg/ml for the
gossypol spectra. The glanded flour fractions exhibited significantly higher UV-visible absorption than the glandless flour
fractions. Even more significant, both of the glanded fractions
exhibited absorption in the 375-400 nm region similar to the 379
nm maximum exhibited by free gossypol. These data again suggest
that gossypol bound to protein is the cause of the intense brown
color in glanded flour-containing biscuits.

Sunflower Seed Flour

Chlorogenic acid (Figure 14) is the phenolic compound, which
according to several research groups (14-17), is responsible for
the discoloration problem observed when sunflower proteins are
used in food systems. The concentration of phenolic compounds in
sunflower seed flour is 3.0 to 3.5% (18). Chlorogenic acid and
caffeic acid constitute about 70% of these phenolics (18). As an
interesting comparison with cottonseed flour studies, the research
of M. A. Sabir and co-workers (14,15) on the chlorogenic acid-
protein interactions in sunflower is reviewed below.

Sabir et al. (14) extracted proteins from sunflower seed
flours with 2.5% sodium chloride at pH 7. They fractionated the
salt-extractable proteins on a Sephadex G-200 column and isolated
five protein fractions (Figure 15). When the undialyzed extracts
(Figure 15a) were used, peak V showed far greater absorption at
280 nm than the high molecular weight fractions. Peak V also
exhibited a UV-visible maximum at 328 nm which was not observed
in the spectra of the other fractions. If the salt extractable
proteins were dialyzed before fractionation, the elution curve
shown in Figure 15b was obtained. Peak V exhibited, in this case,
an absorption at 280 nm more in proportion to the other peaks.
This fraction still exhibited a second absorption maximum at 328
nm. Chlorogenic acid gives a UV-visible maximum at 328 nm and
other phenolic acids and esters absorb between 310 and 335 nm.
Presumably, phenolic components were present in fraction V, some

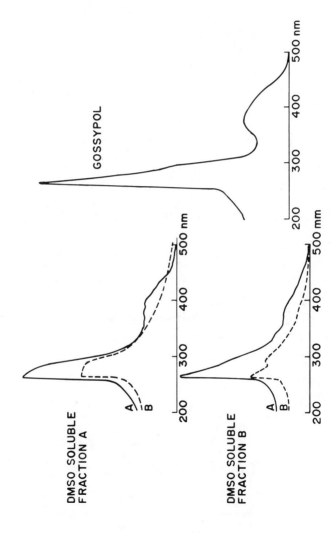

Figure 13.   UV-visible spectra of DMSO-soluble fractions of glanded (A) and glandless (B) cottonseed flours.

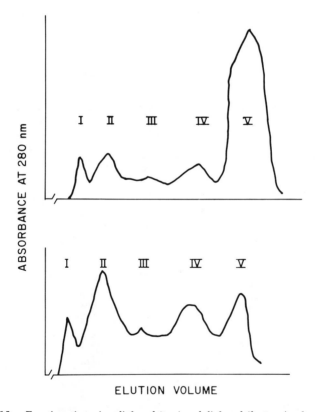

Figure 14.   Structure of chlorogenic acid.

Figure 15.   Fractionation of undialyzed (top) and dialyzed (bottom) salt-extractable sunflower proteins on a Sephadex G–200 column with neutral salt solution. (Reproduced from Ref. 14. Copyright 1973, American Chemical Society.)

of which were removed by dialysis. The phenolics that remain in fraction V after dialysis are strongly bound to this protein fraction.

Sabir and coworkers (15) refractionated fraction V on a Sephadex G-200 column with neutral salt solution and 7 M urea. They reasoned that a strong hydrogen-bonding medium such as 7 M urea would dissociate any phenolics that were hydrogen bonded to the proteins but would not influence any phenolic compounds covalently bound. As shown in Figure 16, they isolated two sub-fractions, V1 and V2. Fraction V1 exhibited a maximum at 280 nm characteristic of proteins but no absorption in the 328 nm region (Figure 17). This subfraction represented 68% of the proteins in fraction V. Fraction V2, on the other hand, showed absorption maxima at 280 and 328 nm. Since 7 M urea did not dissociate these bonds, these researchers concluded that chlorogenic acid was covalently bonded to this low molecular weight protein fraction. This fraction represented 4% of the original salt-extractable proteins.

Conclusions

Current research on sunflower and cottonseed flours indicates that phenolic compounds contribute to discoloration problems when these plant protein products are used in food systems. In the case of cottonseed flours, flavonoids cause a yellow coloration but these phenolics do not interact to any significant extent with the proteins. The phenolic compound gossypol in glanded cotton-seed flour is believed to be covalently bonded to membrane type lipoproteins and cause an intense brown color in food products containing this flour. Both glanded and glandless cottonseed flours contain pigments causing light brown color in foods. These pigments are bound to or are part of the most insoluble flour components. Lignin or condensed tannins may be the moieties responsible for this light brown color.

Sunflower seed protein products also cause marked discoloration in food products. Experimental evidence obtained by Sabir et al. (14, 15) indicates that strong hydrogen bonding and covalent bonding of chlorogenic acid to low molecular weight sunflower proteins occurs and that these interactions are probably responsible for the discoloration problem.

If plant protein products are to be used in food systems, discoloration problems must be solved. A knowledge of the nature of the bonding between the color-causing components and the proteins should lead to development of better methods for prevention and elimination of these problems.

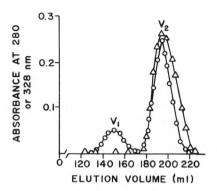

Figure 16.   Refractionation of salt-extractable sunflower protein Fraction V on a Sephadex G–200 column with neutral salt solution and 7 M urea. Key: O–O, 280 nm; and △ – △, 328 nm. (Reproduced from Ref. 15. Copyright 1974, American Chemical Society.)

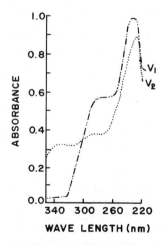

Figure 17.   UV spectra of Subfractions V_1 and V_2 of salt-extractable sunflower proteins. (Reproduced from Ref. 15. Copyright 1974, American Chemical Society.)

Acknowledgement

Names of companies or commercial products are given solely for the purpose of providing specific information; their mention does not imply recommendation or endorsement by the U.S. Department of Agriculture over others not mentioned.

Literature Cited

1. Griffins, L. Biochem. J., 1958, 70, 120.
2. Blouin, F. A.; Zarins, Z. M.; Cherry, J. P. Color, In: "Protein Functionality in Foods"; Cherry, J. P., Ed.; Amer. Chem. Soc. Symposium Series 147: Washington, D. C., 1981; p. 21.
3. Kohler, G. O.; Knuckles, B. E. Food Technol., 1977, 31(5), 191.
4. Lahiry, N. L.; Satterlee, L. D.; Hsu, H. M.; Wallace, G. W. J. Food Sci., 1977, 42, 83.
5. Loomis, W. D.; Battaile, J. Phytochem., 1966, 5, 423.
6. Flett, M. St. C. J. Soc. Dyers Colourists, 1952, 68, 59.
7. Mason, H. S.; Peterson, E. W. Biochim. Biophys. Acta, 1965, 111, 134.
8. Gardner, H. K., Jr.; Hron, R. J., Sr.; Vix, H. L. E. Cereal Chem., 1976, 53, 549.
9. Blouin, F. A.; Zarins, Z. M.; Cherry, J. P. J. Food Sci., 1981, 46, 266.
10. Blouin, F. A.; Cherry, J. P. J. Food Sci., 1980, 45, 953.
11. Pons, W. A., Jr.; Pittman, R. A.; Hoffpauir, C. L. J. Am. Oil Chem. Soc., 1958, 35, 93.
12. Singleton, V. L. Common Plant Phenols Other Than Anthocyanins, Contributions to Coloration and Discoloration, In: "The Chemistry of Plant Pigments"; Chichester, C. A., Ed; Academic Press: N. Y., 1972; p. 162.
13. Gurr, M. I., James, A. T. "Lipid Biochemistry: An Introduction"; Cornell Univ. Press: Ithaca, N. Y., 1972; p. 192.
14. Sabir, M. A.; Sosulski, F. W.; MacKenzie. S. L. J. Agr. Food Chem., 1973, 21, 988.
15. Sabir, M. A.; Sosulski, F. W.; Finlayson, A. J. J. Agr. Food Chem., 1974, 22, 575.
16. Sodini, G.; Canella, M. J. Agric Food Chem., 1977, 25, 822.
17. Cheftel, C.; Cuq, J. L.; Pronansal, M.; Besancon, P. Rev. Francaise des corps Gras, 1976, 23, 9.
18. Sabir, M. A.; Sosulski, F. W.; Kernan, J. A. J. Agr. Food Chem., 1974, 22, 572.

RECEIVED July 16, 1982.

Seed Protein Deterioration by Storage Fungi

JOHN P. CHERRY [1]

United States Department of Agriculture, Southern Regional Research Center,
Agricultural Research Service, New Orleans, LA 70179

Fungi are used to process a variety of seed materials into
various fermented food products, such as: 1) miso--a peanut
butterlike product prepared by fermenting mixtures of rice and
soybeans with Aspergillus oryzae or A. soyae and Saccharomyces
rouxii (1-4); 2) shoyu--a liquid food (soya sauce) prepared by
fermenting soybeans and rice with A. oryzae or A. flavus and
Zygosaccharomyces sp. yeast (5,6,7); 3) tempeh--a material from
soybeans fermented with Rhizopus oligosporus (8,9); 4)
ang-khak--a rice product fermented with Monasus purpureas and
used as a food coloring agent (10); and 5) ontjom--a peanut
presscake fermented by Neurospora sitophila (4,11). Enhanced
nutritive quality and digestibility of these fermented products
have been partially attributed to proteolytic activities of the
various fungi used in the fermentation processes (12-19).
   The aging processes of seeds that produce degradative
changes leading to seed deterioration are, in many cases,
enhanced by microorganisms (20-23). Pathogens produce metabolites that enhance deterioration of their host's cellular
structures (24-27). Cell membrane disruption allows solute
leakage and activation, for example, of autolytic host enzymes
such as proteases (27,28). This is in addition to increases in
the amounts and kinds of enzymes from the fungi (29,30,31).
Biochemical transformations include deletion of some proteins
and enzymes, intensification of others, and/or production of
new components as evidenced by quantitative and qualitative
changes in free amino acids and band patterns in polyacrylamide
electrophoretic disc-gels (32-37).
   This chapter examines the sequence of events initiated by
the inoculation of oilseeds and their products with selected
fungi (A. parasiticus, A. oryzae, A. flavus, R. oligosporus and
N. sitophila), which leads to the hydrolysis of proteins to
small polypeptides and/or aqueous insoluble components, then to
free amino acids. Changes in certain enzymes during the
infection period that are characterized by gel electrophoretic
techniques are also described.

[1]Current address: United States Department of Agriculture, Eastern Regional Research
Center, Philadelphia, PA 19118.

Experimental Procedures

Aspergillus parasiticus, NRRL A-16,462, A. oryzae, NRRL
1988, nine strains of A. flavus Link ex, N. sitophila, NRRL
2884, and R. oligosporus, NRRL 2710, were cultured on potato
dextrose agar slants between 24° and 29°C for 4 to 16 days.
Fungal spores were collected from the surface of the culture
slants with a sterile solution of 0.005% Span 20. After
removal of their skins, the peanut seeds of the cultivar,
Florunner, were soaked in an inoculum of each fungus for 1 min.
They were then placed in Petri dishes set in ventilated
containers lined with water-saturated absorbent cotton and
incubated at 29°C. Control uninoculated seeds were similarly
treated, omitting the fungi in the inoculation step. After
test periods ranging from 2 to 18 days, duplicate samples of
groups of three uninoculated and three of each fungi-infected
seeds were collected. They were individually ground in 7 ml of
sodium phosphate buffer (pH 7.9; I = 0.01) in a mortar with a
pestle, then centrifuged at 43,500 x g for 30 min to separate
soluble and insoluble fractions. Polyacrylamide disc- and
starch slab-gel electrophoresis were used to indicate the
qualitative and semi-quantitative changes that occurred in
proteins and/or selected enzymes in the soluble fractions of
control and inoculated seeds (30). All soluble and insoluble
samples were then lyophilized, ground into meals, and defatted
with diethyl ether. The quantities of proteins in the soluble
and insoluble fractions were determined by the macro-Kjeldahl
technique (N x 5.41). Defatted whole seed meals were used for
determination of free amino acid content. Measurements for
free amino acids were made utilizing the procedure of Young et
al. (38). All tests were conducted twice and labeled as experiments 1 and 2. In instances where there were no statistically
significant differences between experiments, data from both
experiments were combined for presentation in the text.

Results

Quantitative protein changes. Protein quantities decreased
in sodium phosphate buffer-soluble fractions of peanut seeds
during early stages of A. parasiticus, A. oryzae, N. sitophila
and R. oligosporus contamination; time intervals of fungi
growth ranged from 2 to 19 days (Table I; 32, 34-37).
Simultaneously, an increase in protein content occurred in
buffer-insoluble fractions. During longer growth periods,
fungi grew luxuriantly (e.g., A. parasiticus; Figure 1), and
most fungi-inoculated seeds showed a continued decrease in soluble proteins with a leveling off at the later stages of growth,
while others increased in soluble protein (A. oryzae). The
rate of these changes varied significantly between experiments
of A. parasiticus-inoculated seeds.
    The data imply different rates and/or mechanisms of peanut
protein hydrolysis by the fungi used in these experiments. The

Table I. Percentages of protein changes in soluble and insoluble fractions of peanut seeds inoculated with various fungi.

Treatment	Days Inoculated[1]							
	0		2-4		7-9		11-18	
	Soluble	Insoluble	Soluble	Insoluble	Soluble	Insoluble	Soluble	Insoluble
Noninfected	63	34	62	34	58	36	62	33
A. parasiticus:								
Experiment 1	63	34	38	54	26	53	27	49
Experiment 2	63	34	45	51	45	47	39	48
A. oryzae	63	34	45	46	64	39	68	31
N. sitophila	63	34	49	40	35	51	-[2]	-
R. oligosporus	63	34	36	53	31	54	-	-

1/Averages of 2-4, 7-9, and 11-18 days.
2/Not analyzed.

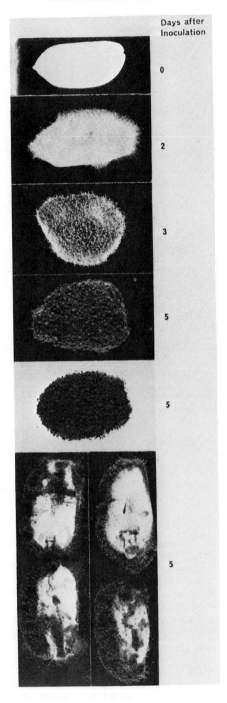

Figure 1. Typical peanut seeds contaminated with A. parasiticus *and then incubated for 0 to 5 days. Included are cross sections of seeds incubated for 5 days to show mold penetration. (Reproduced, with permission, from Ref. 34. Copyright 1974, Physiol. Plant Pathol.)*

organisms are known to exhibit strong and several types of proteolytic activity (13-19,29,30,31,34,39). The sustained increase in amount of proteins in soluble and insoluble fractions during the later fungi growth stages suggests that there is an accumulation of low molecular weight proteins, peptides and amino acids which have differing molecular properties and solubilities in the phosphate buffer. This may be due in part to increased levels of extractable and nonextractable fungal mycelial enzymatic proteins and related metabolites that are present in seed fractions during the later stages of the test periods. A significant contribution to the apparent increase in soluble protein in A. oryzae-contaminated seeds may be due to the accumulation of peanut protein hydrolysates that are not efficiently utilized by this fungus for growth and metabolism.

   Gel electrophoretic properties. Gel electrophoretic patterns of proteins in buffer-soluble fractions showed that each fungus caused specific changes in these storage components during the growth period (Figures 2, 3, and 4). No protein changes were noted in gel patterns of control seeds. In general, protein patterns of seeds contaminated with the various fungi, when compared to those of the noninoculated control, showed new protein components in region 0-1.0 cm and increased mobility and poor resolution of the major dark staining bands in region 1.0-2.0 cm, as fungi growth progressed. At the same time, bands normally located in region 2.0-3.5 cm disappeared, and a new group of polypeptides appeared in region 3.5-7.0 cm. During later stages of growth, many of the proteins became difficult to distinguish in the lower half of the gel patterns.
   The changes in protein properties on electrophoretic gels were more rapid in experiment 1 than experiment 2 with A. parasiticus, as was shown with the quantitative data (cf Table I and Figure 2). With A. oryzae, the major difference between the two experiments was that the smaller molecular weight compounds in region 2.0-6.0 cm seemed to disappear more rapidly in experiment 1 than 2 (Figure 3).
   The two protein bands in region 0.5-2.0 cm of electrophoretic gels shown in Figures 2, 3, and 4 have been characterized as the major storage globulin, arachin, which is present in aleurone grains of peanut seeds (40). These bands were not clearly discernable in the gel patterns after day 3; this was especially noted in experiment 1 with A. parasiticus (Figure 2). Evidently, as portions of the arachin components were hydrolyzed by a combination of proteases from the fungi and the peanut seeds, their sizes, conformations and electrical charges gradually changed resulting in an increase in electrophoretic mobility. Arachin is most likely being hydrolyzed to small polypeptide components and free amino acids.

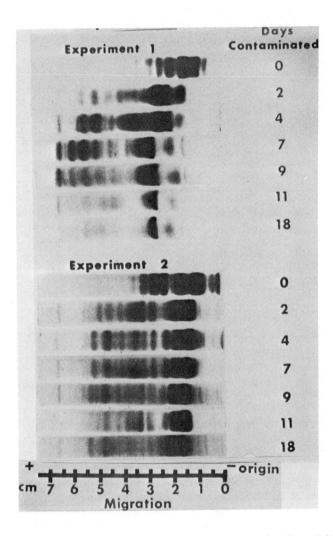

Figure 2. Polyacrylamide disc-gel electrophoretic patterns of buffer-soluble proteins from A. parasiticus-*inoculated peanut seeds of Experiments 1 and 2 after incubation for 0–18 days. (Reproduced, with permission, from Ref. 35. Copyright 1975,* Can. J. Bot.)

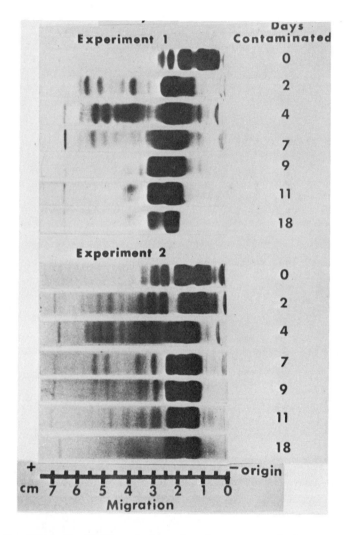

Figure 3. Polyacrylamide disc-gel electrophoretic patterns of buffer-soluble proteins from A. oryzae-inoculated peanut seeds of Experiments 1 and 2 after incubation for 0–18 days. (Reproduced from Ref. 36. Copyright 1976, American Chemical Society.)

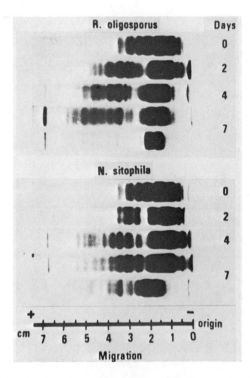

Figure 4. *Polyacrylamide disc-gel electrophoretic patterns of buffer-soluble proteins from* R. oligosporus- *and* N. sitophilia-*inoculated peanut seeds after incubation for 0–7 days. (Reproduced, with permission, from Ref. 32. Copyright 1976, Cereal Chem.)*

Free amino acids. Analysis of free essential amino acid contents of peanut seeds during a 2- to 7-day period of rapid growth of the various fungi used in this study showed that most of these components increased to levels greater than those observed in uninoculated seeds (Table II; 32,35,36,37). Similar observations were noted with many of the nonessential amino acids (32,35,36,37). During later stages, many of the free amino acids decreased in amount, most likely being used as nutrients by the fungi. These observations coincide with the deterioration of proteins to small molecular weight components and their disappearance from gel electrophoretic patterns.

Enzyme analyses. Gel electrophoretic patterns of buffer-soluble extracts from fungi-inoculated peanut seeds showed many biochemical transformations in enzymes (29,30,31,34). Examples of selected enzyme patterns of A. flavus-contaminated seeds are shown in Figure 5 (31). Changes included deletion of some enzymes, intensification of others and/or the production of new multiple molecular forms (31). These enzyme systems cause degradation of storage protein peptides (leucine aminopeptidases), hydrolysis of many ester linkages (esterases, acid and alkaline phosphatases) and oxidation reactions (phosphogluconate, alcohol, malate and glucose-6-phosphate dehydrogenases). As a result of enzyme changes, other studies suggested increased hormonal interaction and/or oxidation of organic substrates with hydrogen peroxide (peroxidases, oxidases) and decomposition of toxic substances such as hydrogen peroxide (catalases) (29,30,34). No doubt, these data on enzyme changes as fungal growth progresses in seeds, make it difficult to prepare vegetable protein products of known composition and stability during handling, storage and processing.

Discussion

The observation that proteins, enzymes and amino acid quantities in various preparations of peanut seeds inoculated with A. parasiticus, A. oryzae, A. flavus, R. oligosporus, or N. sitophila are different from those of noninoculated seeds expands presently known information on the effects of saprophytic organisms on plant tissues. For example, gel electrophoretic data show that water-soluble proteins from fungi-inoculated peanut seeds are hydrolyzed to their structural components during various test periods. New multiple molecular forms of enzymes appear, or if stored in inactive forms, are activated. These data also imply various rates and/or mechanisms of storage protein hydrolysis for the different fungi included in these studies.

Other studies have shown that although certain proteolytic enzymes may be common to different fungi, each species has the capacity to produce proteinases endemic to itself (13-19). Quantitatively, there are decreases and, in some cases, increases, in percentages of protein of soluble extracts during

Table II. Essential free amino acid changes in peanut seeds inoculated for time intervals of 0 to 7 days with various fungi.[1] A: Averages of free amino acid changes within each seed treatment for the entire 7-day test period. B: Free amino acid changes averaged for the five treatments within each time interval of the test period. (37)

Treatments	Threonine	Methionine	Isoleucine	Leucine	Phenylalanine	Lysine	Arginine
			Free Amino Acids. ($nM/100$ mg Fat-Free Meal)				
Noninfected	0.65b	0.51a	0.48b	0.45c	3.47a	0.44a	1.34b
A. parasiticus	0.74b	0.35a	0.59ab	1.01a	1.89b	0.80a	1.68b
A. oryzae	0.92a	0.40a	0.69a	1.10a	2.04b	0.79a	1.61b
R. oligosporus	0.62b	0.29a	0.45b	0.61bc	1.81b	0.59a	1.66b
N. sitophila	0.66b	0.39a	0.58ab	0.89ab	2.07b	0.80a	2.60a
Time intervals (days)							
0	0.35c	0.12b	0.40c	0.34c	2.52a	0.37b	0.90b
2	0.74b	0.58a	0.51bc	0.79b	2.25a	0.72a	2.05a
4	0.83ab	0.37ab	0.56b	0.89b	2.22a	0.78a	2.12a
7	0.95a	0.47a	0.76a	1.23a	2.03a	0.87a	2.00a

(A: applies to the first group of treatments; B: applies to the Time intervals group.)

[1] Valine, histidine and tryptophan are not lsited because they did not show any statistically significant changes within either A or B. Values having no common postscript letter in each amino acid column within A or B separately are significantly different ($P \leq 0.05$) different.

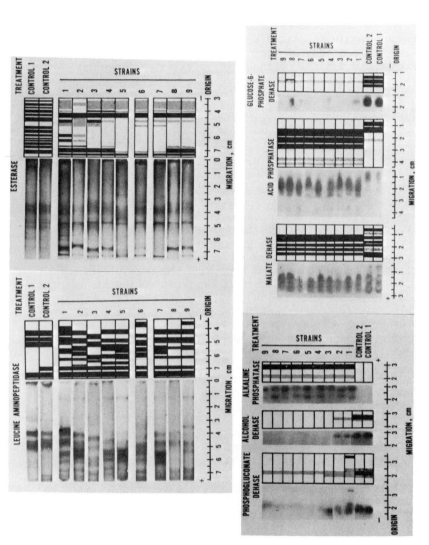

Figure 5.   Polyacrylamide disc- and starch slab-gel electrophoretic patterns of select enzymes from peanut seeds not inoculated (Controls 1 and 2) and inoculated with nine strains of A. flavus (31).

the test periods, regardless of the fungus used, while at the
same time an increase in these constituents occurs in insoluble
fractions. These changes are further confirmed by observations
showing that the total protein amino acid composition of the
soluble and insoluble fractions are continually changing
(32,35,36) and that free amino acid quantities are increasing.
Evidently, in the presence of these fungi the major storage
proteins of fungi-inoculated seeds are converted to free amino
acids and polypeptides of various sizes having different solubility characteristics. Changes in solubility of various
hydrolyzed products of proteins may also be related to their
differential interactions with other degraded constituents
stored in peanut seeds (oils, fatty acids, sugars, etc.; 33).

Previous to these studies on peanut seeds inoculated with
different fungi, most research on this subject was on the proximate composition of finished fermented products compared to the
nonfermented substrates (18,19,41,42,13). Other studies showed
that quantities and proportions of essential amino acids in
certain fermented products were greatly improved over those of
raw substrates (12). These improvements were partly attributed
to fungal digestion of proteins to their structural components,
which yielded more nutritious food products. Thus, while
hydrolyzed protein components in fermenting substrates serve as
primary sources of readily available nutrients for fungal
metabolism and growth, they in turn have the possibility of
improving the nutritional and functional properties of ferments
as foods or feeds.

This chapter showed that techniques normally used to prepare protein extracts from high quality oilseeds will not
necessarily produce fractions similar to those from seeds
inoculated with A. parasiticus, A. oryzae, A. flavus, R.
oligosporus, or N. sitophila. In fact, the resulting extracts
will depend on the species of fungus used and the length of the
growth period. Since these conditions affect the type and
quantity of proteins and amino acids in various peanut
extracts, they should also alter the nutritional and functional
properties of peanut protein products to different forms from
those of quality seeds. For example, fungus-inoculated peanut
seeds have greater quantities of certain essential amino acids
in soluble and insoluble protein fractions than those of
noninoculated seeds. In future studies, these factors need to
be considered in research to expand utilization in foods or
feeds of protein isolates or concentrates from various fungi-
inoculated oilseeds.

Conclusions

Fungi (e.g., A. parasiticus, A. oryzae, A. flavus, R.
oligosporus and N. sitophila) growing on oilseeds and their
products for various times greatly altered the protein's
solubility and gel electrophoretic properties. In aqueous
extracts of inoculated materials, protein quantities changed to

levels much lower or higher than those not inoculated with
fungi. Simultaneously, the amounts of insoluble proteinaceous
compounds changed to quantities greater or lesser than those
contained in soluble fractions. Gel electrophoresis of soluble
extracts from inoculated seeds showed that proteins were hydrolyzed to small molecular-weight components which eventually
disappeared as fungal growth progressed. A corresponding
increase in quantity of most free amino acids coincided with
the substantial alterations of proteins in both soluble and
insoluble fractions. During the interval of protein changes,
electrophoretic patterns of inoculated seeds showed that some
enzymes were deleted, others were intensified, and/or new
molecular forms of these constituents appeared. These data
suggested that inoculation of oilseeds and their products with
fungi initiated a sequence of events whereby enzyme composition
was changed, and proteins were hydrolyzed first to small
polypeptides and/or insoluble components, then to free amino
acids.

Acknowledgment

Use of a company and/or product named by the U.S. Department of
Agriculture does not imply approval or recommendation of the
product to the exclusion of others which may also be suitable.

Literature Cited

1. Shibasaki, K.; Hesseltine, C. W. J. Biochem. Microbiol.
   Technol., 1961, 3, 161.
2. Shibasaki, K.; Hesseltine, C. W. Dev. Ind. Microbiol.,
   1961, 2, 205.
3. Shibasaki, K.; Hesseltine, C. W. Econ. Bot., 1962, 16,
   180.
4. Hesseltine, C. W.; Wang, H. L. Biotechnol. Bioeng., 1967,
   9, 275.
5. Dyson, G. M. Pharmeceut., 1928, 121, 375.
6. Lockwood, L. B. Soybean Digest, 1947, 7(12), 10.
7. Yokotsuka, T. Adv. Food Res., 1960, 10, 75.
8. Steinkraus, K. H.; Yap, B. H.; Van Buren, J. P.;
   Provvidenti, M. I.; Hand, D. B. Food Res., 1960, 25, 777.
9. Djien, K. S.; Hesseltine, C. W. Soybean Digest, 1961,
   22(1), 14.
10. Palo, M. A.; Vidal-Adeva, L.; Maceda, L. Philipp. J.
    Sci., 1961, 89, 1.
11. Gray, W. D. Crit. Rev. Food Technol., 1970, 1, 225.
12. Hesseltine, C. W. Mycologia, 1965, 57, 149.
13. Plating, S. L.; Cherry, J. P. J. Food Sci., 1979, 44,
    1178.
14. Steinkraus, K. H.; Lee, C. Y.; Buck, P. A. Food Technol.,
    1965, 19, 1301.
15. Nakadai, T.; Nasuno, S.; Iguchi, N. Agric. Biol. Chem.,
    1972, 36, 1481.

16. Nakadai, T.; Nasuno, S.; Iguchi, N. Agric. Biol. Chem.,
    1972, 37, 2703.
17. Wang, H. L; Vespa, J. P.; Hesseltine, C. W. Appl.
    Microbiol., 1974, 27, 906.
18. Beuchat, L. R.; Young, C. T.; Cherry, J. P. Can. Inst.
    Food Sci. Technol. J., 1975, 8, 40.
19. Quinn, M. R.; Beuchat, L. R.; Miller, J.; Young, C. T.;
    Worthington, R. E. J. Food Sci., 1975, 40, 470.
20. Barton, L. V. "Seed Preservation and Longevity." Leonard
    Hill: London, England, 1961; 216 pp.
21. Christensen, C. M. Seed Sci. Technol., 1973, 1, 547.
22. Bothast, R. J. Fungal deterioration and related phenomena
    in cereals, legumes and oilseeds, In: "Postharvest
    Biology and Biotechnology"; Hultin, H. O.; Milner, M.,
    Eds.; Food and Nutrition Press, Inc.: Westport. Conn.,
    1978; p. 210.
23. Cherry, J. P. Phytopath., 1982, In press.
24. Harman, G. E.; Granett, A. L. Physiol. Plant Pathol.,
    1972, 2, 271.
25. Koostra, P. Seed Sci. Technol., 1973, 1, 417.
26. Roberts, E. H. Seed Sci. Technol., 1973, 1, 529.
27. Villiers, T. A. Ageing and the longevity of seeds in
    field conditions, In: "Seed Ecology"; Heydecker, W.,
    Ed.; The Pennsyl. State Univ. Press, Page Bros. (Norwich)
    Ltd.: Norwich, England, 1973; p. 265.
28. Ryan, C. A. Ann. Rev. Plant Physiol., 1973, 24, 173.
29. Cherry, J. P. Oilseed enzymes as biological indicators
    for food uses and applications, In: "Enzymes in Food and
    Beverage Processing"; Ory, R. L.; St. Angelo, A. J., Eds.;
    ACS Symposium Series No. 47, American Chemical Society:
    Washington, D.C., 1977; p. 209.
30. Cherry, J. P. Enzymes as quality indicators in edible
    plant tissues, In: "Postharvest Biology and Biotechnol-
    ogy"; Hultin, H. O.; Milner, M., Eds.; Food and Nutrition
    Press, Inc.: Westport, Conn., 1978; p. 370.
31. Cherry, J. P.; Beuchat, L. R.; Koehler, P. E. J. Agric.
    Food Chem., 1978, 26, 242.
32. Cherry, J. P.; Beuchat, L. R. Cereal Chem., 1976, 53, 750.
33. Cherry, J. P.; Beuchat, L. R. J. Amer. Oil Chem. Soc.,
    1976, 53, 551.
34. Cherry, J. P.; Mayne, R. Y.; Ory, L. R. Physiol. Plant
    Pathol., 1974, 4, 425.
35. Cherry, J. P.; Young, C. T.; Beuchat, L. R. Can. J. Bot.,
    1975, 53, 2639.
36. Cherry, J. P.; Beuchat, L. R.; Young, C. T. J. Agric.
    Food Chem., 1976, 24, 79.
37. Cherry, J. P.; Young, C. T.; Beuchat, L. R. Proc. Am.
    Peanut Res. Educ. Assoc., Inc., 1976, 8, 3.
38. Young, C. T.; Matlock, R. S.; Mason, M. E.; Waller, G. R.
    J. Am. Oil Chem. Soc., 1974, 51, 269.
39. Nakadai, T.; Nasumo. S.; Iguchi, N. Agric. Biol. Chem.,
    1972, 36, 261.

40. Basha, S. M. M.; Cherry, J. P. J. Agric. Food Chem.,
      1976, 24, 359.
41. Beuchat, L. R.; Worthington, R. E. J. Agric. Food Chem.,
      1974, 22, 509.
42. van Veen, A. G.; Graham, D. C. W.; Steinkraus, E. H.
      Cereal Sci. Today, 1968, 13, 96.

RECEIVED September 3, 1982.

Behavior of Proteins at Low Temperatures

OWEN FENNEMA

University of Wisconsin—Madison, Department of Food Science,
Madison, WI 53706

Protein behavior at below ambient temperatures has received
little attention compared to that of protein behavior at physiological temperatures or above. This is, of course, understand-
able, but nonetheless unfortunate, since proteins are believed by
many authorities to be involved in significant, if not substantial, ways with the ability or inability of living matter and
foods to tolerate a variety of low temperature situations.

*The ability of some animals to hibernate and others to
acclimate when exposed to a low, nonfreezing environment.
*The ability of some plants to "winter harden" and thereby
tolerate freezing conditions.
*The ability of some fish to avoid freezing at water tempera-
tures normally low enough to cause freezing.
*The ability of some microbial cultures and other small
biological specimens to survive freezing, frozen storage
and thawing.
*The inability of humans to tolerate low nonfreezing body
temperatures.
*The inability of large biological specimens such as whole
organs and small, whole animals to survive preservation by
freezing.
*The inability of plant and animal food tissue to withstand
commercial freeze preservation procedures without undergoing detrimental changes, particularly with respect to
water holding capacity and texture.

In addition, it should be noted that all too many researchers
pay insufficient attention to the effects of low temperature
storage on proteinaceous samples, assuming simply, and often
falsely, that the lower the temperature the better.

As will be documented in this Chapter, the behavior of
proteins at low temperatures often deviates from intuitive
expectations. Functional properties, especially enzyme activity,
can change nonlinearly as the temperature is lowered and this
unexpected behavior becomes even more erratic when cellular
systems are involved and/or freezing occurs. Changes in enzyme

0097-6156/82/0206-0109$07.75/0

activity resulting from a lowering of temperature can be reversible, partially reversible or irreversible depending on factors such as the kind of enzyme, the nature of the sample and the time-temperature scheme employed. From what has already been said, a reduction in temperature also would be expected to cause significant intra- and intermolecular changes in proteins, and these alterations do in fact occur.

The behavior of proteins in noncellular and cellular systems will be considered at temperatures covering above and below freezing conditions. Attention will be given to both changes in protein functionality (mainly enzyme activity) and changes in protein structure.

Experimental Procedures

One obvious reason for the limited amount of research being generated on proteins at low temperatures relates to the difficulties of conducting this kind of research, particularly in the presence of ice. Optical procedures that are relied on so heavily in unfrozen liquid samples are generally useless when an ice phase is present. Furthermore, gross nonhomogeneity exists in frozen samples and this nonhomogeneity changes with frozen storage. Procedures to accurately assess the characteristics of samples of this nature simply do not exist. Thus, many researchers evaluate the effects of freezing and frozen storage after thawing. Furthermore, all too many investigators fail to design their studies so that the effects of freezing rate can be isolated from the effects of freezing depth (nadir).

One approach to studying enzyme-catalyzed reactions at very low temperatures (down to about -100°C) involves the use of aqueous-organic solvents with freezing points sufficiently low to avoid ice formation (1, 2). This technique was developed primarily to study reaction intermediates and has proven reasonably successful. Some concern exists, however, that high concentrations of organic solvents might influence reaction pathways.

Factors Influencing Enzyme Activity at Low Temperatures

In this section, enzyme activity in simple, noncellular systems will be considered first and this will be followed by a discussion of enzyme activity in cellular systems.

Enzymes in noncellular, simple systems at low temperatures. Factors considered to be of significance in governing the activity of enzymes at low temperatures include nature of the enzyme, temperature dependence, cooling-freezing procedures, and composition of the medium.

1) Nature of the enzyme. It is well known that enzymes differ greatly in their response to a variety of conditions so it should be no surprise that marked differences in the behavior of

various enzymes are observed during cooling and freezing. Some
enzymes, for example, exhibit great tolerance to freeze-thaw
treatments, whereas others do not (3, 4, 5); and some exhibit
significant activity in partially frozen systems (e.g. oxidases
and lipases) whereas others do not (e.g. some proteinases) (6-10).
Based largely on a review of the Japanese and Russian literature,
Lozina-Lozinskii (11) concluded that freezing causes greater
changes in fibrillar proteins than in globular proteins. The
ramifications of this conclusion, of course, involve more than
just enzyme activity.

   2) Temperature dependence. Temperature is, of course,
a factor of primary importance in determining enzyme activity and
reaction rates. The relationship between the reaction rate constant and absolute temperature is expressed by the well known
Arrhenius equation ($k = se^{-\Delta Ha/RT}$), and it is normally expected
that Arrhenius plots (log k vs. 1/T) will be linear over reasonably small (20-30°C) temperature spans. This expectation is not
normally met, however, for enzyme suspensions at low or even moderately low temperatures (12, 13, 14). For example, nonlinear
Arrhenius plots have been observed at slightly above-freezing
temperatures for pyruvate carboxylase (15), glutamate carboxylase
(16), acetyl CoA-carboxylase (17), pyruvate orthophosphate
dikinase (18), lipase (10) and for phosphatase and peroxidase
(9). The behavior of lipase is shown in Figure 1. It has been
suggested that these data conform equally well to a smooth curve
and that more acceptable interpretations are possible in this
form (14). It also should be emphasized that the plot in Figure 1
is obtained both in the presence and absence of ice. Thus, the
independent effect of decreasing temperature can cause enzyme-
catalyzed reactions in simple systems to depart in a significant
negative fashion from the Arrhenius equation.

   It should not be assumed, however, that the effects of
temperature lowering and the effects of temperature lowering plus
the formation of ice always produce Arrhenius plots that are
superimposable. Freezing introduces many more complexities
(changes in pH, ionic strength, solute concentration, etc.) than
temperature lowering alone and the effects of freezing are therefore not always consistent with the pattern in Figure 1.
Although rare, freezing can, for example, actually cause an
increase in the rate constant for some enzyme-catalyzed reactions
(19, 20).

   The general pattern of a negative deviation from the
Arrhenius relationship as temperature is lowered (with or without
freezing) probably involves conformational changes in the enzyme
and/or an association-dissociation type transition that leads to
enzyme inactivity. In this case, one of the assumptions under-
lying the applicability to the Arrhenius relationship is violated
and linearity would not be expected.

   A second property of enzymes that is sometimes temperature
dependent is their affinity for substrate. Under appropriate

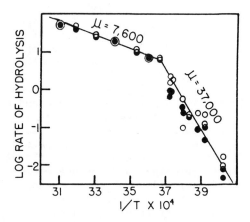

*Figure 1. Log rate of hydrolysis (i.e., log reciprocal of time required for 1%
hydrolysis) of tributyrin by lipase as influenced by temperature. (Reproduced, with
permission, from Ref. 10. Copyright 1942, Institute of Food Technologists.)*

circumstances, enzyme-substrate affinity is indicated by the
magnitude of the Michaelis constant (Km), and reports of a strong
temperature dependence of this value have been observed for
several enzymes including lactate dehydrogenases (21, 22), phosphophenol pyruvate carboxylase (23) and phenylalanine ammonia
lyase (24). The relationship between Km and temperature for
lactate dehydrogenases from various sources is shown in Figure 2.
In these examples, maximum apparent enzyme-substrate affinity
(minimum Km) corresponds very closely to the habitat or
acclimation temperature of the host from which the enzyme is
derived. A reversible change in conformation near the enzyme
active site could be responsible for this behavior (23). Since
substrate concentrations in cells are normally low (usually 1 mM
or less) and seldom, if ever, attain saturation levels for most
enzymes, the temperature dependence of Km values in living
systems is considered of far greater importance than the temperature dependence of V_{max} values (22).

Still another temperature-dependent property of enzyme-
catalyzed reactions is worthy of mention. This concerns the
ultimate extent to which these reactions proceed. It is commonly
observed that the ultimate accumulation of reaction products tends
to decrease as the subfreezing temperature is decreased. For
example, this result has been observed for denaturation of
chymotrypsinogen (25) and for reactions catalyzed by lipase (26),
invertase (27) and lipoxygenase (28), with data illustrating the
latter example appearing in Figure 3. Possible explanations for
this behavior include:

   *Reversible changes in enzyme activity stemming from
    reversible changes in enzyme conformation and/or associa-
    tion. These changes could occur because of the direct
    effects of temperature or the indirect effects of freezing
    (pH change, increase in ionic strength, etc.).
   *Restraints on diffusion of substrate, reaction products
    and/or enzyme. This could occur because of entrapment in
    ice and/or slow diffusion in the concentrated, high-
    viscosity, unfrozen phase. In liquid systems, viscosity is
    known to decrease as a function of 1/T, but the consequences
    on enzyme activity, even at very low temperatures, is not
    considered important (29). However, once ice formation
    occurs, viscosity increases abruptly to a very high level
    and the effect on enzyme activity, although not well known,
    is likely to be substantial.

      3) Cooling and freezing procedures. Cooling and
freezing procedures, especially the latter, can have pronounced
effects on enzyme activity. Factors such as rates of cooling,
freezing and warming, temperature nadir, and the time and temperature of frozen storage all are significant (3, 8, 30, 31). In
evaluating these effects, one should be aware that most investigators have measured enzyme activity following thawing rather
than during frozen storage, and have confounded the effects of
freezing rate and temperature nadir.

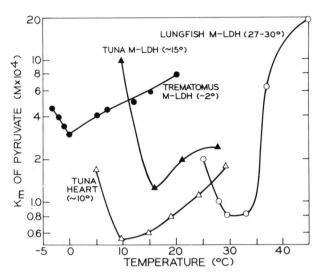

Figure 2. Temperature dependence of Michaelis constant (K_m) of pyruvate in reactions catalyzed by lactate dehydrogenases. Temperatures indicate average tissue temperatures. (Reproduced, with permission, from Ref. 21. Copyright 1968, Pergamon Press.)

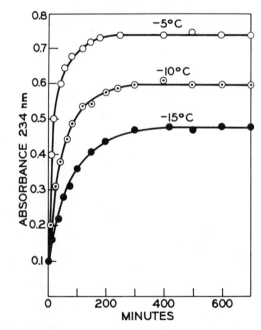

Figure 3. Effect of subfreezing storage temperatures on the ultimate accumulation of oxidation products of linolenic acid. (All samples were rapidly frozen to −78.5 °C, then stored at temperatures indicated.) Each value is a mean of four determinations. (Reproduced, with permission, from Ref. 28. Copyright 1980, Academic Press.)

4) <u>Composition of the sample</u>. Enzyme purity and concentration, pH of the medium, the kind and quantity of electrolytes present and the absence or presence of protective agents (such as proteins, glycerol, sugars, dimethyl sulfoxide, etc.) can greatly influence enzyme activity at low temperatures. For example, retention of activity of lactate dehydrogenase in a simple solution following thawing is favored when enzyme concentration is high, salt concentration is low, chlorides rather than phosphates are present, other proteins or glycerol are present, and pH is at the upper end of the range 6.0-7.8 (<u>31</u>, <u>32</u>). During freezing, the concentration of solutes in the unfrozen phase always increases and, as a result, properties such as pH (<u>31</u>, <u>33</u>, <u>34</u>), ionic strength, surface tension, concentration of dissolved gases and viscosity undergo marked changes. All of these properties have important influences on enzyme activity.

<u>Enzymes in cellular systems at low temperatures</u>. Enzyme activity in cellular systems is influenced by all of the factors previously discussed, with additional complications arising because of the cellular structure. Two complications of particular importance involve the behavior of enzymes during chilling injury of plants, and delocalization of enzymes during freezing of both plant and animal tissues.

It has long been recognized that plants of tropical and subtropical origins undergo detrimental physiological changes, known as "chilling injury," when exposed to low nonfreezing temperatures. Two occurrences associated with chilling injury are 1) alteration of membrane structure in organelles (<u>35</u>, <u>36</u>) and 2) discontinuities in, or at least a sharp change in the curvature of, Arrhenius plots for reactions catalyzed by membrane-bound enzymes. These discontinuities generally coincide well with the maximum temperature for the onset of membrane damage and chilling injury (<u>35</u>, <u>37</u>). The latter occurrence exemplifies protein behavior that is apparently unique to cellular systems and is worthy of further consideration.

Numerous enzymes, typically nonsoluble (e.g. associated with membrane systems or starch grains) and from chilling-sensitive plants, exhibit nonlinear Arrhenius plots at above-freezing temperatures. An example involving cytochrome oxidase activity in intact mitochondria from tomato is shown in Figure 4. The most attractive explanation for this behavior is that the enzyme is associated with a lipid, and it is temperature-induced changes in this association, perhaps changes in fluidity of the lipid and/or changes in the geometry of the lipid molecules, that produce a conformational change in the enzyme and reversible inactivation.

Although most reports of the anamolous behavior mentioned above involve enzymes that are attached to membranes or other cellular particles, attachment is not a prerequisite for this behavior. Enzymes experimentally detached from membranes of

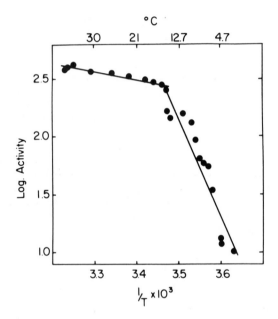

Figure 4. Arrhenius plot of cytochrome oxidase activity in mitochondria. (Reproduced, with permission, from Ref. 40. Copyright 1979, Academic Press.)

chilling-sensitive plants will exhibit this behavior (39) and
there is at least one report of a soluble enzyme, phosphophenol
pyruvate carboxylase from tomato (23), that exhibits this behavior. Since treatment of the enzyme with a detergent, or some
other substance that will remove lipids, will often (but not
always) cause the enzyme to yield linear Arrhenius plots, it is
the lipid-protein interaction, and perhaps the nature of the
lipid itself, that is believed responsible for the nonlinear
Arrhenius plots (38, 39, 40).

Although most investigators believe that proteins are not
directly involved in chilling injury (41) this view is not
universally held (23).

The behavior of enzymes during freeze-preservation can vary
greatly depending on whether the enzyme is located in a cellular
or noncellular environment. Many enzyme-catalyzed reactions in
cellular systems actually increase in rate during freezing. Examples include enzyme-catalyzed degradation of glycogen and/or accumulation of lactate in frog, fish, beef or poultry muscle (42-46);
enzyme-catalyzed degradation of high energy phosphates in fish,
beef and poultry muscle (42, 47-52); enzyme-catalyzed hydrolysis
of phospholipids in cod muscle (53); enzymic decomposition of
peroxides in rapidly frozen potatoes and in slowly frozen peas
(54); and oxidation of L-ascorbic acid in rose hips (55), straw-
berries (30) and Brussels sprouts (56). In noncellular systems,
this behavior is very uncommon, and has been observed only when
the sample is extremely dilute prior to freezing.

The increase in solute concentration that occurs in the
unfrozen phase during freezing is often suggested as being
responsible for instances of freeze-induced increases in reaction
rates. For enzyme-catalyzed reactions in cellular systems, this
explanation is inappropriate because:

*Enhanced reaction rates in frozen cellular systems often do
not decrease upon thawing (57, 58, 59). This would not be
so if freeze-induced concentrations of solutes were the
cause since thawing would reverse the concentration
effect.

*Enzyme-catalyzed reactions in noncellular systems rarely
accelerate during freezing, even though the freeze-concen-
tration phenomena is fully operative.

The most reasonable explanation for the freeze-induced rate
enhancement of enzyme-catalyzed reactions in cellular systems
involves membrane damage. It is well known that freezing can
disrupt membranes and thereby cause release of enzymes.
Documented instances include release of cytochrome oxidase from
mitochondria in beef and trout muscle and in chicken liver (60);
release of acid lipases from lysosomes from rainbow trout (61);
and release of malate and glutamate dehydrogenases, glutamate
pyruvate transaminase, glutamate or aloacetate transaminase,
aconitase and fumarase from mitochondria of bovine and porcine
muscle (62). It is also reasonable to expect that not only

enzymes would be delocalized by freezing, but also substrates and
other constituents that influence rates of enzyme reactions (8).
    Finally, it should be noted that chilling, or freezing and
frozen storage, can also alter the binding states (ionically
bound vs. soluble), and the relative concentrations of isoenzymes
in cellular systems (5, 63, 64).

Effect of Low Temperature on Protein Structure

    Although it is commonly believed that proteins become more
stable as the temperature is lowered, it has already been shown
that many proteins exhibit instability, as measured by partial
loss of functionality, at low, especially subfreezing, temperatures. In this section, attention will be directed to molecular
changes that proteins can undergo at low temperatures. These are
important because they are no doubt responsible for observed
alterations in protein functionality at low temperatures.

    Classification of molecular changes in proteins at low
temperatures.
                1) Dissociation of oligomers. According to Taborsky
(65) and Lozina-Lozinskii (11), dissociation of protein oligomers
into subunits is a primary response of enzymes to freezing.
Examples of proteins behaving in this manner include lactate
dehydrogenase (66), α-chymotrypsin (67), glutamate dehydrogenase
(68), an apoprotein of high-density serum lipoproteins (69),
glucose-6-phosphate dehydrogenase (70), pyruvate orthophosphate
dikinase (18), albumins and globulins (71) and deoxyribonucleo-
proteins (72).
                2) Rearrangement of subunits within oligomers. An
example of this kind involves lactate dehydrogenase (3).
                3) Aggregation. Low temperature association (a specific
form of aggregation usually involving subunits of oligomers and
specific bonding sites) of protein subunits is comparatively
rare. One example involves a transition of monomers to trimers
in a membrane proteinase from Streptococcus lactis (73). Fink
(2) also has found that association of enzyme subunits can occur
at low temperatures in aqueous-organic solvents.
    Gelation of egg yolk is routinely observed during freezing
and thawing (74, 75). Low temperature aggregation has also been
reported for 17β-hydroxysteroid dehydrogenase (76), urease (77)
and myosin (78). Casein micelles in milk can also aggregate
during frozen storage. In the early stages, a flocculent
precipitate forms, and in the more advanced stages, a gel (79).
                4) Changes in conformation. Reports of conformational
changes in proteins at low temperatures are reasonably abundant
(11, 12, 80). Examples include phosvitin (81), actomyosin (82),
myosin B (83), virions of southern bean mosaic virus (84), snake
venom L-amino oxidase (85) and numerous instances of denaturation
cited by Brandts (12).

Discussion of changes in protein structure at low temperature. The types of changes cited above can be best understood if
two premises, that appear to have general applicability to these
situations, are accepted. First, it is generally agreed that the
tertiary native structure of protein is maintained by several
types of interactions, namely hydrophobic, hydrogen-bonding,
electrostatic and van der Waals forces (86). Among these,
hydrophobic interactions are regarded as being of primary
importance in most proteins (65, 80, 87) and the strength of
these interactions is weakened by lowering the temperature.
Therefore, lowering the temperature will tend to produce
structural changes in those protein molecules that are dependent
on hydrophobic interactions for maintenance of native structure
(65, 87). It should also be recognized that this statement
applies equally well to both association-dissociation transitions
and to changes in conformation (88). Although this clearly
applies to the situations being considered here, it should nevertheless be recognized that the simple relationship between hydrophobic interactions and temperature is not capable of accounting
for all protein structural changes that are observed at low
temperatures. For example, factors other than temperature have
profound effects on the structure of proteins, and some of these
factors (pH, ionic strength, surface tension, protein concentration, concentration of nonprotein solutes) change substantially
during freezing.

Second, a thermodynamic approach to the molecular behavior
of proteins at low temperatures provides information of great
importance and must therefore occupy a position of central
importance in this discussion. All of the protein alterations
mentioned previously involve the transfer of some portion of a
protein molecule from an aqueous environment to an essentially
organic environment (to the interior of a folded protein) or the
reverse. Dissociation and unfolding involve increased exposure
of protein surface to an aqueous environment (increased protein-
water interaction). Whether dissociation and/or unfolding will
actually occur depends solely on the change in total free energy.
If the change in free energy for increased interaction with water
is negative, the dissociation-unfolding type transitions will
occur spontaneously, but nothing will be known about rate. If
the change in free energy of increased water interaction is
positive, the association-folding-aggregation type transitions
will occur spontaneously, but again, nothing will be known about
rate. These statements, of course, apply to the overall behavior
of the sample, not to individual molecules. Furthermore, changes
in temperature can alter whether the transitions just mentioned
will result in negative or positive changes in free energy, so at
one temperature, increased water interaction may be favored and
at another temperature, the reverse may be true.

It is now instructive to examine some quantitative data for
thermodynamic properties of globular proteins exposed to low

temperatures. At the outset, it should be emphasized that data
is not abundant since the thermodynamic prerequisite of reaction
reversibility severely limits the conditions that can be used
experimentally. Furthermore, the data presented here were
developed on the assumption that the transitions under consideration (association⇄dissociation; native⇄denatured) involve only
two states. Reaction intermediates, if they occur, must not be
stable under the conditions used. This assumption is considered
reasonable for many proteins, including those considered here,
but not necessarily for all proteins (12, 89, 90, 91).

Shown in Figure 5 is the temperature dependence of the
standard free energy ($\Delta G°$ or $\Delta F°$) of denaturation for chymotrypsinogen at three different pH values. Conditions were such that
association did not occur and observations therefore pertain to
native⇄unfolded transitions of monomers. Since accurate values
of $\Delta G°$ could be directly determined only within the range of
±2,000 calories, the remainder of the data were calculated from
a four-term power series involving absolute temperature (92).
The data in Figure 5 clearly indicate that $\Delta G°$ is not a linear
function of temperature as had been frequently assumed prior to
publication of these data. Instead, the relationship is parabolic
in nature, with a temperature of maximum stability (T_{max}) occurring at about 10-12°C, i.e. raising or lowering the temperature
from T_{max} can result in denaturation (unfolding). Thus, low
temperature denaturation is clearly possible.

Lowering the pH of the sample causes the entire curve to
shift to a more negative value of $\Delta G°$ and thereby reduce the
temperature at which a value of $\Delta G° = 0$ (for the high-temperature
leg of the graph) is attained. According to Brandts (12), the
essential features of these data are undoubtedly typical of
denaturation reactions for most proteins.

The data in Figures 6 and 7 provide clear evidence that the
conclusions drawn from Figure 5 are valid. The data in Figure 6
relate to chymotrypsin at pH 1.47 and it is evident that this
protein does, in fact, exhibit maximum stability at about 12°C,
with increased denaturation occurring at higher or lower temperatures. T_{max} for chymotrypsin is therefore almost identical to
T_{max} for chymotrypsinogen in Figure 5.

Figure 7 is a van't Hoff plot (log equilibrium constant, K =
unfolded/native vs. 1/T) for β-lactoglobulin A showing T_{max} at
about 35°C (90). A sufficient number of other investigators have
reported incidences of low temperature denaturation to verify
that this behavior is a reasonably common occurrence (12, 25).

From the slopes of the curves in Figure 7 it is possible to
calculate the enthalpy of transition from the native to the
denatured state, assuming a two-state process at all temperatures.
From such calculations, it is found that the unfolding reaction
for β-lactoglobulin A is exothermic below 35°C and endothermic
above this temperature (90).

Calculated thermodynamic parameters for the α-chymotrypsino-

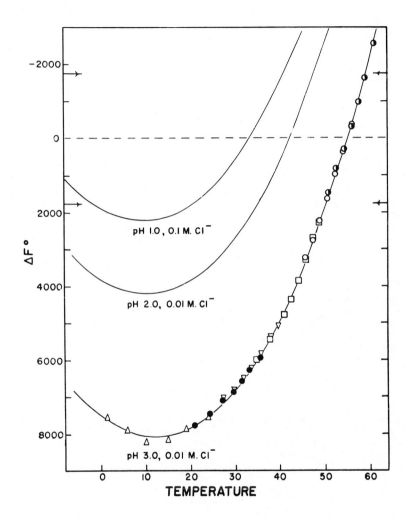

Figure 5.   Temperature dependence of the free energy of denaturation of chymo-trypsinogen at different pH values and ionic strengths. (Reproduced from Ref. 92. Copyright 1964, American Chemical Society.)

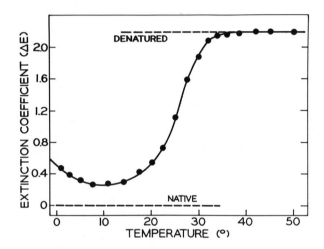

Figure 6. Temperature dependence of the extinction coefficient at 293 nm for chymotrypsin at pH 1.47. The dashed lines at the top and bottom of the plot indicate the extinction coefficients for the denatured and native states, respectively. (Reproduced, with permission, from Ref. 12. Copyright 1967, Academic Press.)

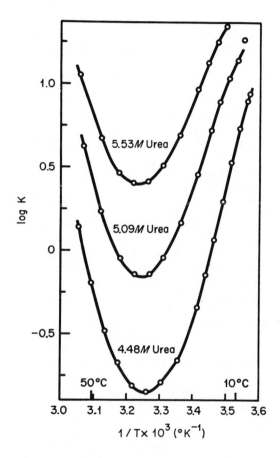

Figure 7. Van't Hoff plots (logarithm of the equilibrium constant, K = $\frac{unfolded}{native}$, vs. reciprocal of absolute temperature) for β-lactoglobulin A in urea at various concentrations. (Reproduced from Ref. 90. Copyright 1968, American Chemical Society.)

gen data (native⇌unfolded) in Figure 5 (pH 3.0) are shown in
Figure 8 (92). Values for ΔH° and ΔS° are highly temperature
dependent, both exhibiting large positive values at 65°C,
decreasing to values of zero at about 10°C and becoming negative
below 10°C.

The change in heat capacity (ΔCp) between the native and
unfolded forms at a given temperature is large and positive for
all temperatures in Figure 8. The large positive values of ΔCp
are responsible for the temperature dependency of both

$$\Delta H° \ (\frac{d(\Delta H)}{dT} = \Delta Cp) \ \text{and} \ \Delta S° \ (\Delta S = \int_{T_1}^{T_2} \frac{nCp}{T} \ dT)$$

and for the occurrence of a temperature of maximum protein
stability (Figures 5 and 7) (88, 90, 92, 93). Although ΔCp is
temperature dependent in Figure 8, this is not especially
important nor is this characteristic true for all denaturation
reactions (88).

The large heat capacity values apparently arise from exposure
of apolar groups to water during the denaturation process. In
water, apolar groups exhibit a partial molal heat capacity that
is about three times greater than their heat capacity in an
organic environment, such as in the interior of the native
protein. Consequently, considerable energy must be expended
during temperature increases to partially disrupt the structural
water existing around apolar groups at low temperature (25).

The general importance of the heat capacity effect at
atmospheric pressure was emphasized by Edelhock and Osborne (80)
in their statement that ". . .most, if not all, protein reactions
in which nonpolar groups are exposed to water, i.e., denaturation,
ligand dissociation, the dissociation of macromolecules or
microscopic structures composed of a large number of subunits,
are controlled, in large part, by a positive heat capacity change
which appears to increase with temperature."

Dissociation of oligomeric proteins into subunits at low
temperatures probably occurs more easily than simple unfolding
since the binding forces involved in dissociation are not as
strong as those involved in unfolding. In Figure 9 the temperature dependence of heats of dissociation (enthalpies, ΔH°) are
shown for glutamate dehydrogenase at pH 7 (68), α-chymotrypsin
at pH 4.12 (67) and apo A-II protein at pH 7.4 from high-density
serum lipoprotein (69). The enthalpies for glutamate dehydrogenase and α-chymotrypsin are negative (exothermic) at low
temperatures and positive (endothermic) at high temperatures.
The enthalpy of dissociation for carboxymethylated (Cm) apo A-II
is negative below, and positive above, 23°C, and contrary to
glutamate dehydrogenase and α-chymotrypsin, which yield globular
molecules upon dissociation, Cm apo A-II dissociates into
molecules that are randomly coiled. Thus, dissociation in this
instance is accompanied by major changes in secondary and
tertiary structure.

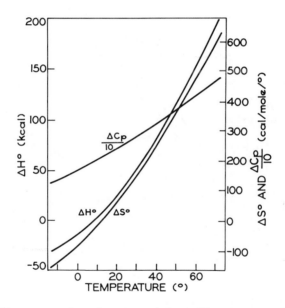

Figure 8. Temperature dependence of enthalpy ($\Delta H°$), entropy ($\Delta S°$), and heat capacity (ΔC_p) pertaining to denaturation of chymotrypsinogen. (Reproduced from Ref. 92. Copyright 1964, American Chemical Society.)

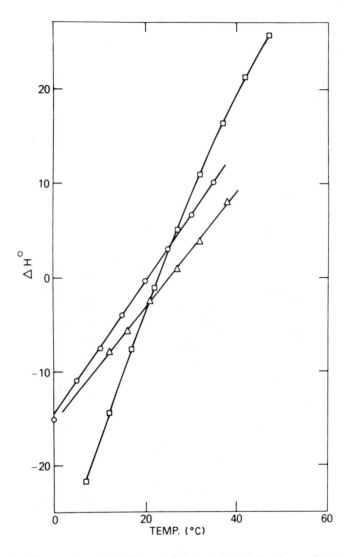

*Figure 9. Temperature dependence of heats of dissociation for α-chymotrypsin
(○) (67), glutamate dehydrogenase (△) (68), and reduced and carboxymethylated
apo A-II protein (□) (69). (Reproduced, with permission, from Ref. 80. Copyright
1976, Academic Press.)*

As mentioned at the beginning of this section, examples do exist for both association and dissociation type reactions in proteins exposed to low temperatures, and the type of transition favored will be that which exhibits a decrease in free energy. From the expression $\Delta G = \Delta H - T\Delta S$ (constant T and P) it is evident that ΔG will become more negative as ΔH becomes more negative (exothermic) or ΔS becomes more positive (greater disorder or greater probability of existence). Situations do exist where ΔH is negative and ΔS is positive, but it is more common for both terms to have the same sign (94).

Based on the data presented, the thermodynamic properties of unfolding and dissociation reactions for proteins can be summarized as shown in Table I. The two assumptions mentioned earlier apply to these data. Since these assumptions are considered valid for many proteins (12), these data, and the associated behavior depicted in Figures 5-7, can be considered reasonably typical.

Omitted from Table I are the association type transitions that some proteins undergo at low temperatures. In these instances, interactions (electrostatic, hydrogen bonding, disulfide interchange) other than hydrophobic may play decisive roles, or hydrophobic interactions may be influenced significantly by factors other than temperature (pH, ionic strength, etc.). Thus, the data in Table I are an oversimplification, but they nevertheless are believed to represent, reasonably well, the behavior of many proteins at low temperatures.

Beginning with a discussion of the data at intermediate to high temperatures (row 1 of Table I), it is evident that the enthalpies and entropies of transition to the undissociated or unfolded state are both positive. Either of these transitions are favored ($-\Delta G$) when the temperature is high enough and ΔS is large enough so that $T\Delta S > \Delta H$. The entropy term, in all instances in Table I is composed of two components, one involving protein structure and the other water structure. Unfolding or dissociation would involve a positive change in entropy for the protein and a negative change in entropy for water, the latter occurring because groups exposed to water during unfolding or dissociation are predominantly hydrophobic, thus promoting increased water structure (88). As the temperature is raised in the region above the temperature of maximum stability (T_{max}) for the protein, water exhibits a reduced tendency to associate in an ordered manner around hydrophobic groups, and more hydrophobic groups become exposed. Thus, the change in protein structure becomes an increasingly dominant component of ΔS, and ΔS assumes a larger positive value, and ΔG a larger negative value, as the temperature is raised above T_{max}.

At T_{max}, the values of ΔH and ΔS both decrease to near zero (row 2, Table I and Figure 8). Since ΔG assumes the largest possible positive value at this temperature, ΔH and ΔS cannot both have values of zero. Small changes in either ΔH or ΔS near

Table I. Thermodynamics of protein unfolding or dissociation
when hydrophobic interactions are of major importance[1]

Temperature	Thermodynamic data				
	ΔG	ΔH	ΔS	$\|T\Delta S/\Delta H\|$	ΔCp[2]
Sufficiently high to cause disordering	−	+	+	>1	~+4 @55°C
Temperature of maximum stability	+	~0	small −	>1	~+2 @10°C
Sufficiently low to cause signif- icant disordering	small +	−	−	>1	~+1.5 @−10°C
Still lower temperature with more disordering	−	−	−	<1	?

[1] Based on data from Edelhoch and Osborne (80), Tanford (88), and Brandts (92).

[2] Applies to chymotrypsinogen unfolding (92). See also Figure 8.

a value of zero can result in various combinations of the two properties that will yield a positive ΔG. In Table I, a positive ΔG is achieved by setting ΔH equal to zero and assigning a slight negative value to ΔS. This combination appears to be consistent with the temperature-dependent trends in Table I.

As the temperature is lowered below T_{max}, ΔH and ΔS become increasingly negative (row 3, Table I; Figure 8) and ΔG becomes less positive (Figure 5). As long as ΔG remains positive on the low temperature side of T_{max}, this must result from $|T\Delta S| > |\Delta H|$, with the ΔS term being dominated by the change in water structure. At still lower temperatures (row 4, Table I) ΔG presumably becomes negative (log K > 0, where $\Delta G = -RT \ln K$; Figure 7), in which case the absolute value of $\Delta H(-)$ must exceed the absolute value of $T\Delta S$ (-).

Conclusions

Possible adverse effects of low temperatures on proteins include delocalization of proteins in cellular systems and a loss of protein solubility and functionality. Factors influencing the degree of damage during freezing include the type of protein, the complexity of the system (cellular vs. noncellular), storage temperature (related to water activity), storage time, and the conditions of freezing and thawing. Proteins at low temperature may undergo changes in conformation and the degree of association, with the degree of reversibility depending on the conditions. Causative factors influencing protein damage during freezing include changes in salt concentration, changes in pH, mechanical effects of ice on membranes and dehydration effects.

Different proteins differ greatly in their responses to low temperatures, and the behavior of a single protein at a given low temperature can be unusual and difficult to predict. The latter is true because low temperatures can have a direct influence on protein structure and functionality, and because low temperatures, especially subfreezing temperatures, can induce dramatic changes in the protein's environment. A proper perspective is perhaps best achieved by recognizing that a great deal remains to be learned about the much-studied behavior of proteins at ambient to high temperatures, and that far more remains to be learned about the little-studied behavior of proteins at low temperatures.

Acknowledgments

Research supported by the College of Agricultural and Life Sciences, University of Wisconsin-Madison.

Literature Cited

1. Douzou, P. Adv. Enzymol., 1980, 51, 1.
2. Fink, A. L. In "Proteins at Low Temperatures"; Fennema, O., Ed.; ACS Advances in Chem. Series: Washington, D.C., 1979; p. 35.
3. Chilson, O. P.; Costello, L. A.; Kaplan, N. D. Fed. Proc., 1965, 24 (2), Suppl. 15, S-55.
4. Duke, S. H.; Koukkari, W. L.; Soulen, T. B. Physiol. Plant., 1975, 34, 8.
5. Krasnuk, M.; Jung, G. A.; Witham, F. H. Cryobiology, 1976, 13, 375.
6. Balls, A. K.; Tucker, T. W. Ind. Eng. Chem., 1938, 30, 415.
7. Balls, A. K.; Lineweaver, H. Food Res., 1938, 3, 57.
8. Fennema, O. In "Dry Biological Systems"; Crowe, J. H.; Clegg, J. S., Eds.; Academic Press: New York, 1978; p. 297.
9. Maier, V. P.; Tappel, A. L.; Volman, D. H. J. Amer. Chem. Soc., 1955, 77, 1278.
10. Sizer, I. W.; Josephson, E. S. Food Res., 1942, 7, 201.
11. Lozina-Lozinskii, L. K. "Studies in Cryobiology"; John Wiley & Sons: New York, 1974; p. 202, 208, 209.
12. Brandts, J. F. In "Thermobiology"; Rose, A. H., Ed.; Academic Press: New York, 1967; p. 25.
13. Han, M. H. J. Theor. Biol., 1972, 35, 543.
14. Kavanau, J. L. J. Gen. Physiol., 1950, 34, 193.
15. Scrutton, M. C.; Utter, M. F. J. Biol. Chem., 1965, 240, 1.
16. Shukuya, R.; Schwert, G. J. Biol. Chem., 1960, 235, 1658.
17. Numa, S.; Ringelmann, E. Biochem. Z., 1965, 343, 258.
18. Shirahashi, K.; Hayakawa, S.; Sugiyama, T. Plant Physiol., 1978, 62, 826.
19. Grant, N. H.; Alburn, H. E. Nature (London), 1966, 212, 194.
20. Tong, M-M.; Pincock, R. E. Biochemistry, 1969, 8, 908.
21. Hochachka, P. W.; Somero, G. N. Comp. Biochem. Physiol., 1968, 27, 659.
22. Somero, G. N. Amer. Naturalist, 1969, 103, 517.
23. Graham, D.; Hockley, D. G.; Patterson, B. D. In "Low Temperature Stress in Crop Plants"; Lyons, J. M.; Graham, D.; Raison, J. K., Eds.; Academic Press: New York, 1979; p. 453.
24. Uritani, I. In "Postharvest Biology and Biotechnology"; Hultin, H. O.; Milner, M., Eds.; Food and Nutrition Press: Westport, CT, 1978; p. 136.
25. Brandts, J. F.; Fu, J.; Nordin, J. H. In "The Frozen Cell"; Wolstenholme, G. E. W.; O'Connor, M., Eds.; J. and A. Churchill: London, 1970; p. 189.
26. Parducci, L. G.; Fennema, O. Cryobiology, 1978, 15, 199.
27. Parducci, L. G.; Fennema, O. Cryobiology, 1979, 16, 578.
28. Fennema, O.; Sung, J. C. Cryobiology, 1980, 17, 500.

29. Douzou, P. Mol. Cell. Biochem., 1973, 1, 15.
30. Fennema, O. In "Water Relations in Foods"; Duckworth, R. B., Ed.; Academic Press: New York, 1975; p. 397.
31. Soliman, F. S.; van den Berg, L. Cryobiology, 1971, 8, 73.
32. Greiff, D.; Kelly, R. T. Cryobiology, 1966, 2, 335.
33. van den Berg, L. Arch. Biochem. Biophys., 1959, 84, 305.
34. van den Berg, L.; Rose, D. Arch. Biochem. Biophys., 1959, 81, 319.
35. Raison, J. K. J. Bioenerg., 1973, 4, 285.
36. Raison, J. K.; Chapman, E. A. Austr. J. Plant Physiol., 1976, 3, 291.
37. Zeylemaker, W. P.; Jansen, H.; Veeger, C.; Slater, E. C. Biochim. Biophys. Acta., 1971, 242, 14.
38. Downton, W. J. S.; Hawker, J. S. Phytochemistry, 1975, 14, 1259.
39. Duke, S. H.; Schrader, L. E.; Miller, M. G. Plant Physiol., 1977, 60, 716.
40. Waring, A.; Glatz, P. In "Low Temperature Stress in Crop Plants"; Lyons, J. M.; Graham, D.; Raison, J. K., Eds.; Academic Press: New York, 1979; p. 365.
41. Lyons, J. M.; Graham, D.; Raison, J. K., Eds. "Low Temperature Stress in Crop Plants"; Academic Press: New York, 1979.
42. Behnke, J. R.; Fennema, O.; Cassens, R. G. J. Agric. Food Chem., 1973, 21, 5.
43. Sharp, J. G. Proc. Roy. Soc., 1934, Ser B, 114, 506.
44. Sharp, J. G. Biochem. J., 1935, 29, 850.
45. Smith, E. C. Proc. Roy. Soc., 1929, Ser B, 105, 198.
46. Tomlinson, N.; Jonas, R. E. E.; Geiger, S. E. J. Fish. Res. Bd. Can., 1963, 20, 1145.
47. Bito, M.; Amano, K. Bull. Tokai Reg. Fish. Res. Lab., 1962, 32, 149. (As cited by Tomlinson et al., 1963, Ref. #46.)
48. Partmann, W. Z. Ernahr. Wiss., 1961, 2, 70.
49. Partmann, W. J. Food Sci., 1963, 28, 15.
50. Saito, T.; Arai, K. Bull. Jap. Soc. Scient. Fish., 1957, 22, 569. (As cited by Tomlinson et al., 1963, Ref. #46.)
51. Saito, T.; Arai, K. Bull. Jap. Soc. Scient. Fish., 1957, 23, 265. (As cited by Tomlinson et al., 1963, Ref. #46.)
52. Saito, T.; Arai, D. Arch. Biochem. Biophys., 1958, 73, 315.
53. Lovern, J. A.; Olley, J. J. Food Sci., 1962, 27, 551.
54. Kiermeier, F. Biochem. Z., 1949, 319, 463.
55. Mapson, L. W.; Tomalin, A. W. J. Sci. Food Agric., 1958, 9, 424.
56. Suhonen, I. J. Sci. Agric. Soc. Finland, 1967, 39, 99.
57. Fishbein, W. N.; Stowell, R. E. Cryobiology, 1966, 6, 227.
58. Rhodes, D. N. J. Sci. Food Agric., 1961, 12, 224.
59. Tappel, A. L. In "Cryobiology"; Meryman, H. T., Ed.; Academic Press: New York, 1966; p. 163.
60. Barbagli, C.; Crescenzi, G. S. J. Food Sci., 1981, 46, 491.

61. Geromel, E. J.; Montgomery, M. W. J. Food Sci., 1980, 45,
     412.
62. Hamm, R. In "Proteins at Low Temperatures"; Fennema, O.,
     Ed.; ACS Advances in Chem. Series: Washington, D.C.,
     1979; p. 191.
63. Gkinis, A. M.; Fennema, O. R. J. Food Sci., 1978, 43, 527.
64. Gusta, L. V.; Weiser, C. J. Plant Physiol., 1972, 49, 91.
65. Taborsky, G. In "Proteins at Low Temperatures"; Fennema, O.,
     Ed.; ACS Advances in Chem. Series: Washington, D.C.,
     1979; p. 1.
66. Markert, C. L. Science., 1963, 140, 1329.
67. Aune, K. C.; Goldsmith, L. C.; Timasheff, S. N.
     Biochemistry, 1971, 10, 1617.
68. Reisler, E.; Eisenberg, H. Biochemistry, 1971, 10, 2659.
69. Osborne, J. C.; Palumbo, G.; Brewer, H. B.; Edelhoch, H.
     Biochemistry, 1975, 14, 3741.
70. Kirkman, H. N.; Hendrickson, E. M. J. Biol. Chem., 1962,
     237, 2371.
71. Pallavicini, C. Sci. Tecnol. Degli Alimenti., 1973, 3, 35.
72. Khenokh, M. A.; Pershina, V. P.; Lapinskaya, E. M.
     Tsitologiya., 1966, 8, 769-772; In Russian (as cited by
     Lozina-Lozinskii, 1974, Ref. #11).
73. Cowman, R. A.; Speck, M. L. Cryobiology, 1969, 5, 291.
74. Moran, T. Proc. Roy. Soc., 1925, Ser B, 98, 436.
75. Thomas, A. W.; Bailey, M. I. Ind. Eng. Chem., 1933, 6, 669.
76. Jarabak, J.; Seeds, Jr., A. E.; Talalay, P. Biochemistry,
     1966, 5, 1269.
77. Hofstee, B. F. J. J. Gen. Physiol., 1948, 32, 339.
78. Connell, J. J. Nature, 1959, 183, 664.
79. Saito, Z.; Niki, R.; Hashimoto, Y. J. Fac. Agr. Hokkaido
     Univ., 1963, 53, 200.
80. Edelhoch, H.; Osborne, Jr., J. C. Adv. Protein Chem.,
     1976, 30, 183.
81. Taborsky, G. J. Biol. Chem., 1970, 24, 1054.
82. Kenokh, M. A.; Aleksandrova, V. N. "Reaktsiya Kletok i ikh
     belkovykh Komponentov na ekstremal'nye vozdeistviya";
     Leningrad, 1966; p. 27. (As cited by Lozina-Lozinskii,
     1974, Ref. #11.)
83. Hanafusa, N. Contr. Inst. Low Temp. Sci., 1972, Ser B (17),
     1.
84. Sehgal, O. P.; Das, P. D. Virology, 1975, 64, 180.
85. Curti, B.; Massey, V.; Zmudka, M. J. Biol. Chem., 1968,
     243, 2306.
86. Wynn, C. H. "The Structure and Function of Enzymes"; Edward
     Arnold: London, 1973; p. 18.
87. Oakenfull, D.; Fenwick, D. E. Aust. J. Chem., 1977, 30, 741.
88. Tanford, C. Adv. Protein Chem., 1968, 23, 121.
89. Eaglund, D. In "Water--A Comprehensive Treatise"; Vol. 4;
     Franks, F., Ed.; Plenum Press: New York, 1975; p. 305.
90. Pace, N. C.; Tanford, C. Biochemistry, 1968, 7, 198.

91. Privalov, P. L.; Khechinashvili, N. N. J. Mol. Biol., 1974, 86, 665.
92. Brandts, J. F. J. Amer. Chem. Soc., 1964, 86, 4291.
93. Brandts, J. F. J. Amer. Chem. Soc., 1964, 86, 4302.
94. Mahan, B. H. "Elementary Chemical Thermodynamics"; W. A. Benjamin: New York, 1963.

RECEIVED May 17, 1982.

Structural Changes and Metabolism of Proteins Following Heat Denaturation

J. N. NEUCERE and J. P. CHERRY [1]

United States Department of Agriculture, Southern Regional Research Center,
Agricultural Research Service, New Orleans, LA 70179

Heat is perhaps the primary factor that enhances the safety, nutritional quality, and palatability of most foods. Mild or moderate temperatures can be beneficial in achieving these goals, whereas, extreme temperatures often are deleterious in terms of biological value and toxicity. The complex reactions that occur during extreme heat treatments, which usually involve simple and intricate molecules and ions, modify the physicochemical properties and metabolic efficiency of proteins. A key concept in protein chemistry is the effect heat has on active centers or sites within these molecules. These centers, involving any number of functional groups, are defined as the regions of space where the binding of other molecules and ions occur. Some of the most reactive sites in proteins are amine, sulfhydryl, tyrosyl, and imidizole groups. Heat-induced changes in any or all of these functional groups are directly related to physicochemical modifications such as solubility, enzymatic activity, antigenic reactivity, electrophoretic migration and association-dissociation constraints.

Anti-nutritional parameters and toxicological aspects of products from protein heated in the presence of carbohydrates are well recognized from the many studies via the Amadori and Maillard reactions. The formation of diverse isopeptides induced by heat are also critical factors that can adversely affect the biological value of proteins. Measuring toxicity or harmful effects, for example, of such complicated reaction products is not a simple matter. Adversity could involve either a modified intact protein or the presence of intricate amino acids and peptides that have formed. This chapter concentrates on several empirical views of structural changes and anti-nutritional aspects of proteins under various conditions of heat treatments.

Experimental Procedures

The reader is referred to the following publications on techniques of gel electrophoresis (1), immunochemistry (2),

[1] Current address: United States Department of Agriculture, Eastern Regional Research Center, Philadelphia, PA 19118.

functionality (3) and nutrition (4). Studies on the
determination of multiple forms of enzymes and their use as
biological indicators for food uses and applications were
reviewed by Cherry (5,6).

Whole seeds, full-fat and fat-free meals, concentrates
and/or isolates are used in research to determine the effects of
moist and dry heat on protein properties. For example, peanut
kernels are moist-heated in water in a temperature-controlled
steam retort kept at various temperatures for different time
intervals (7). After moist heat treatment, the seeds are lyophilized and then ground into a meal for protein analyses. Dry heat
treatments may be conducted by placing seeds in a forced-draft
oven at various temperatures and time intervals (8); purified
proteins such as peanut α-arachin are similarly heat-treated (9).
Seeds can be made to imbibe varying amounts of water to 40% and
then dry heated (8). Microwave heating of seeds containing only
their innate moisture or after they have been placed in water is
conducted at operational frequencies as high as 2,480 MHz (10).

Evidence of Protein Structural Changes Induced by Heat

Solubility. Perhaps the most obvious evidence of conformational changes in protein structure are the alterations in protein
solubility. In general, protein solubility decreases with time
and temperature of heat treatment; and considerable variations in
the degree of change exist among proteins. Interactions with
several functional groups (e.g., SH- or tyrosyl-residues) become
more prevalent following unfolding of protein chains, resulting
in diverse and complicated precipitation reactions. The exact
mechanisms leading to precipitation are most difficult to study
after heat treatments because the process is usually irreversible.
It is well known that the interactions of water molecules with
ionic and polar groups of proteins strongly influence the folded
conformations of proteins (11). Consequently, moist heat usually
has a more complex effect on the solubility properties of proteins than does dry heat, and is strongly influenced by ionic
strength and pH.

An example of solubility changes of peanut proteins induced
by heat is shown in Figure 1. In this study (8), seeds were
allowed to imbibe distilled water for 16 hr at 25°C to a final
moisture content of 40% and were heated for 1 hr from 110 to
155°C. Other seed samples were heated on an "as is" basis (5%
moisture) under identical conditions. Solubility of total protein decreased almost linearly with temperature for the dry-heated
seeds. For the wet-heated samples, the relationship was quite
different; a plot of solubility versus temperature showed a
sigmoid-like curve with a minimum at 120°C decreasing sharply
after 145°C. Compared to the 110°C wet- and dry-heated samples,
the control had an intermediate degree of solubility. This
information implies that the initial steps of denaturation in the

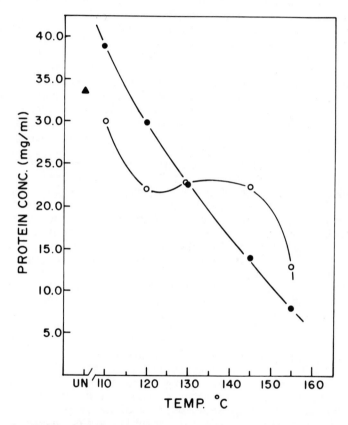

Figure 1.   Relative protein solubilities of intact peanuts, after moist (○) and dry (●) heat. Triangles indicate native seed or control. (Reproduced, from Ref. 9. Copyright 1974, American Chemical Society.)

imbibed seed involve a gradual change of the water-protein system within the seed between 110 and 120°C. Beyond this point, a more complex water-protein system appears to develop possibly involving protein fragments and other macromolecules. At low moisture content, a continuous change in the water-protein system seems to occur.

Cherry et al. (7) showed that the percentages of water-soluble protein in peanut seeds wet-heated (temperature-controlled steam retort) at 100°C remained constant between 0 and 45 min, then declined from 55.7 to 21.8% between 45 and 210 min (Figure 2). During this same time interval, percentages of water insoluble proteins increased from 38.9 to 52.2%. In this same study, the percentages of water-soluble proteins in peanut seeds wet-heated at 120°C were shown to decline very rapidly to 10.9% in 45 min, then increase to 32.8% as heating was extended to 180 min; the protein content in the insoluble fraction gradually increased and then leveled off during this heating period.

Ultraviolet circular dichroic spectra of peanut seed α-arachin (major globulin stored in aleurone grains; 12) that was heated at various temperatures showed the occurrence of many changes in its conformation modes (13). These changes were related to increased content of unordered structures within the globulin from decreased amounts of helical and pleated sheet modes. The unordered structures allow the formation of protein-protein and protein-nonprotein component interactions that are water-insoluble. The increase in solubilization during long heating periods at high temperature is probably caused by disruption of these interacted components (7,14,15).

Theoretical studies dealing with the binding of water vapor by purified proteins have been reported (16,17). The solubility of purified egg albumin as a function of water content and heat is described in Figure 3 (16). In this study, hydrated egg albumin samples were allowed to equilibrate with reference solutions at 25°C. Test samples were heated in a water bath for 10 min at 70, 76.5, and 80°C. Clearly, the protein samples at 25°C suffered no irreversible changes in solubility up to 0.4 g of water per g of protein. At elevated temperatures, however, decreases in solubility were linear up to about 0.275 g of water per g of protein; thereafter, a gradual decrease in solubility was observed. Close observation of the plot shows that the change in solubility at 76.5 and 80°C is almost identical. Other data from this study showed that 6 moles of water are complexed with each mole of hydrophilic amino acid residue when water content was about 0.3 g of water per g of protein at a temperature of 23°C. This corresponds to the point of maximum insolubility at 25°C on the plot.

In establishing protein functionality, it is apparent that the effects of heat on solubility become a prime factor of consideration. Properties such as emulsification, whippability, texturization and water and oil absorptivity, for example, are

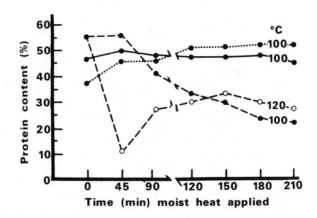

Figure 2.   Percentages (Kjeldahl N × 6.25) of protein in whole seed (——) and water soluble (● – – and ○ – –), and insoluble (● - - -) extracts of peanut seeds after moist-heat treatment at 100 and 120°C for intervals of 0–210 min. (Reproduced, with permission, from Ref. 7. Copyright 1975, Institute of Food Technologists.)

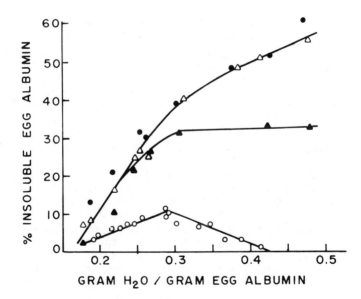

Figure 3.   Solubility of egg albumin as a function of water content. Key: ○, 25°C; ▲, 70°C; △, 76.5°C; and ●, 80°C. (Reproduced, with permission, from Ref. 16. Copyright 1968, Academic Press.)

all interrelated to solubility in the production of fabricated
foods from flours, concentrates, and isolates (18,19).
    Gel electrophoretic properties. Gel electrophoretic tech-
niques are excellent examples of modern technology developed to
qualitatively detect proteins (5,6). Basically, the methods
separate partially to highly purified fractions of proteins by an
electric charge in aqueous extracts in a gel matrix such as polyacrylamide or starch. Protein mobility through electrophoretic
media depends upon a combination of protein-related factors,
including net charge, molecular size, and conformation.
    Heating aqueous extracts of peanut seed proteins to 85°C did
not affect their mobility in electrophoretic gels (Figure 4; 23,
24). At 95°C, however, α-arachin (Figure 4; arrows 1 and 2)
decreased in staining density with the greatest reduction
occurring in band 2. Simultaneously, two new bands (arrows 5 and
6) appeared anodal to α-arachin. At 100°C, sites of α-arachin
had further decreased in staining intensity and were accompanied
by the appearance of an additional band (arrow 7). Thus, as the
apparent concentration of α-arachin decreased (arrows 3 and 4 compared to arrows 1 and 2), three new bands appeared (arrows 5 to
7). Furthermore, a weak band at R_f 1.0 (arrow 8; bromophenol
marker front) became more obvious at 95 and 100°C as α-arachin
and the heated products migrated anodally to different sites in
the gel.
    The proteins in the soluble fractions of wet-heated (100°C)
peanut seeds (Figure 2) were characterized by polyacrylamide gel
electrophoresis (Figure 5; 7,19). Alpha-arachin (region 0.5-1.5
cm) in the gel patterns of peanut seeds wet-heated at 100°C for
15 to 90 min remained unchanged. The gel patterns showed that
the quality of proteins of nonarachin fractions (region 2.0-4.0
cm) decreased in extracts from peanut seeds wet-heated at 100°C
for 30 to 75 min. The quantity of α-arachin declined in the gel
patterns between 105 and 180 min. Between 90 and 180 min, the
number of protein bands greatly increased in region 3.0-6.5 cm of
the gels.
    Electrophoretically detected changes in the soluble proteins
of peanut seeds wet-heated at 120°C were clearly shown at 15 min
(Figure 6). This gel pattern resembled those of peanut seeds
wet-heated at 100°C for 90 to 105 min; i.e., a decrease in the
quantity of nonarachin proteins and a simultaneous increase in
the number of components that migrated to region 3.0-6.5 cm of
the gels. Between 30 and 210 min, α-arachin and the proteins in
region 3.0-6.5 cm became difficult to discern and at the same
time, a broad diffuse band formed through the gel (region 1.0-5.0
cm).
    These studies showed that in addition to denaturing proteins
to insoluble forms in aqueous solutions, moist heating of peanut
seeds at high temperatures sequentially alters them to subunit
polypeptide forms or fragments, then to aggregates, and finally
to insoluble components. Continued heating causes an increase in

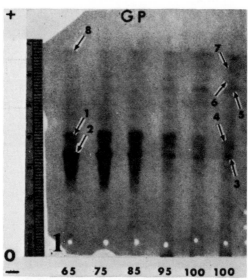

Figure 4. Starch gel stained for proteins in heated aqueous extracts from dormant peanut cotyledons. Numbers represent temperatures (°C) of heat treatment. (GP = general soluble proteins.) (Reproduced from Ref. 8. Copyright 1972, American Chemical Society.)

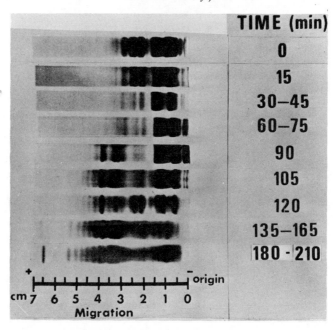

Figure 5. Disc polyacrylamide gel electrophoretic patterns of water-soluble proteins from 1973 peanut seeds moist-heated at 100°C for periods of 15–210 min. (Reproduced, with permission, from Ref. 7. Copyright 1975, Institute of Food Technologists.)

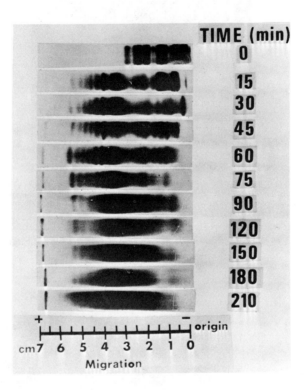

Figure 6.   Disc polyacrylamide gel electrophoretic patterns of water-soluble proteins from peanut seeds moist-heated at 120°C for periods of 15–210 min. (Reproduced, with permission, from Ref. 7. Copyright 1975, Institute of Food Technologists.)

solubility of the proteins and the formation of a major diffuse
component or aggregate that is detectable by gel electrophoresis.
Similar observations have been noted with soybean proteins (5,6,
7,19). It appears that during heating of peanut and soybean seed
products the extent of unfolding of the structural components of
proteins, and their interactions with other proteins (or subunits) and/or various types of constituents occurs in a stepwise
fashion and is dependent on heating temperature and the time
interval heat is applied.

Cherry and McWatters (19) showed that the degree of denaturation of proteins from 1974 harvested peanut seeds moist-heated at
100°C for various time intervals differed from those harvested in
1973 (Figure 7). The increase in the number of bands that
migrated to regions 1.5-2.5 and 4.5-6.5 cm in electrophoretic gels
from extracts of heated 1973 seeds were not shown in those of the
1974 seeds. Instead, changes in aqueous extracts of 1974 seeds
heated for 75 to 210 min were mainly the formation of two major
proteins that migrated in the gels to region 3.0-4.0 cm. These
proteins were noted mainly in the gels of 1973 seeds heated for
90 to 105 min.

These data showed that peanut seeds grown during different
crop years and/or stored for various time intervals are affected
differently by moist-heat processing. This work presented
evidence that environmental, agronomic, and handling practices
can contribute to variations in peanut seed proteins during
moist-heat processing. It is necessary, therefore, to consider
these factors in characterizing the functionality of peanut seeds
and their vegetable protein products prior to their use in food
formulations.

Inactivation of enzymes. The catalytic activity of enzymes
depends on the maintenance of three-dimensional structures which
are stabilized by hydrogen bonds and hydrophobic interactions.
Ultimate stability, however, depends on the primary structures of
polypeptide chains. Measurement of deactivation rates conducted
by heat are used extensively in distinguishing between multiple
forms of enzymes. Individual members of a family of multiple
forms of enzymes can have a wide range of variation in heat
stabilities (5,6). Both close similarities and wide differences
in rates of inactivation by heat can co-exist among multiple
forms of a single enzyme. This has been shown, for example, on
studies of human alkaline phosphatase (21,22). Alkaline
phosphatase extracted from placental tissue is stable to heat up
to 70°C, whereas the enzyme prepared from bone loses half of its
activity in less than 10 min at 55°C. Rates of inactivation of
multiple forms of enzymes by heat are also generally affected by
changes in other factors such as pH, protein concentrations, and
the amounts of substrates and co-factors present. It has been
shown that certain enzymes have increased heat-stability in the
presence of their substrates. Lactate dehydrogenase multiple
forms in human serum, for example, showed increased

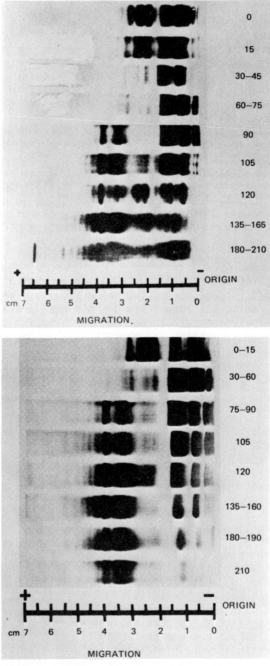

Figure 7. Disc polyacrylamide gel electrophoretic patterns of water-soluble proteins from 1973 (top) and 1974 (bottom) peanut seeds moist-heated at 100°C for periods of 15–210 min. (Reproduced, with permission, from Ref. 19. Copyright 1975, Institute of Food Technologists.)

heat-stability in the presence of coenzyme nicotinamide adenine dinucleotide (NAD) (23).

Separation procedures such as electrophoresis and chromatography, as approaches to multiple enzyme analysis, have proved to be powerful tools (5,6,11,14). The distribution patterns on starch gels, of multiple molecular forms of peanut seed enzymes before and after heat treatments have been reproduced (5,6,20,24). Examples of these results, on heated extracts from dormant peanut colyledons, are shown in Figure 8 (1-5).

Malate dehydrogenase (MDH) of nonheated dormant peanut cotyledons exhibited five multiple forms, one band stained very lightly (Figure 8-1; arrow). The MDHs in Figure 8-2 were allowed to react longer with substrate then those in Figure 8-1 to intensify the banding pattern. No changes were apparent in the distribution pattern of MDH between 25 and 60°C (Figure 8-1,2); at 61°C, only one band remained. This band persisted through 63°C and disappeared at 65°C (cf Figures, 8-1 and 8-2).

Three anodal bands of nonspecific α-esterases (α-EST) present in the nonheated extract (Figure 8-3; arrows 1-3) began exhibiting thermal sensitivity between 45 and 50°C; only a trace of activity occurred at the latter temperature. However, a single band in the cathodal section indicated that enzymatic activity at this site was evident through 55°C but not present at 65°C.

The nonheated extract exhibited five sites of leucine aminopeptidase (LAP) activity (Figure 8-4; arrows 1-5). The activities of the five enzyme forms were unaffected through 45°C. At 55°C, only one band was easily discernible but traces of two bands were also present (Figure 8-4; arrows 3,4). The most heat stable band was barely visible at 65°C (Figure 8-4; arrow 5).

Glutamate dehydrogenase (GDH) displayed three distinct bands which were unaffected by heat through 75°C; at 85°C, no banding occurred (Figure 8-5).

Another study of plant enzymes after dry-roasting whole seeds showed definitive banding at 110°C with weak responses at 130°C (24). This same study also showed that imbibed seeds heated in situ displayed no enzyme activity above 100°C.

It is known that a number of factors affect the process of thermal inactivation of enzymes. Among these are pH and the presence of certain ions and various compounds in different molecular environments. The specific cell conditions in which enzymes normally function are difficult to define. Obviously, in vitro conditions are not the same as those in situ. One may assume that the cellular environment affords better heat protection for enzymes than buffered solutions. However, the amount of moisture in the cell and the state of metabolic activity are also significant factors.

Quantification of the activity of multiple forms of enzymes can be made by semi-logarithmic plots of decay of activity at a constant temperature (25). This method of analysis has been shown

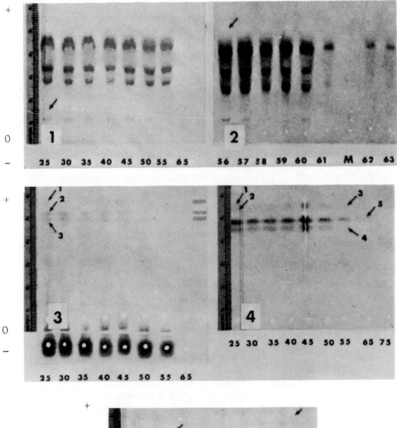

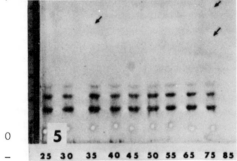

Figure 8.   Photographs of starch gels stained for enzymes in heated extracts from dormant peanut cotyledons. Basal numbers represent temperatures (°C) of heat treatment. Key: 1 and 2, malate dehydrogenase; 3, nonspecific α-esterase (stipled bands are on the gel of the nonheated sample); 4, leucine aminopeptidase; and 5, glutamate dehydrogenase. (Reproduced from Ref. 24. Copyright 1973, American Chemical Society.)

to be useful for two-component systems, e.g., liver and bone alkaline phosphatases (Figure 9). The linear components represent a decline in activity as a function of time for the two enzymes. Note that complete loss of activity for bone alkaline phosphatase occurs after heating for about 10 min at 56°C and that liver alkaline phosphatase remains active up to 30 min at this temperature. Provided that qualitative differences in stabilities between multiple forms are maintained from one specimen to the next, slight variations in incubation temperatures do not affect comparability of results.

Inhibition of antigenic reactivities. Differences in antigenicity offer a wide range of immunochemical analyses for both qualitative and quantitative work. Because the antigenicity of individual globular proteins is due mostly to the three-dimensional conformation of the molecule, any type of steric change induced by heat usually affects the antigenic determinants of the native protein. Changes in the immunochemical profile of peanut proteins heated in situ have been reported (8). The immunoelectrophoretic profile of the major peanut proteins after different heat treatments is shown in Figure 10. The profile of native seeds in example 1 of Figure 10 is based on the original work of Daussant et al. (26).

The precipitin lines corresponding to α-arachin (designated B) and one other antigen were detected throughout the series of wet and dry heat treatments. The antigenic sites on all of the other proteins were inactivated at a relatively low temperature (110°C) at high moisture content (samples 2 to 6). Conversely, dry heat (samples 7 to 11) induced sequential antigenic degradation with increased temperature; α2-conarachin (C) remained antigenic up to 130°C. Note also, that the leading edge of the precipitin line for α-arachin in the imbibed seed (2 to 6) extended closer to the anode than the corresponding lines from the dry heated samples (7 to 11). This slight change in mobility implies a probable change in shape and/or conformation of this protein molecule, that exposes masked negatively charged functional groups.

Further immunochemical analyses of the effects of wet and dry heat on "isolated" proteins from peanuts have been described (9). In this study, isolated α-arachin was heated in solution and in the dry state from 80°C to 195°C for one hour. The immunoelectrophoretic analysis of the dry-heated samples is shown in Figure 11. Up to 130° (sample 4), essentially no precipitin variation from that of the control was observed. However, the minor antigenic protein or the so-called α-arachin contaminant (arrow) was inactivated at higher temperatures. At 155°C (sample 6) the precipitin line of α-arachin appeared diffused with a slight anodic shift. Analysis after heating at 175°C showed a wide indistinct line that formed a double arc (arrow in sample 7), and all antigenic activity was destroyed at 195°C.

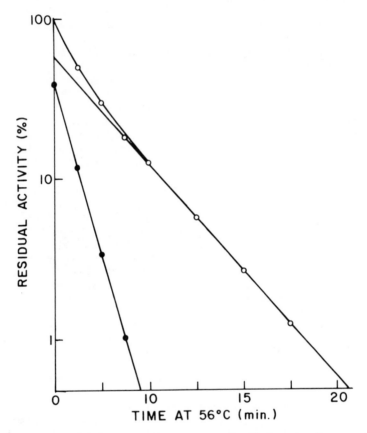

Figure 9.   The decline of liver (○) and bone (●) alkaline phosphatase activity with time of incubation at 56°C resolved into two linear components of different stabilities. (Reproduced, with permission, from Ref. 25. Copyright 1979, Royal Society of Chemistry.)

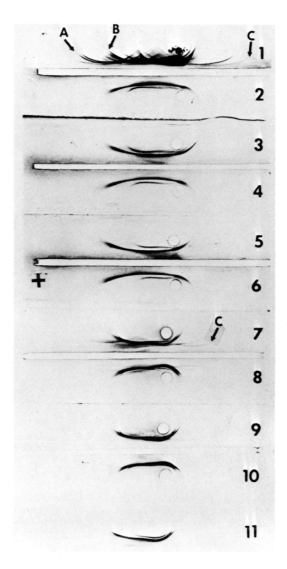

Figure 10. Immunoelectrophoretograms of peanut proteins. Key: A, α_1-conarachin; B, α-arachin; C, α_2-conarachin; 1, control; 2–6, proteins extracted from seeds imbibed 16 h; and 7–11, proteins extracted from seeds not imbibed. The members of each series (2,3,4,5,6, and 7,8,9,10,11) were heated to 110, 120, 130, 145, and 155°C, respectively. (Reproduced from Ref. 8. Copyright 1972, American Chemical Society.)

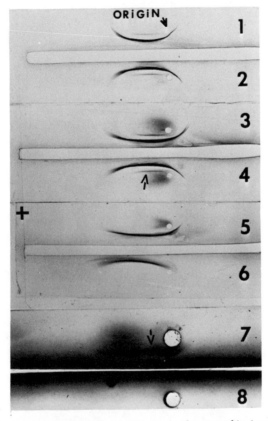

Figure 11.   Immunoelectrophoresis of proteins in the α-arachin fraction after dry heating. Key: 1, control; 2–8, samples heated at 110, 120, 130, 145, 155, 175, and 195°C, respectively. (Reproduced from Ref. 9. Copyright, American Chemical Society.)

Analyses by antibody-in-gel electrophoresis (semi-quantita-
tion) of samples containing equal protein concentrations heated
in solution are shown in Figure 12. Part A shows the relative
changes in electrophoretic migration of the precipitin complexes.
At 80°C, a slight increase in migration of the conical peaks were
observed for both α-arachin (a) and the so-called arachin
contaminant (ac). However, at 90°C the reverse was observed for
α-arachin and the contaminant was completely inactivated; only a
trace of activity was observed at 100°C and no reaction occurred
at 110°C. The increase in migration could be due to either an
increase in charge of the antigens or a reduction of active
determinant groups on the antigens. On the other hand, reduction
in migration reflects decreased overall molecular charge or
possibly the exposure of additional antigenic sites. Double diffusion of the same samples in part B showed similar results. The
reaction of partial fusion after heating at 100°C (well 4 in B)
suggested major molecular disorganization without complete
destruction of determinant groups. Note also the slight spur
(dashed arrow) which formed between the control and the sample
heated at 80°C (wells 1 and 2 in B) indicating a reaction of
partial identity.

For most globular proteins, partial or complete loss of antigenic specificity is generally due to modifications of sequential
and/or conformational determinants. A sequential determinant is
maintained by a distinct amino acid sequence in random coil form,
whereas conformational specificity depends on disulfide bonds
and/or side chain polar groups that maintain the steric structure
of the antigen (27). A study of heat denaturation of ovalbumin
and serum albumin in solution showed that even if the secondary
structure of the native antigen was modified, some determinant
groups reacted weakly with the native antibody (28). However,
the reactions of native antigen with its denatured antibody were
stronger than that of the denatured antigen. Hence, the
antigenic behavior of denatured protein depends on its specific
structure and generally cannot be described. The lack of
cross-reaction or any reaction changing from the "normal"
precipitin reaction must be due to modification(s) of the antigen
molecule (27).

Changes in Absorption Spectra. Small changes occur in the
ultraviolet spectra of protein in response to small temperature
changes in the predenaturation range. These changes are characteristic mainly for the perturbation of tyrosine and tryptophan
residues in aqueous solution. Nicola and Leach (29) showed with
model compounds and proteins the possible types of contributions
to the thermal difference spectra. In proteins, the environment
of tyrosine and tryptophan residues within the molecule can be
distinguished by their thermal difference spectra. Residues on
the protein surface produce a unique and recognizable difference
spectrum. On the other hand, residues in the protein interior
make a small contribution to the thermal difference spectrum of a

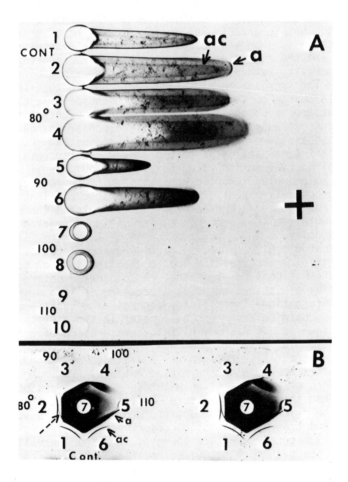

Figure 12. Antibody-in-gel (A) and double diffusion (B) analyses of samples heated in solution for 1 h (cont = control).

In A, the odd- and even-numbered wells contain 30 and 60 µg of protein, respectively; 2% immune serum against the α-arachin fraction was mixed in the agar. In B, all antigen wells on the left contained 300 µg of protein, those on the right, 600 µg. The antibody wells (7) were filled three and six times, respectively, with the same immune serum. (Reproduced from Ref. 9. Copyright 1974, American Chemical Society.)

protein. This study pointed out the uniqueness of water in the
heating difference spectrum of tyrosine and tryptophan. With
increasing temperature, a red shift was observed in the absorption curve, which was attributed to preferential disruption of
hydrogen bonds where tyrosine or tryptophan act as proton
acceptors.

Changes in infrared absorption and circular dichroic spectra
of peanut globulin that were induced by heat were reported (13).
Infrared analysis of isolated α-arachin heated dry at several
temperatures are shown in Figure 13. Absorbancies assigned to
amide I, II, and V bands occurred at 1645 to 1655 cm-1, about
1530 cm-1, and at approximately 710 cm-1, respectively (Figure
13-I, II). According to standard analyses on model systems
(30,31), conformation of α-arachin in each sample was principally
from pleated sheet and unordered structures. Apparently,
conformational changes sufficient to affect the spectral
locations of amide I, II, and V were not thermally induced in
α-arachin. Examination of the amide IV and VI bands (Figure
13-III) showed a gradual shift from amide VI to amide IV as
α-arachin was heated. The significance of the change from
carbonyl out-plane bending to in-plane bending relative to
conformational modification was not investigated further.
Circular dichroic spectra of solubilized native and heated
samples in this study indicated that native α-arachin was
composed of 14.6% α-helical, 27.0% pleated sheet, and 58.4%
unordered structures. Computations from ultraviolet circular
dichroic spectra of native and heated α-arachin are shown in
Table I. These data indicate that changes of conformational

Table I.   Contents of α-helical, pleated sheet and unordered
           structures in conformations of heated and unheated
           arachin. (13)

Treatment	Conformational modes (%)		
	α-Helical	Pleated sheet	Unordered
Unheated (control)	14.6	17.0	58.4
Heated (150°C)	9.6	29.4	61.0
Heated (170°C)	5.0	22.8	72.2
Heated (190°C)	4.3	19.7	76.0

modes were thermally induced in α-arachin. Heat treatments
lessened the contents of α-helical structures, and increased
pleated sheet structures. The amount of unordered structures
increased during the heat treatment. Since the antigenicity of
α-arachin also decreased as the contents of α-helical structures
decreased and pleated sheet structural modes increased, the
results from circular dichroism suggest that antigenic determinant groups in native α-arachin reside in these conformational

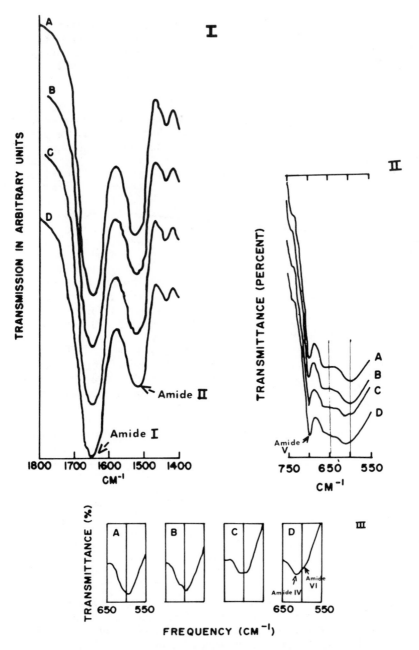

Figure 13.   IR absorbancies of α-arachin after dry heating. Amide I, II, and V bands are shown in Parts I and II. An expanded scale of amide bands IV and VI is shown in Part III D. A is native α-arachin, and B, C, and D are α-arachin heated at 150, 170, and 190°C, respectively. (Reproduced, with permission, from Ref. 13. Copyright 1975, Munksgaard International Publishers, Ltd.)

modes. Furthermore, heat treatments induced interchanges among
the modes of secondary structures in α-arachin, resulting in
altered α-helical and pleated sheet structures, with concomitant
changes and losses in antigenic reactivity.

Anti-Nutritional Effects Of Heat

Inactivation and destruction of essential amino acids. The
ultimate index in measuring amino acid availability to any organism is digestibility. The decrease in biological availability of
amino acids is usually caused by their binding to compounds which
cannot be disrupted by digestive enzymes. Excessive heat
treatment is known to impair digestibility of proteins by causing
them to undergo many complex reactions. Lysine, for example
undergoes destruction and is also rendered unavailable by
Maillard reactions (32). Cystine, arginine, tryptophan, histidine and serine are either partially destroyed or inactivated as
a result of excessive heating (33). A recent study showed that
microwave cooking produced a greater loss of amino acids in peas
than did conventional cooking (Table II; 34). Microwave

Table II.   Effect of cooking on amino acid content of Colossus
Peas[1]. (34)

Amino acid	Concentration, g/100g dry wt		
	Raw	Cooked	
		Conventional	Microwave
Lysine	1.82	1.72(5.5)	1.68(7.7)
Histidine	0.86	0.80(7.0)	0.75(12.8)
Arginine	1.41	1.35(4.2)	1.28(9.2)
Asparagine	2.94	1.76(6.1)	1.67(9.2)
Glutamine	4.46	4.20(5.8)	4.05(9.2)
Threonine	1.04	0.98(5.8)	0.94(9.6)
Serine	1.41	1.35(4.2)	1.28(9.2)
Proline	1.09	1.05(3.7)	1.01(7.3)
Glycine	0.93	0.89(4.2)	0.83(10.7)
Alanine	1.09	1.04(4.6)	1.00(8.2)
Cystine	0.11	0.10(9.1)	0.09(18.2)
Valine	1.31	1.26(3.8)	1.20(8.4)
Metionine	0.23	0.22(4.3)	0.20(13.0)
Isoleucine	1.16	1.09(6.0)	1.07(7.8)
Leucine	2.13	1.01(5.6)	1.98(7.0)
Tyrosine	0.74	0.70(5.4)	0.68(8.1)
Phenylalanine	1.53	1.47(3.9)	1.40(8.5)

[1]Values in parenthesis are percent losses.

processing was especially destructive to histidine, glycine,
cystine and methionine. This finding suggested that microwave

energy may be promoting chemical alterations in the side chains
on these particular amino acids.
     Formation of isopeptides and related reactions. During heat
treatment, cross-linkages can form in proteins that reduce their
digestability (35,36). One of the most likely mechanisms of this
heat damage seems to be the formation of isopeptide bonds or
reaction of the ϵ-amino group of lysine with either the carboxyl
group of aspartic or glutamic acid (37), or amide group of
glutamine or asparagine (38). Other cross-linkages may result
from the degradation of cystine. Dehydroalanine may condense
with cysteine to form lanthionine, or with the ϵ-amino group of
lysine to form lysinoalanine (39). The lysine-isopeptides are
acid labile and have been determined as residues after enzymatic
digestion of the peptide linkages (40).
     The fate of lysine-containing isopeptides in the rat's diges-
tive tract has been studied (35). The feces and undigested ileal
contents from rats fed on diets containing chicken muscle,
unheated or heated for 8 or 27 h at 121°, and a nitrogen-free
diet, were analysed, to determine the digestibility values for
dietary protein and lysine components. These values, and the net
protein ratio, (NPR), net digestible protein ratio (NDPR), and
floro-dinitrobenzene-reactive- (FDNB-) or available-lysine
content, of the chicken muscle are presented in Table III. With

Table III. Protein quality and digestibility values for unheated
            and heated chicken muscle fed to rats. (35)

Treatment	NPR	NDPR	FDNB-reactive lysine[1]	CID[2]/ N	CFD[3]/ N
Unheated (control)	5·5	6·2	93	0·88	0·98
Heated: 8 h at 121°C	3·5	4·9	70	0.88	0·94
	(0·64)[4]	(0·79)	(0·81)	(0·81)	(0·96)
27 h at 121°C	1·9	3·3	63	0·57	0·80)
	(0·35)	(0·53)	(0·68)	(0·65)	(0·82)

[1]/Measured as mg/g protein.
[2]/Coefficient of ileal digestibility.
[3]/Coefficient of fecal digestibility.
[4]/Values in parentheses represent ratios of heated to unheated
    samples.

heated samples, the digestibility of the protein was decreased,
as were NPR-, NDPR- and FDNB-reactive lysine values. NPR
decreased to a much greater extent than either FDNB-reactive
lysine values or ileal nitrogen digestibility. According to the
NDPR results, the growth-promoting value of the heated protein
that was digested and absorbed by the rats was greatly reduced.
Ileal nitrogen digestibilites of heated samples were lower than

those of fecal digestibilities. Feces from rats given chicken muscle heated for 27 h at 121°C contained 37.0 mg FDNB-reactive lysine/g fecal protein. This corresponded to about half the value for unheated chicken muscle (63 mg/g protein). Other data showed that lysine-isopeptide digestibilities in both the ileal and the feces were higher than that of corresponding nitrogen digestibility. It was postulated that some of the isopeptides were destroyed either by the bacteria or by the additional enzymes detected in the ileal and fecal samples containing the heated samples. These data suggested that cross-linkages are formed in proteins during severe heat treatment which reduce the rate of protein digestion, possibly by preventing enzyme penetration, or by masking the sites of enzymatic attack. However, the lysine-isopeptides appeared to be at least as digestible as the rest of the protein molecule.

A possible chemical change during the heating of proteins in the presence of water is hydrolytic cleavage of peptide chains. Experiments with bovine plasma albumin (38) as a model system, showed that the number of free amino groups, based on dinitrophenol measurements, increased about 0.05% (dry weight) after heating the sample for 7 h at 115°C and 145°C.

It has also been reported that when heat-damaged protein is fed to rats, an accumulation of "unavailable small peptides" occurs in the intestine of the animals (41). This may be due to the hindrance of amino acid absorption across the mucosal barrier. Experiments on the influence of unavailable peptides isolated from an enzyme digest of heat-damaged cod fillet on the uptake of leucine from rat small intestine were reported (Table IV; 42).

Table IV. Uptake of leucine and glucose and production of lactate alone everted sacs of rat small intestine and in the presence of "unavailable peptides" isolated from enzymatic digests of heat-damaged cod fillet. (42)

	Peptides absent	Peptides present
Uptake of leucine (μg/ml):		
Determined microbiologically	21·7	8·0
Determined by radioactive counts	24·3	12·7
Loss of glucose from mucosal fluid (μg/ml)	348	354
Lactate concentration in serosal fluid (μg/ml)	269	251
Serosal:mucosal leucine	2.0	1.3

Absorption of leucine from intestinal sacs was expressed in terms
of the value for final serosal concentration:final mucosal concentration of leucine. This value is an indication of the ability
of the gut wall to concentrate leucine against a concentration
gradient. In the absence of unavailable peptides the value for
serosal:mucoscal uptake for leucine was 2.0. The initial concentration for glucose in the mucosal and serosal fluids was 2000
μg/ml. After incubation for 30 min, 348 μg/ml had disappeared
from the mucosal fluid while an increase of 269 μg/ml of lactate
was noted in the serosal fluid. The value for serosal:mucosal
leucine decreased significantly from 2.0 to 1.3 in the presence
of unavailable peptide. Based on radioactive counts, the amount
of leucine lost from the mucosal fluid in the absence of unavailable peptides was 24.7 μg/ml. In the presence of unavailable
peptides the corresponding value was 12.7 μg/ml. Clearly, the
presence of the unavailable peptide had a marked effect on the
uptake of the leucine. These peptides had no effect on the
uptake of glucose or its metabolism to lactate. The reduction in
leucine uptake is not easily explained. Perhaps the unavailable
peptides bind adjacent to sites for amino acid absorption or they
interfere, upon accumulation, with the normal function of mucosal
cells (43).

   Toxicological implications in heated protein. Poor performance of animals given heat-damaged foods could be due in part to
Maillard compounds of free amino acids and sugars that have antinutritional or even toxic characteristics (44). However, this
might not be typical of heated proteins because the Maillard reaction products of free amino acids and proteins behave differently
in the digestive tract. For example, the fructose-lysine complex
was excreted in the urine of rats, whereas fructose-lysine bound
to protein remained primarily in the intestinal tract (45).
Protein heated in alkaline solution has been shown to induce
lesions in kidneys when fed to experimental rats (46). Such
histopathological changes, however, are not necessarily due to
the presence of a "toxic" substance in the diet. Calcification
in the kidneys of rats, for example, which was provoked by
feeding ammonia heated yeast was dramatically reduced by a
moderate excess of methionine in the diet (49). Consequently,
induction of some type of nutritional imbalance must always be
taken into consideration when evaluating heat processed foods.

   Cereals and legumes are known to contain toxic factors which
are both growth depressants and lethal (47-49). Some of these
factors are heat-labile. In South American countries, Phaseolus
vulgaris is the most consumed leguminous seed in human diets
(50). Feeding tests of this raw bean flour and its different
fractions with rats killed them between the third and 23rd day of
the experiment. All materials, except the water-insoluble solids
and the globulins, killed all of the rats by the 10th day of the
diets. For the water-insoluble solids and the globulins, the
lethal periods were between 10 and 14 and 12 and 23 days,

respectively. Similar toxicity has been reported for the black bean and the navy bean (51,52).

Other experiments showed that there was a correlation between protein solubility, trypsin inhibitor activity, and hemagglutination activity after water-soaked beans were heated at 97°C (50; Figure 14). The data showed that heating water-soaked beans in boiling water for 5 and 10 min inactivated the trypsin inhibitor and the hemagglutinin, respectively. And at the end of 15 min, less than 25% of the total protein remained in solution. Feeding studies showed that the optimum protein efficiency ratio occurred for samples held 10 min in boiling water at 97°C. These data suggest that the toxic substances in the bean are either destroyed or react with some other cell materials that inactivate them during mild heat treatment.

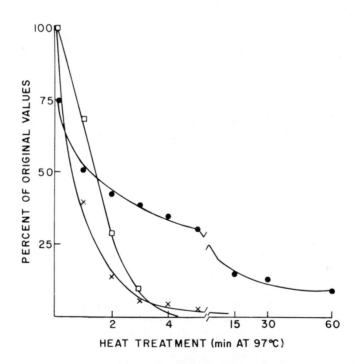

Figure 14. Effect of heating water-soaked beans (Phaseolus vulgaris) *on protein extractability and on trypsin inhibitor and phytohemagglutinin activities. Key: □, activity curve for trypsin inhibitors; ×, hemagglutinin; and ●, protein solubility curve. (Reproduced from Ref. 50. Copyright 1980, American Chemical Society.)*

Conclusions

Heat-induced denaturation of protein in foods are presumably
limited by the number of polypeptides and the availability of
their side-chain functional groups. The diversity of ions and
macromolecules (such as simple and complex lipids and carbohydrates) present in foods, however, allows for any number of
unique reactions that can be triggered by heat. Analytical tests
are made on a comparative basis to establish deviations from a
normal standard profile. The most important parameters to evaluate in predicting conformational changes include: solubility
characteristics, enzymic deactivation, electrophoretic migration,
antigen reactivity, ultraviolet absorption, chromatographic properties, amino acid destruction, and reactions with other
macromolecules. The effects of heat, both beneficial and adverse,
range from flavor enhancement to severe limitation of essential
amino acid utilization. Both in vitro and in vivo analyses show
wide variations in biological values of proteins after heating at
different levels of moisture. Although the chemical mechanisms
of heat damage are not fully understood, evidence shows that
enzyme-resistant linkages form during heat treatments. The new
bonds referred to as "isopeptide" linkages reportedly involve
primarily Σ-amino groups of lysine reacting with carboxyl groups
of glutamic and aspartic acids and with amide groups of glutamine
and asparagine. Carbohydrates are also involved in these
reactions. Fission of disulfide bonds occurs during heating,
causing the depolymerization of polypeptide subunits that lead to
other complex reactions.

LITERATURE CITED

1. Bietz, J. A. Cereal Foods World, 1979, 24, 199.
2. Catsimpoolas, N., Ed.; "Immunological Aspects of Foods;" The
   AVI Publishing Co., Inc.: Westport, CT, 1977; 400pp.
3. Cherry, J. P., Ed.; "Protein Functionality in Foods;" ACS
   Symposium Series 147: Washington, D. C., 1981; 332pp.
4. Friedman, M., Ed.; "Nutritional Improvement of Food and Feed
   Proteins;" Plenum Press: New York, NY, 1978; 882pp.
5. Cherry, J. P. In "Enzymes in Food and Beverage Processing;"
   Ory, R. L.; St. Angelo, A. J., Eds.; ACS Symposium Series;
   American Chemical Society: Washington, D. C., 1977; p. 209.
6. Cherry, J. P. In "Postharvest Biology and Biotechnology;"
   Hultin, H. O.; Milner, M., Eds.; Food and Nutrition Press,
   Inc.: Westport, CT, 1978; p. 370.
7. Cherry, J. P.; McWatters, K. H.; Holmes, M. R. J. Food
   Sci., 1975, 40, 1199.
8. Neucere, N. J. J. Agric. Food Chem., 1972, 20, 252.
9. Neucere, N. J. J. Agric. Food Chem., 1974, 22, 146.

10. Pour-El, A.; Nelson, S. O.; Peck, E. E.; Tjhio, B.; Stetson, L. E. J. Food Sci., 1981, 46, 880.
11. Edelhock, H.; Osborne, J. C., Jr. In "Advance in Protein Chemistry;" Anfinsen, C. B.; Edsall, J. T.; Richards, F. M., Eds.; Academic Press: New York, NY, 1976; p. 183.
12. Basha, S. M. M.; Cherry, J. P. J. Agric. Food Chem., 1976, 24, 359.
13. Jacks, T. J.; Neucere, N. J.; McCall, E. R. Int. J. Peptide Protein Res., 1975, 7, 155.
14. Wolf, W. J. J. Agri. Food Chem., 1970, 18, 969.
15. Wolf, W. J.; Cowan, J. C. CRC Crit. Rev. Food Tech., 1971, 2, 81.
16. Bull, H. B.; Bresse, K. Arch. Biochem. Biophys., 1968, 128, 488.
17. Bull, H. B.; Bresse, K. Arch. Biochem. Biophys., 1968, 128, 497.
18. Morr, C. V. Food Technology, 1976, 30, 18.
19. Cherry, J. P.; McWatters, K. H. J. Food Sci., 1975, 40, 1257.
20. Thomas, D. L.; Bright, J. E. Can. J. Bot., 1973, 51, 1191.
21. Moss, D. W.; King E. J. Biochemical J., 1962, 84, 192.
22. Neale, F. G.; Clubb, J. S.; Hotchkis D.; Posen, S. J. Clin Path., 1965, 18, 359.
23. Vesell, E. S.; Yielding, K. L. Ann N. Y. Acad. Sci., 1968, 151, 678.
24. Thomas, D. L.; Neucere, N. J. J. Agric. Food Chem., 1973, 21, 479.
25. Moss, D. W. "Isoenzyme Analysis"; Analytical Sciences Monographs, The Chemical Society Burlington House: London, 1979; p. 83.
26. Daussant, J.; Neucere, N. J.; Yatsu, L. Plant Physiol., 1969, 44, 471.
27. Sela, M.; Schechter, B.; Schechter, I.; Borex, F. Cold Spring Harbor Symp. Quant. Biol., 1967, 32, 537.
28. Arat, F. Nature, 1966, 212, 848.
29. Nicola, N. A.; Leach, S. J. Int. J. Peptide Protein Res., 1976, 8, 393.
30. Miyazawa, T.; Blout, E. R. J. Am. Chem. Soc., 1961, 83, 712.
31. Miyazawa, T.; Masuda, Y.; Fukushima, K. J. Polymer Sci., 1962, 62, 562.
32. Taira, H.; Sugimura, K.; Saknrai, Y. Agric. Biol. Chem. (Tokyo), 1965, 29, 1074.
33. Liener, I. E. In "Processed Plant Protein Foodstuffs"; Altschul, A. M., Ed.; Academic Press: New York, NY, 1958; p. 79.
34. Chung, S. Y.; Morr, C. U.; Hen, J. J. J. Food Sci., 1981, 46, 272.
35. Hurrell, R. F.; Carpenter, K. J. British J. Nutr., 1976, 35, 383.

36. Mauron, J. In "International Encyclopedia of Food and Nutrition." Vol. 2; Bigwood, E. J., Ed.; Pergamon Press: Oxford, 1972; p. 417.
37. Asquith, R. S.; Otterburn, M.S. J. Text. Inst., 1969, 60, 208.
38. Bearnason, J.; Carpenter, K. J. British J. Nutr., 1970, 24, 313.
39. Ziegler, K. J. Biol. Chem., 1964, 239, 2713.
40 Asquith, R. S.; Otterburn, M. S.; Gardner, K. L. Experientia, 1971, 27, 1388.
41. Buraczewski, S.; Buraczewski, L.; Ford, J. E. Acta Biochem. Pol., 1967, 14, 21.
42. Shorrock, C.; Ford, U. E. British J. Nutr., 1978, 40, 185.
43. Orlowski, M.; Meister, A. Proc. Nat'l Acad. Sci., 1970, 67, 1248.
44. Adrian, J. World Rev. Nutrition Dietetics, 1974, 19, 71.
45. Mauron, J. In "Nestle' Research News". Boella, C., Ed.; Nestle' Products Technical Assistance Co., Ltd: Switzerland, 1978-79; p. 56.
46. Woodward, J. C.; Short, D. P. J. Nutr., 1973, 103, 569.
47. Liener, I. E.; Kakade, M. L. In "Toxic Constituents of Plant Food Stuffs"; Liener, I. E, Ed.; Academic Press: New York, NY, 1980; p. 7.
48. Wall, D. Series Seminars No. SE, Centro Internacional de Agriculture Tropical: Cali, Colombia, 1973.
49. Jaffe', W. G. Arch. Latinoam Nutr., 1977, 27, Suppl. No. 2.
50. Antunes, P. L.; Sgarbieri, V. C. J. Agric. Food Chem., 1980, 28, 935.
51. Kakade, M. L.; Evans, R. J. British J. Nutr., 1963, 17, 69.
52. Bressani, R.; Elias, L. G.; Valiente, J. British J. Nutr., 1963, 17, 69.

RECEIVED September 9, 1982.

Heat–Stir Denaturation of Cottonseed Proteins: Texturization and Gelation

JOHN P. CHERRY[1] and LEAH C. BERARDI

United States Department of Agriculture, Southern Regional Research Center, Agricultural Research Service, New Orleans, LA 70179

The developments of glandless, or gossypol-free, cotton (1, 2, 3,) and the liquid cyclone process (LCP) (4, 5) to mechanically extract gossypol-containing glands make it possible to produce edible vegetable protein products such as flours, concentrates and isolates from cottonseeds (6-11). An enriched bread containing toasted glandless kernels is presently marketed in selected areas of the United States under the trade name of "Proteina" (12).

Spinning and thermoplastic extrusion processes were applied to further expand the potential utilization of cottonseed as texturized protein products (13-22). Berardi and Cherry (23) developed a simple heat-stir procedure to produce texturized products and self-sustaining gels from glandless cottonseed protein isolates. Water suspensions of classical and storage protein isolates at pH's of 4.5 to 9.0, were heated to 90°C with stirring. Texturized products having the mouthfeel and chewiness of cooked meat were formed.

Although methods are available for texturizing vegetable proteins, there is limited information on the changes that proteins undergo when they are converted to texturized or gelled products (22, 24, 25). During the spinning and thermoplastic extrusion processes, moistened vegetable proteins are fed into an extruder, mixed and heated. Data suggest that during these processes, the polypeptides in the proteins hydrate, gradually unravel, become stretched by the shearing action of the rotating screw flites and form new cross-linkages which result in fibrous structures. The heated plasticized mass is forced through a die at the extruder discharge to form expanded texturized strands of protein with meat-like characteristics.

Rhee et al. (24) showed that texturized fibers formed from raw ingredients at acid pHs had stronger physical properties, rougher surfaces and smaller diameters than those made at alkaline pH. Proteolytic enzyme modified proteins containing mainly low molecular weight polypeptides could not maintain textural integrity upon hydration and retort heat treatment.

[1] Current address: United States Department of Agriculture, Eastern Regional Research Center, Philadelphia, PA 19118.

High NaCl concentrations interfered with, while $CaCl_2$ enhanced, the texturization processes. An understanding of the mechanisms involved in the changes of protein produced consistencies (or relative viscosities) during the formation of textured or gelled products at varying protein concentrations, temperatures and pHs is important in advancing the formulation and processing of aqueous food systems.

This chapter examines the consistency changes, i.e., thixotropism, gelation and texturization, of various aqueous suspensions of cottonseed protein products during the heat-stir process (23) under varying conditions of concentration, temperature and pH. The C. W. Brabender Visco-Amylo-Graph, developed, as one of its functions, to continuously measure the consistency or relative textural changes of starch pastes during cooking and cooling (21, 26, 27) was extended to studies of similar changes with cottonseed protein products in aqueous suspensions.

Experimental Procedures

Preparation of basic protein materials. The protein products tested or evaluated included LCP, glandless and air-classified hexane-defatted cottonseed flours, concentrates and isolates. Glandless cottonseeds from the varieties Acala, Gregg-25-V and Watson GL-16-A were flaked, hexane-defatted and desolventized under laboratory conditions that included no heat treatments. Watson GL-16-B and Watson GL-16-C flours were prepared by commercial pilot plant operations which included heat in the desolventization step. The Watson GL-16-C flour was heated further to reduce its bacterial count and moisture content (28).

The Acala flour was further processed by air-classification into five fractions with a Pillsbury Laboratory Model No. 1 Unit (29). The first and second separations, which contained the smallest sized particles, were combined and labeled as the fines fraction. Separations three and four were combined as the coarse fraction, and five, consisting mainly of debris, was discarded.

Two glandless cottonseed concentrates were prepared as follows: 1) double extraction of the Acala flour with water (H_2O, H_2O-extracted concentrate), and 2) extraction first with 0.008 M $CaCl_2$, and then with water ($CaCl_2$, H_2O-extracted concentrate). The resulting concentrates were dried by lyophilization. The classical isolate (CI) (30) containing both nonstorage proteins (NSP) and storage proteins (SP) was precipitated at pH 5.0 from a 0.034 N NaOH extraction (pH 10.5) of Acala flour, washed at pH 5.0 and spray-dried at pH 7.0. The NSP- and SP-isolates were prepared by the selective precipitation procedure (11). In summary, the pH of the 0.034 N NaOH extract (pH 10.5) was lowered to 6.8 to precipitate SP which was then water-washed and spray-dried at pH 6.8. The remaining whey was then adjusted to pH 4.0 to precipitate NSP, which was then water-washed at pH 4.8 and

spray-dried at pH 5.1. All of the dry protein products were
ground to flour consistencies in an Alpine 160 Z Kolloplex pin
mill.

Proximate compositions of the cottonseed protein products
are presented in Table I.

Analysis methods. Proximate and amino acid compositions of
cottonseed protein products were determined with established
procedures (31, 32, 33). Meat loaf products containing 0 to 50%
textured CI and SP-isolate were prepared by a basic recipe as
outlined by Berardi and Cherry (23). A 10-member taste-panel
compared cooked meat-loaf products with and without textured
cottonseed proteins on a scale of 1 to 4; the all-meat product
was arbitrarily assigned a value of 4.0 (23).

Chemical modifications and additives. The LCP-flour in
water was chemically modified by succinylation and acetylation
according to the method of Beuchat et al. (34). The anhydrides
were added to the protein suspensions in equivalent amounts
varying from 2.5% to 80% of the protein weight in the flour.

At 1% by weight concentration, $CaSO_4$, soybean flour and
concentrate, soluble starch, Toruway -49, -42 and -30 (yeast
proteins), sodium alginate, wheat gluten, casein, cysteine-HCl,
raffinose, sucrose, glucose, gossypol, NSP and crystalline
cellulose were added alone or in various combinations to selected
cottonseed protein suspensions prior to the texturization
studies. Other tests included the addition of 0.3% NaCl.

Slow heat-stir treatment. The C. W. Brabender Visco-Amylo-
Graph was used to study consistency (or relative viscosity)
changes of thixotropism, gelation and texturization that occur
during slow heat-stir (45 min to raise the temperature from 25°C
to 92.5°C) processing of aqueous cottonseed protein suspensions
(4% to 30% by weight of cottonseed products). The viscograph has
a rotating sample cup that contains stirring pins surrounded by
heating elements. The cup is sealed with a lid that contains a
second set of pins surrounded with cooling elements. Both sets
of pins are sensitive to changes in the consistency of the
aqueous protein suspension in the cup which are graphically
recorded for easy conversion to centipoise (cps) units. The
mechanical thermoregulator, operating in conjunction with the
heating and cooling system, permits the sample to be heated and
cooled through a programmed temperature-time cycle, and sample
consistency is recorded continuously. From the chart, the peak
or highest consistency (relative viscosity) during the heating
cycle, the apparent viscosity at any temperature during the
heating or cooling cycle, the viscosity stability of the hot
sample and the degree of setback on cooling can be determined.

The cup, containing 460 g of aqueous protein suspension and
rotating at a rate of 70 rpm, was 1) maintained for 3 min at 25°C

Table I. Proximate compositions of cottonseed protein products

Cottonseed Protein Product	Composition--% (Moisture-Free Basis)					
	Nitrogen	Protein[1]	Crude fiber	Lipid	Ash	Carbo-hydrates[2]
Glanded cottonseed						
LCP-flour	10.89	68.1	2.2	0.3	7.8	23.8
Glandless cottonseed						
Acala flour	10.61	66.3	2.4	1.0	8.7	24.0
Gregg-25-V flour	10.29	64.3	3.2	1.7	7.7	26.3
Watson GL-16-A, -B, -C flours[3]	9.17	57.3	4.2	2.7	9.2	30.8
Air-classified fine fraction[3]	11.86	74.3	1.7	0.5	10.1	15.1
Air-classified coarse fraction	9.59	59.9	3.2	1.0	8.0	31.1
H2O, H2O-extracted concentrate	12.15	75.9	3.7	2.6	8.5	13.0
0.008 M CaCl2, H2O-extracted concentrate	11.27	69.4	4.7	1.5	10.5	18.6
Classical protein isolate	15.30	95.6	<0.3	1.8	2.3	0.3
Nonstorage protein isolate	13.56	84.8	0.2	6.4	5.9	2.9
Storage protein isolate	16.67	104.2[4]	<0.2	0.5	2.4	0.2

1/ Crude protein = % N x 6.25.
2/ Includes crude fiber. On moisture-free basis, % carbohydrate = 100% - %(protein + lipid + ash).
3/ Contains more than 70% protein because air-classification has concentrated the protein by removal of some residual lipid, crude fiber and carbohydrate fractions and thus could be labeled as a concentrate.
4/ % N x 6.25 = 104.2% protein. Based on other components of SP-isolate; protein content is about 96.9%.

to establish equilibrium, 2) heated for 45 min to 92.5°C, 3) held for 10 min at 92.5°C, and 4) cooled for 45 min to 25°C. The viscograph was programmed to heat and cool at a temperature change rate of 1.5°C per min.

Rapid heat-stir treatment. The rapid heat-stir method simply involved 1) placing a 100 g protein suspension in a 400 ml beaker that was set in a silicone oil bath on a magnetic stirrer-heater and, 2) raising the temperature of the stirring suspension from 25°C to 95°C in 18-20 min (23). The final product was cooled to room temperature and examined for texturized or gelled properties.

Consistency Changes During Slow Heat-Stir Processing

Varietal effects. The viscograph consistency patterns of 18% Acala, Gregg-25-V and Watson GL-16-A laboratory prepared flour suspensions (at natural pHs of approximately 6.3) are presented in Figure 1. At approximately 72-79°C, or 31-36 min of heating, the consistencies of all three suspensions increased sharply. A little more heat was required to initiate the consistency increase of the Watson Gl-16-A flour suspension than the other two samples. Microscopic examination of the three suspensions showed that at the time of the consistency increases, the protein bodies agglomerated and remained this way throughout the remaining test periods. During the 10 min hold period at 92.5°C, the consistencies of the three suspensions remained unchanged, then as cooling progressed to 25°C, they slowly increased. Although all suspensions had increased relative viscosities after all of the viscograph treatments, they remained pourable.

Lengthening the hold period at 92.5°C, from 10 min to 20 min, caused slightly higher increases in consistencies during the cooling period. The relative viscosity increases during cooling are most likely due to the occurrence of small amounts of molecular interactions (hydrophobic, hydrophilic, Van der Waals, etc.) between proteins and nonprotein constituents that are induced as a result of heat denaturation/renaturation (22, 35).

The small differences in consistencies among suspensions of the three flours were obtained repeatedly and thus may be variety-related. These data suggest that the consistency patterns may be used to identify varietal sources of cottonseed flours that have been defatted and desolventized by similar processes; they may also be used to distinguish cottonseed products from those of other oilseeds.

Concentration effects. At different concentrations of 20%, 18%, and 15%, the initial rises in consistencies of Gregg-25-V flour suspensions occurred at 72°C, 79°C and 81°C, respectively (Figure 2). The 10% suspension had a very small rise in

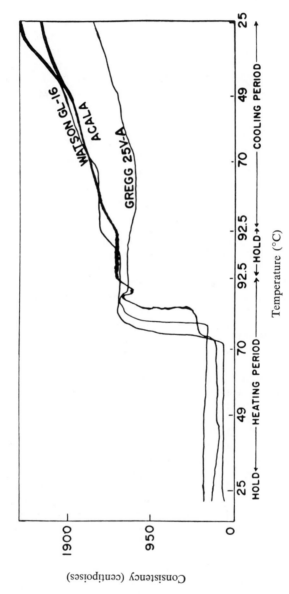

Figure 1. Viscograph consistency patterns of 18% suspensions of cottonseed flours from different cultivars.

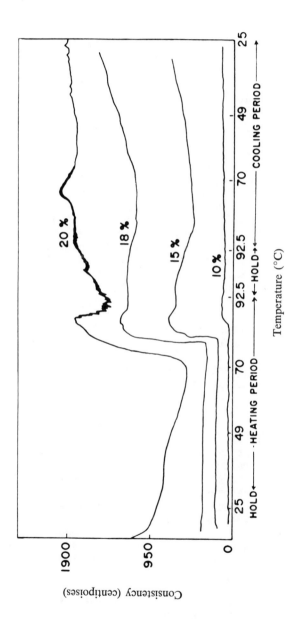

Figure 2.   Viscograph consistency patterns of 10–20% suspensions of Gregg-25-V flour.

consistency at 85°C. These data showed that the higher the concentration of flour in the suspension, the lower the temperature
required for an initial rise in consistency. The higher
concentration suspensions, with more flour constituents to react,
had greater consistencies at the completion of the experiments
supporting the idea that molecular interactions are occurring
(37); at higher concentrations, more reactions occur between
constituents. None of these suspensions exhibited gelling or
texturing properties. Similar observations were made with the
Acala and Watson GL-16-A flours.

pH effects. Figure 3 illustrates the type of changes that
occur in the consistency patterns of Acala flour suspensions at
pH's 3.5, 6.3 (natural pH of the suspension), 8.5 and 9.5. The
pH 6.3 suspension's consistency exhibited the typical changes
described for the 18% suspension of Gregg-25-V flour in Figure 2.
Throughout the treatment, the pH 3.5 suspension rose only slightly
between 70°C and 90°C, remaining below 950 cps throughout the
experiment. The pH 8.5 and 9.5 suspensions displayed thixotropic
properties between 30°C and 70°C. After decreases to consistencies similar to those of the low pH samples at 70°C, the
consistencies of these suspensions increased similarly to high
values found at 92.5°C, remained relatively constant during the
holding period, and then increased further during the cooling
period. However, no gelling or texturing properties were noted
in these experiments. Under conditions of heat and alkaline pH,
the protein bodies ruptured releasing the SP to interact with
each other, NSP and nonprotein components. The degree of protein
body rupture and the amount of protein and nonprotein component
solubilization may account for the final higher consistency
values by the pH 8.5 suspension than those at pH 9.5 (30) at the
end of the cooling period.

Processing effects. The consistency patterns of differently
processed flours, Watson GL-16-A, -B, and -C are shown in Figure
4. The consistency patterns of Watson GL-16-B and -C flour
suspensions which were processed under commercial conditions that
included heat treatment(s) never attained the high level of consistency of the laboratory-prepared Watson GL-16-A flour. More
heat was required to produce the initial consistency rises of the
-B and -C flour suspensions. Evidently, heat processing of
cottonseed flours affects their ability in aqueous suspension to
become more viscous during the viscograph procedure. None of
these flour suspensions, however, attained gelled or texturized
properties.

Air-classified product changes. The 18% suspension of the
air-classified fine (or protein concentrate) fraction was
converted by the viscograph treatment to a distinct dual-phase
system of liquid and beads of agglomerated protein bodies. This

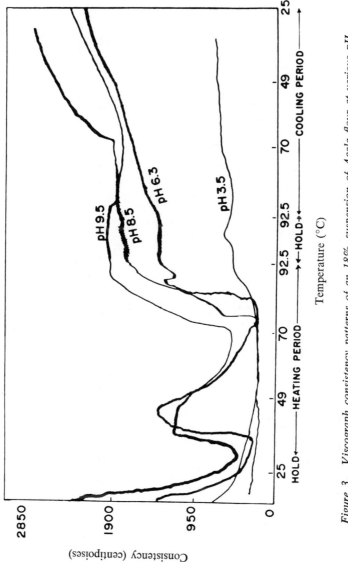

Figure 3. Viscograph consistency patterns of an 18% suspension of Acala flour at various pH values.

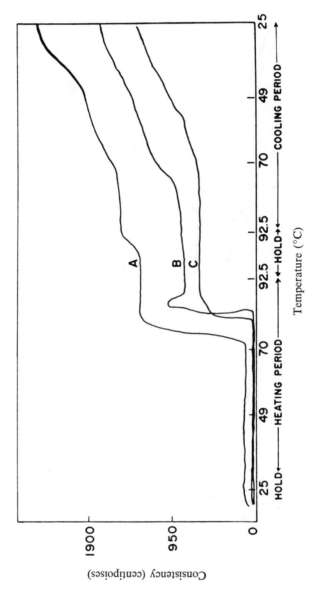

Figure 4. Viscograph consistency patterns of 18% suspensions of Watson GL-16 flours prepared by different procedures (see Experimental Procedures).

occurred during the rise in consistency at approximately 85°C
which was followed by a decrease at 92.5°C and no further changes
during the holding and cooling stages (Figure 5). The protein
beads ranged in size from just visible to about 3.0 mm in diameter
and suggested that the membranes of protein bodies had become
resistant to rupture. During the holding and cooling stages, the
protein bodies only rolled across the revolving vertical rods in
the cup and thus exerted no measurable additional contribution to
consistency.

The 18% suspension of the air-classified coarse fraction
which was low in proteins and high in cell wall fragments and carbohydrates (Table I) reached high consistencies of 2700-2900 cps
during the holding and cooling periods (Figure 5). The final suspension had smooth, pudding-like and self-sustaining properties.
Increasing the concentration of the air-classified coarse fraction in the suspension improved the self-sustaining properties of
the pudding-like product. Evidently, many charged groups are
associated with the membrane and cell wall fragments, SP and NSP
which form intermolecular interactions during the heating period
and develop a partially gelled system during cooling.
Catsimpoolas and Meyer (36) noted this occurrence with soybean
proteins and described the process in terms of a progel state,
i.e., a fluid state characterized by high viscosity.

Adjusting the pH of the air-classified concentrate or fine
fraction suspension to 3 improved its ability to become more
viscous between 75°C and 92.5°C than that at pH 6.3 (Figure 6).
This increase continued throughout the holding period, gradually
decreased during cooling to 70°C, then remained relatively
unchanged to 25°C. During the cooling period, the pH 3 suspension did not separate into two phases until the temperature
dropped to below 70°C. The two phase system of the pH 6.3
suspension formed shortly after the rapid rise in consistency at
approximately 90°C. At pH 8.5, the suspension demonstrated the
same consistency increase at 47°C as noted for the Acala flour
suspensions at pH 8.5 and 9.5 (Figure 3). The rest of the consistency pattern of this product was similar to that of the pH
6.3 suspension except that values remained approximately two
times higher during the holding and cooling periods.

At pH 3.0, the suspension of the air-classified coarse
fraction exhibited a consistency pattern that was high (3800 cps)
during the initial holding period and then dropped to between 950
cps and 1900 cps during the heating, holding and cooling periods
(Figure 7). The pH 8.5 suspension also dropped from 3300 cps to
950 cps between 25°C and 70°C, but increased rapidly to approximately 3300 cps at 92.5°C. This high consistency remained until
the end of the cooling period. The pH 6.3 suspension stayed
below 950 cps between 25°C and 70°C, then increased similarly to
that of the pH 8.5 suspension to 2850 cps at 92.5°C and remained
at this consistency throughout the experiment. Evidently, the
mechanical forces of the viscograph were able to overcome the gel

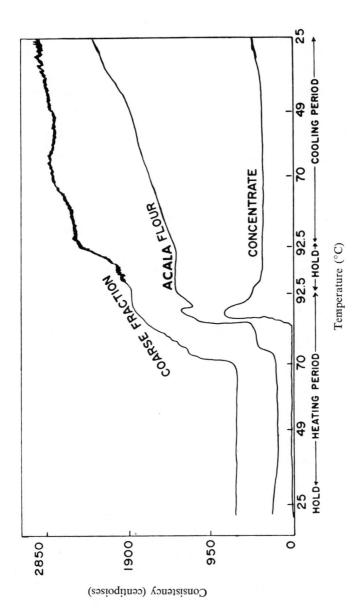

Figure 5. Viscograph consistency patterns of 18% suspensions of Acala flour and the fine (concentrate) and coarse fractions of air classification.

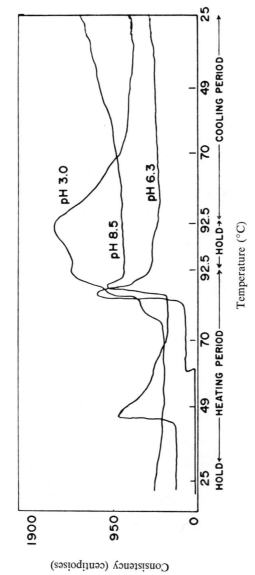

Figure 6. Viscograph consistency patterns of 20% suspensions of air classified protein concentrate (fine fraction) at various pH values.

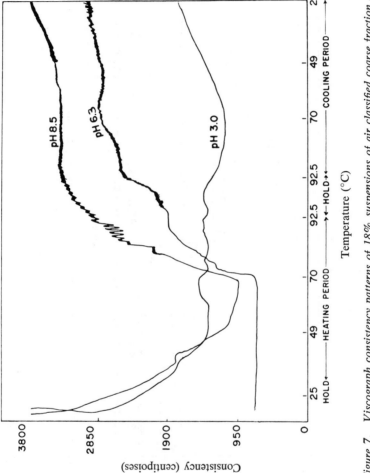

Figure 7. Viscograph consistency patterns of 18% suspensions of air classified coarse fraction at various pH values.

forming properties of the pH 6.3 and 8.5 suspensions between 25°C
and 70°C. The suspensions formed final products that had pudding-
like properties.

Concentrate changes. The consistency patterns of the water-
extracted concentrate and 0.008 M $CaCl_2$, H_2O double extracted
concentrate changed similarly during the viscograph treatment;
i.e., initial thixotropic properties and then an increase in consistency during the heating, holding and cooling periods (Figure
8). The only difference was that the double water (H_2O-H_2O)-
extracted concentrate had higher consistencies (500 to 1000 cps)
than the $CaCl_2$-H_2O-concentrate during the holding and cooling
periods. The $CaCl_2$ evidently hindered protein-protein interactions and prevented the formation of a highly viscous, pudding-
like product similar to that of the H_2O-H_2O-concentrate. This is
contrary to observations noted in spinning and extrusion
processes where $CaCl_2$ decreased the bulk density, but increased
the strengths of the texturized proteins (24).

Isolate changes. The consistency patterns of 20% CI
suspensions at pH's 5.5, 7.0 and 8.5 are presented in Figure 9.
The pH 5.5 suspension demonstrated little change in consistency
during the heating, holding and cooling periods. This was not
unexpected since at pH 5.5, cottonseed NSP and SP have low
solubilities. Adjusting the pH to 7.0 increased NSP solubility
and caused a rapid rise in consistency to occur, due to the SP,
as the temperature was increased to 83°C. The consistency of
this suspension declined during the cooling period and produced a
thickened, but pourable, final product. The pH 8.5 suspension,
which contains solubilized NSP and SP, was thixotropic between
25°C and 77°C, then showed a great increase in viscosity from
less than 950 cps to 4700 cps at 92.5°C. The suspension became
highly thixotropic during the holding period dropping in
consistency to approximately 1900 cps, but then increased
significantly during the cooling period. The final product was
thick, but pourable.

Reducing the amount of CI in the pH 8.5 suspension to 18%
lowered the degree of rise at 83°C to approximately 950 cps
(Figure 10). The suspension's consistency decreased slightly
during the hold period and then remained relatively unchanged
during the cooling period. The pH 7.0 18% suspension increased
to 3800 cps at 92.5°C, dropped to 2850 cps during the hold period,
and remained at this level during the cooling period. These
values were higher than those of the 20% suspension. The pH 5.5
18% suspension showed no changes during the study.

Suspensions containing 18% NSP-isolate, adjusted to pH's 5.8
and 8.5 showed no consistency changes during the heating, holding
and cooling treatments (Figure 11). The pH 5.5 suspension's
consistency did not change during the heating and holding

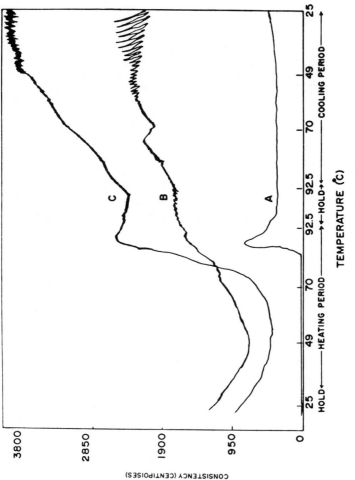

Figure 8. Viscograph consistency patterns of 18% suspensions of air classified concentrate (fine fraction) (A), CaCl₂–H₂O extracted air classified concentrate (B), and H₂O–H₂O extracted air classified concentrate (C).

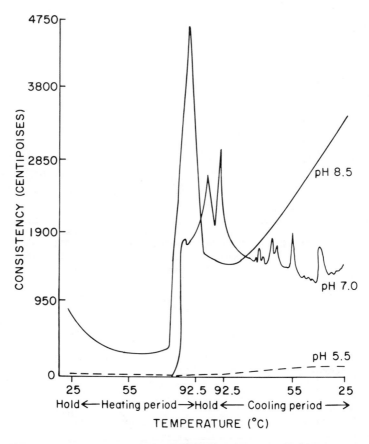

Figure 9.   Viscograph consistency patterns of 20% suspensions of classical protein isolate (CI) at various pH values.

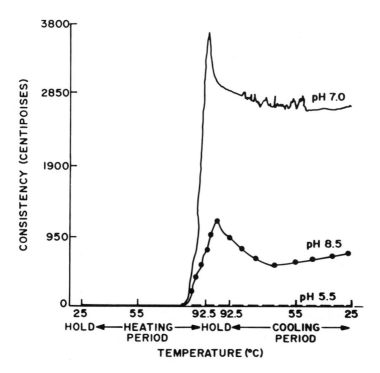

Figure 10. Viscograph consistency patterns of 18% suspensions of CI at various pH values.

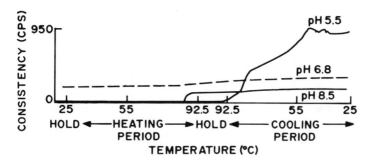

Figure 11. Viscograph consistency patterns of 18% suspensions of nonstorage protein (NSP) isolate at various pH values.

periods, but did increase gradually to 950 cps during the cooling period to form a slightly thickened but pourable product.

Figure 12 shows the consistency patterns of 12% SP-isolate suspensions at various pH values. At pH's 4.5 and 7.0, no consistency changes were noted thoughout the experiment; the isoelectric pH of SP is 6.8 and its solubility remains low at pH 4.5. The pH 6.0 and 8.0 suspensions showed small consistency increases during the heating period at about 90-92.5°C. Suspensions at pH's 5.0, 5.5 and 8.5, however, were converted to textured protein products when the temperature reached 85-90°C. It is unclear why texturization of SP occurred at pH's 5.0 and 5.5 where these globulins are highly insoluble. Apparently, certain structural changes do occur in these globulins at these pHs that enable them to undergo the interactions necessary for texturization. This texturization could be done with suspensions containing as low as 7% SP-isolate in a pH 5.5 suspension (Figure 13). At pH 8.5, textured products were obtained with 12% or higher SP-isolate suspensions (Figure 14). The textured products formed at pH's 5.5 and 8.5 looked like and had the mouthfeel and chewiness of cooked meat (Figure 15). Addition of a beef bouillon cube to the suspension followed by texturization yielded products that tasted like cooked hamburger.

Microscopic examination of an SP-isolate suspension during the heating stage up to texturization showed the following sequence of changes. The initial suspension contained small spheres of spray-dried SP. At 45°C, these sphere-like particles began to expand indicating that they were becoming hydrated. This increased hydration continued until the temperature of the suspension reached 72°C. The hydrated spheres then became glassy-like, undergoing a melt stage, then immediately converted to fiber-like particles that were in parallel alignment. The fiber-like particles were probably disrupted polypeptides of the globulin proteins. This was followed by the production of a mass of entwined fibers which suggested intermolecular interactions similar to those that occur during spinning and thermoplastic extrusion processes were also occurring in the heat-stir process (24).

Kinsella (25) reported that texturization during the extrusion process involved hydration of globular proteins in protein bodies which was followed by unravelling and stretching of the proteins by the shearing action of the rotating screw-filter. The proteins became aligned in sheaths. While passing through the extruder die, the proteins compressed further, laminated longitudinally and denatured. Changes occurring while the proteins emerged from the die resulted in production of air-space vacuoles within the laminated extrudate and rapid thermo-setting of the stretched protein fibers. Similar molecular changes and interactions could be occurring during the slow heat-stir process without the compression step used in extrusion processes.

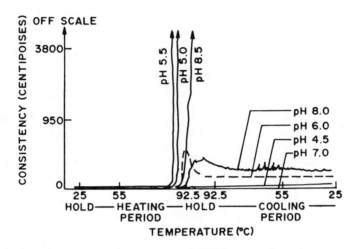

Figure 12.   Viscograph consistency patterns of 12% suspensions of storage protein (SP) isolate at various pH values.

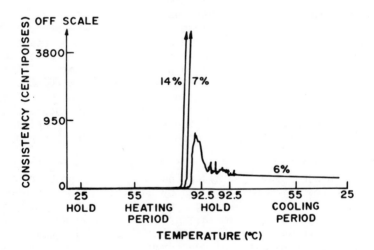

Figure 13.   Viscograph consistency patterns of 6, 7, and 14% suspensions of SP-isolate at pH 5.5.

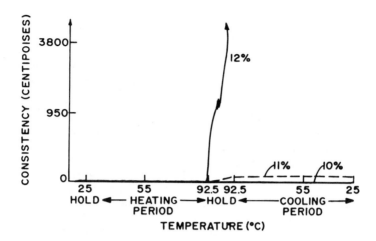

Figure 14.   Viscograph consistency patterns of 10, 11, and 12% suspensions of SP-isolate at pH 8.5.

Figure 15.   Texturized cottonseed SP.

Addition of 1% or more of NSP-isolate to a 12% SP-isolate suspension at pH 5.5 hindered the texturing process (Figure 16). Non-storage proteins or any number of non-protein constituents in these products could be interferring with the texturization mechanism of the SP-isolate.

Figure 17 shows a comparison of the consistency changes during slow heat-stir treatment of 18% suspensions of SP-isolate without and with added 1% NSP-isolate to those occurring in suspensions of the H_2O-H_2O-concentrate and the air-classified fine concentrate at pH 5.5. The H_2O-H_2O-concentrate was relatively free of NSP. It contained mainly protein bodies with SP and any carbohydrate moities not extracted with water. The air-classified fraction contained maninly NSP, protein bodies with their SP contents and carbohydrates, and yielded a dual-phase system of a liquid and beads of agglomerated protein bodies upon treatment. However, the H_2O-H_2O-concentrate increased in consistency during the cooling period. The final pudding-like consistency may have been due to structural changes in SP content and especially in the fibrous and pectinaceous components under conditions of the slow heat-stir treatment. On the other hand, the changes in the latter's consistency may be the result of the slow heating rate that permitted some protein denaturation before heat-setting of proteins into the fibrous networks of the final product.

Consistency Changes During Rapid Heat-Stir Processing

Twenty percent SP-isolate and 16% CI suspensions at various pHs of 3.0 to 10.0 were subjected to a rapid heat-stir (20 min to raise the temperature of a stirring protein suspension from 25°C to 95-98°C) to determine their texturability by a method other than that used with the C. W. Brabender Visco-Amylo-Graph (see Experimental Procedures). Table II compares the results of these experiments. Texturization of SP-isolate by the viscograph method only occurred in suspensions at pH's 5.0, 5.5 and 8.5. The rapid heat-stir method produced texturized products in all suspensions at pH's 5.0 to 9.0, and gelled products at pH's 3.0, 4.0 and 10.0. Gels made at pH's 3 and 10 were dark colored and translucent. At pH 4.0, they were light cream-colored and opaque. The gels were self-sustaining, i.e., they maintained their shape at room temperatures. No gelled products were produced with the slow heat-stir process. The rapid heat-stir method converted pH 3.0, 4.0 and 10.0 suspensions of CI to gelled products. Like the suspensions of the slow heat-stir process, those at the other pH values became only slightly thickened pourable suspensions. Addition of 0.3% NaCl to the CI suspensions caused those at pHs between 4.0 and 9.0 to texturize during the rapid heat-stir process. Possibly, the salt reduced the solubility of NSP or another component(s) and thus reduced interference with the texturization properties of SP. Figure 18 shows the textural

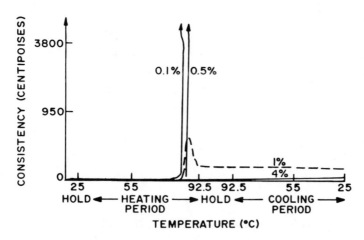

Figure 16. Viscograph consistency patterns of 18% suspensions of SP-isolate at pH 5.5 after addition of various percentages of NSP-isolate.

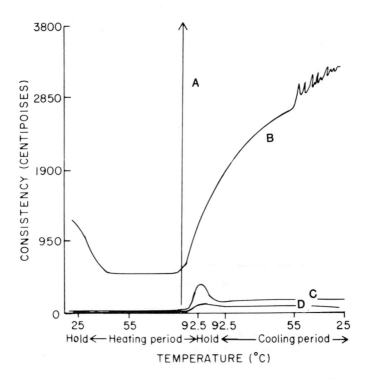

Figure 17. Viscograph consistency patterns of 18% suspensions of SP-isolate at pH 5.5 with and without 1% NSP-isolate, air classified concentrate (fine fraction), and H₂O–H₂O extracted concentrate. Key: A, SP-isolate; B, H₂O–H₂O extracted concentrate; C, air classified fine fraction; and D, SP-isolate with 1% NSP.

Table II. Effect of rate of heating, pH, and added NaCl on texture changes[1] in 16% CI and 20% SP-isolate suspensions (23)

Suspension pH	16% CI Suspension			20% SP-Isolate Suspension	
	Slow Heating[2]	Rapid Heating[3]		Slow Heating[2]	Rapid Heating[3]
	w/o Salt	w/o Salt	w/Salt[4]	w/o Salt	w/o Salt or w/Salt[4]
3.0	NVC	Gelled	Gelled	NVC	Gelled
4.0	NVC	Gelled	Textured	NVC	Gelled
5.0	NVC	Thicker	Textured	Textured	Textured
5.5	NVC	Thicker	Textured	Textured	Textured
6.0	NVC	Thicker	Textured	Sl. thicker	Textured
7.0	Beady	Thicker	Textured	Sl. thicker	Textured
8.0	Thicker	Thicker	Textured	Thicker	Textured
8.5	Thicker	Thicker	Textured	Textured	Textured
9.0	Thicker	Thicker	Textured	Sl. thicker	Textured
10.0	Thicker	Gelled	Gelled	Sl. thicker	Gelled

1/ Where NVC = no visible change; sl. thicker = slightly thickened, pourable suspension; thicker = more viscous but pourable suspension; gelled = self-sustaining gel; textured = definite fibrous structure resembling cooked meat; beady = two-phase system of beads (coagulated protein) and liquor.
2/ 45 min heating to raise temperature from 25° to 92.5°C.
3/ 20 min heating to raise temperature from 25° to 95.0°-98°C.
4/ 0.3% NaCl added to original protein suspension.

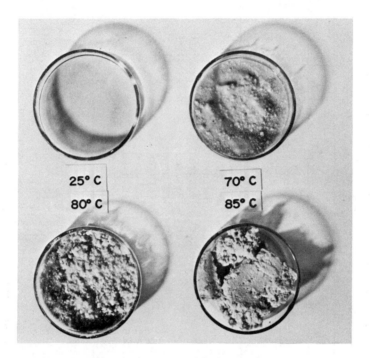

Figure 18.   Texture changes during rapid heating of a 20% SP-isolate suspension at pH 5.5. (Reproduced from Ref. 23.)

changes that occur during the rapid heat-stir processing of 20% SP-isolate at pH 5.5. Texturization of SP-isolate or CI gave products demonstrating 160% to 250% increases in weight.

None of the cottonseed flour suspensions, with and without 0.3% NaCl and at various pHs, textured or gelled during the rapid heat-stir method as was the case with the viscograph process. Suspensions of flour containing 0.3% NaCl heated first to 60°C for 30 min, cooled and then rapid-heat-stirred, formed weakly texturized products; i.e., the texturized structures could be disrupted in running water. A pH 5.5 suspension of H_2O-H_2O-concentrate with 0.3% NaCl formed a similar texturized product. This was also noted with a pH 10.5 extract of NSP and SP from cottonseed flour that had been adjusted to neutral pH and lyophilized. All of these textured samples had less firm structures than the textured SP-isolate and CI (with added NaCl) prepared by the rapid heat-stir method. In all cases, suspensions had to contain 7% or more protein to form any type of texturized product.

Various additives and protein modifications were examined to determine their effects on the texturing and gelling properties of SP-isolate and CI by the rapid heat-stir process. Of the additives listed in <u>Experimental Procedures,</u> soluble starch at a 1% level enhanced texturization and gellation. Addition of 1% crystalline cellulose increased the yield of textured protein, probably because of an increase in water uptake. Raffinose (1%) inhibited texturization. The other compounds did not affect the texturization process. Succinylation and acetylation of the proteins in LCP-flour improved their ability to texturize, but the products were disrupted in running water.

The texturized protein products can be dehydrated for prolonged shelf-life and then rehydrated as needed by letting them stand for 30 min in warm water. Table III presents the composition of dehydrated and rehydrated texturized SP-isolate. Also presented is the composition of dehydrated and rehydrated texturized SP-isolate plus added 7% cottonseed oil. The rehydrated products without and with oil showed percentages of weight increases of 168 and 102, respectively.

Meat loaves with various percentages of textured SP-isolate and CI, and 0.3% NaCl that were prepared at different pHs are shown in Figures 19 and 20. Similar products with and without added oil are shown in Figure 21. Incorporation of texturized protein into meat loaves increased their protein content and decreased lipid, crude fiber and ash content (Table IV). Essential amino acid content of the cooked meat loaves was improved by the addition of texturized proteins (Table V).

Taste panelists reported no significant differences in initial flavor, general flavor, aftertaste, chewiness, color, texture and moisture of the 90% meat-10% texturized SP-isolate-meat loaf and the standard reference all-meat loaf (Table VI). The 20% texturized SP-isolate-meat loaf was similar to the all-

Table III. Characteristics of dry and rehydrated textured SP-isolate ± salad oil (23)

Parameter	Textured SP-Isolate			
	14% SP-Isolate in pH 5.5 Suspension[1]		14% SP-Isolate and 7% Oil in pH 5.5 Suspension[1]	
	Lyophilized	Rehydrated	Lyophilized	Rehydrated
Weight increase (%)	-	168.0	-	102.0
Composition (%)				
Moisture	4.0	62.7	3.1	50.5
Protein[2]	97.8	36.4	69.2	35.8
Lipid	0.4	0.2	26.4	13.1
Crude fiber	0.4	0.2	0.4	0.2
Ash	1.4	0.5	0.9	0.4

1/ Suspension subjected to rapid heat-stir treatment.
2/ % Crude protein = % N x 6.25.

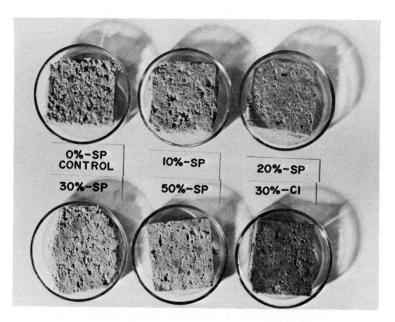

Figure 19. Meat loaves containing varying amounts of texturized SP-isolate and CI. (Reproduced from Ref. 23.)

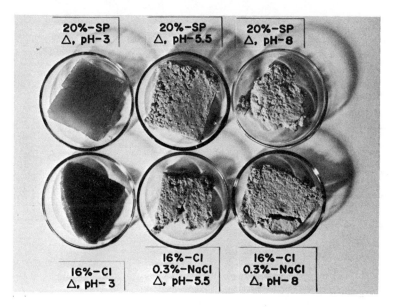

Figure 20. Gelled products (20% SP, pH 3; 16% CI, pH 3) and meat loaves containing 20% texturized SP-isolate and 16% texturized CI plus 0.3% NaCl.

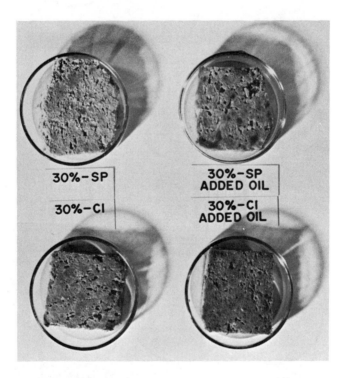

Figure 21.   Meat loaves containing 30% texturized SP-isolate and CI, with and without cottonseed oil.

Table IV. Proximate compositions of textured cottonseed proteins, raw meat, and cooked meat loaves (23)

Sample	Composition – %(Moisture-Free Basis)			
	Protein	Lipid	Ash	Crude Fiber
Textured SP-isolate	97.3	0.2	2.2	<0.1
Textured CI	92.3	1.6	6.0	<0.1
Raw meat	50.7	46.4	2.4	0.3
Cooked meat loaves				
100% meat	63.1	31.7	4.4	0.8
90% meat, 10% TSPI1/	72.3	23.8	3.1	0.5
80% meat, 20% TSPI	70.7	28.3	3.5	0.3
70% meat, 30% TSPI	73.4	21.8	3.3	0.2
70% meat, 30% TSPI + oil	69.7	29.1	3.0	0.2
70% meat, 30% TSPI + oil, flavor	70.8	24.7	4.3	0.2
50% meat, 50% TSPI	82.1	14.5	2.9	0.2
70% meat, 30% TCI1/	71.9	23.7	3.8	<0.2
70% meat, 30% TCI + oil	66.5	28.7	4.3	<0.2
70% meat, 30% TCI + oil, flavor	69.7	24.1	6.0	<0.2

1/ TSPI = texturized storage protein isolate; TCI = texturized classical isolate.

Table V. Essential amino acid composition of textured SP-isolate, textured CI, raw meat, and cooked meat loaves (23)

Amino Acid [1/]	Textured SPI [2/]	Textured CI	Raw Meat	100% Meat	90% Meat, 10% SPI	80% Meat, 20% SPI	70% Meat, 30% SPI	50% Meat, 50% SPI	70% Meat, 30% SPI, Oil	70% Meat, 30% SPI, Oil, Flavor	70% Meat, 30% CI	70% Meat, 30% CI, Oil	70% Meat, 30% CI, Oil, Flavor
Valine	3.67	2.81	0.82	1.14	1.17	1.15	1.53	1.53	1.39	1.73	1.76	1.48	1.92
Isoleucine	2.46	2.78	0.71	0.99	0.95	0.84	1.15	1.17	1.05	1.27	1.18	1.05	1.44
Leucine	5.89	5.27	1.28	1.66	1.73	1.71	2.21	1.99	1.89	2.17	3.56	2.58	3.40
Threonine	3.01	2.61	0.73	0.88	0.98	0.95	1.16	0.99	1.01	1.05	1.86	1.78	1.83
Methionine	0.72	0.93	0.37	0.48	0.52	0.43	0.42	0.45	0.39	0.51	0.89	0.61	0.86
Phenylalanine	6.18	4.91	0.66	0.90	1.30	1.45	1.94	1.73	1.69	1.90	2.69	2.09	2.47
Tyrosine	3.04	2.33	0.51	0.65	0.73	0.77	1.12	0.91	0.95	1.00	1.83	1.23	1.49
Lysine	3.38	3.30	1.54	1.66	1.57	1.32	1.51	1.46	1.36	1.52	2.78	1.91	2.73
1/2-Cystine	0.86	1.04	0.23	0.29	0.33	0.32	0.41	0.38	0.35	0.37	0.66	0.49	0.60

1/ Amino acid (g/100 g sample).
2/ SPI = storage protein isolate.

Table VI. Sensory evaluation[1] of cooked meat loaves containing various amounts of textured SP-isolate (23)

| Quality Attributes | % Meat Replaced With Texturized SP-Isolate | | | | |
	0%[2]	10%	20%	30%	50%
Aroma	4.0c	2.8b	2.6b	1.5a	2.4ab
Initial flavor	4.0c	3.1bc	2.7b	1.2a	2.3b
General flavor	4.0d	3.2cd	2.0bc	1.0a	1.8ab
Aftertaste	4.0c	3.1bc	3.2bc	1.9a	2.4ab
Juiciness	4.0c	2.8b	3.0b	1.2a	2.0a
Chewiness	4.0d	3.1cd	3.0bc	1.6a	2.1ab
Color	4.0c	3.4bc	3.2bc	2.1a	2.9ab
Texture	4.0c	3.0bc	3.3c	1.5a	2.0ab
Moisture	4.0c	3.1bc	3.5c	1.6a	2.5ab
Overall quality	4.0d	2.9c	3.0c	1.2a	1.7b

1/ The panelists used a scale of 1 to 4 to score meat loaves. Taste panel evaluations subjected to one-way analysis of variance and ranking of means. Mean values having a common letter are not significantly different (P>0.05).
2/ Standard all-meat loaf assigned an arbitrary value of 4.0.

meat loaf in aftertaste, color, texture and moisture. None of
the meat loaves containing texturized SP-isolate had aroma,
juiciness, and overall quality as good as the reference loaf.
Meat loaves containing 30% texturized SP-isolate and added oil,
with or without added flavor, scored higher on general flavor
characteristics than the 30% SP-isolate-meat loaf (Table VII).
      Meat loaves prepared with 30% texturized CI plus oil, or oil
and added flavoring, were scored higher in initial flavor, juiciness and moisture than the 30% texturized CI-meat loaves without
the oil and flavor. However, quality attributes of these products were not as good as those of the all-meat loaf (Table
VIII).
      In general, meat loaves formulated with heavy beef and added
beef flavor and lipid to replace those properties lost due to the
substitution of meat with the bland, fat-free, texturized SP-isolate and CI should become more acceptable with higher levels of
texturized cottonseed proteins.

Conclusions

      Increasing the temperature of a constantly stirring aqueous
suspension of cottonseed SP-isolate from 25°C to 95-98°C in 20
min (rapid heat-stir method), or to 92.5°C in 45 min (slow heat-
stir method) produced texturized products with mouth-feel and
chewiness similar to cooked chopped meat. The texturized
products were comparable to those formed by the spinning and thermoplastic extrusion processes. The cottonseed CI, which contains
both NSP and SP, requires suspension in a 0.3% NaCl solution
before it texturizes. Formation of the textured products was
pH-related, occurring mainly when the suspensions were between
pH's 5.0 and 9.0. At the other pH values, suspensions form
heat-stable, self-sustaining, gelatin-like products or thickened
pourable suspensions. The suspensions should contain 7% or more
protein isolate to obtain quality textured products or gels.
Texturization of SP-isolate and CI provides products demonstrating
160% and 250% weight increases, respectively. Addition of 1%
soluble starch produces similar results as 0.3% NaCl, in that it
enhances the texturability of the protein isolates. Cottonseed
flour undergoes texturization in the presence of 0.3% NaCl by the
rapid heat-stir method after its proteins have been preheat-
treated at 60°C for 30 min and cooled. The texturization is
improved after the flour's proteins have been modified by
succinic or acetic anhydride. Although the degree of texturing
of these treated flours was not similar to the optimum texturability of the SP-isolate and CI, the likelihood exists for
improving their functionality by further protein modification(s).
The texturized proteins can be dehydrated to provide stable
storage products and rehydrated to their original form when
needed to formulate fabricated meat-like foods and feeds.

Table VII. Sensory evaluation[1] of cooked meat loaves containing 30% textured SP-isolate and with or without added oil and flavor (23)

Quality Attribute	Cooked Meat Loaves			
	Std Ref[2]	30% SPI[3]	30% SPI, Oil, Flavor	30% SPI, Oil
Aroma	4.0b	1.5a	2.0a	2.0a
Initial flavor	4.0b	1.2a	2.0a	2.1a
General flavor	4.0c	1.0a	2.1b	2.1b
Aftertaste	4.0b	1.9a	2.4a	1.8a
Juiciness	4.0b	1.2a	2.0a	2.0a
Chewiness	4.0b	1.6a	1.8a	1.8a
Color	4.0b	2.1a	2.4a	2.1a
Texture	4.0b	1.5a	1.8a	1.6a
Moisture	4.0b	1.6a	2.2a	1.7a
Overall quality	4.0b	1.2a	1.7a	1.8a

1/ The panelists used a scale of 1 to 4 to score meat loaves. Taste panel results subjected to one-way analysis of variance and ranking order of means. Mean values having a common letter are not signifi- cantly different (P>0.05).

2/ Std Ref = standard reference all-meat loaf assigned arbitrary value of 4.0.

3/ SPI = storage protein isolate.

Table VIII. Sensory evaluation[1]/ of cooked meat loaves containing 30% textured CI and with or without added oil and flavor (23)

Quality Attribute	Cooked Meat Loaves			
	Std ref[2]/	30% CI	30% CI, Oil	30% CI, Oil, Flavor
Aroma	4.0b	1.5a	1.4a	2.1a
Initial flavor	4.0c	1.7a	1.9ab	2.7b
General flavor	4.0b	1.9a	1.8a	2.6a
Aftertaste	4.0b	2.2a	2.1a	2.4a
Juiciness	4.0c	2.4a	3.1b	3.0ab
Chewiness	4.0b	2.2a	2.4a	2.8a
Color	4.0b	2.2a	2.5a	2.9a
Texture	4.0b	2.2a	2.5a	2.7a
Moisture	4.0c	1.9a	2.8b	2.6b
Overall quality	4.0b	2.0a	1.9a	2.5a

[1]/ The panelists used a scale of 1 to 4 to score meat loaves. Taste panel results subjected to one-way analysis of variance and ranking order of means. Mean values having a common letter are not significantly different (P>0.05).

[2]/ Std Ref = standard reference all-meat loaf assigned arbitrary value of 4.0.

Acknowledgments

Use of a company and/or product named by the U.S. Department of Agriculture does not imply approval or recommendation of the product to the exclusion of others which may also be suitable.

Literature Cited

1. Miravalle, R. J. Proc. Conf. Protein-Rich Food Prod. from Oilseeds; ARS 72-71, U.S. Department of Agriculture, 1969; p. 68.
2. Agricultural Research Service, U.S. Department of Agriculture and National Cottonseed Products Association, Inc. Glandless Cotton: Its Significance, Status and Prospects, Proc. of a Conf., Dec. 13-14, 1977, Dallas, Texas.; 184 pp.
3. Hess, D. C. Cereal Foods World, 1977, 22, 98.
4. Vix, H. L. E.; Eaves, P. H.; Gardner, H. K.; Lambou, M. G. J. Am. Oil Chemists' Soc., 1971, 48, 611.
5. Gardner, H. K.; Hron, R. J.; Vix, H. L. E. Cereal Chem., 1976, 53, 549.
6. Olsen, R. J. Oil Mill Gazett., 1973, 77(9), 7.
7. Cherry, J. P.; Berardi, L. C.; Zarins, Z. M.; Wadsworth, J. I.; Vinnett, C. H. Cottonseed protein derivatives as nutritional and functional supplements in food formulations, In: "Nutritional Improvement of Food and Feed Proteins"; Friedman, M., Ed.; Plenum Publ. Corp.: New York, N.Y., 1978; p. 767.
8. Cherry, J. P.; Berardi, L. C. Cottonseed, In: "Handbook of Processing and Utilization in Agriculture; Wolff, I. A., Ed.; CRC Press, Inc.: Boca Raton, Fla., 1982; in press.
9. Berardi, L. C.; Cherry, J. P. Cotton Gin and Oil Mill Press, 1979, 80(9), 14.
10. Lawhon, J. T.; Cater, C. M.; Mattil, K. F. Food Technol., 1970, 24(6), 77.
11. Martinez, W. H.; Berardi, L. C.; Goldblatt, L. A. 3rd Internat. Congr. Food Sci. Technol., 1970, p. 248.
12. Pigg, D. Texas Agric. Prog., 1980, 26(1), 20.
13. Inglett, G. E. "Fabricated Foods"; AVI Publ. Co., Inc.: Westport, Conn., 1975; 215 pp.
14. Noyes, R. "Protein Food Supplements"; Noyes Development Corp.: Park Ridge, N.J., 1969; 412 pp.
15. Taranto, M. V.; Meinke, W. W.; Cater, C. M.; Mattil, K. F. J. Food Sci., 1975, 40, 1264.
16. Taranto, M. V.; Cegla, G. F.; Bell, K. R.; Rhee, K. C. J. Food Sci., 1978, 43, 767.
17. Taranto, M. V.; Cegla, G. F.; Rhee, K. C. J. Food Sci., 1978, 43, 973.
18. Cegla, C. F.; Taranto, M. V.; Bell, K. R.; Rhee, K. C. J. Food Sci., 1978, 43, 775.

19. Wilding, M. D. J. Am. Oil Chemists' Soc., 1974, 51, 128A.
20. Szczesniak, A. S.; Kleyn, D. H. Food Technol., 1963, 17, 74.
21. Elder, A. L.; Smith, R. J. Food Technol., 1969, 23(5), 629.
22. Harper, J. M. "Extrusion of Foods, Volume II"; CRC Press, Inc.: Boca Raton, Fla., 1981; 240 pp.
23. Berardi, L. C.; Cherry, J. P. J. Food Sci., 1980, 45, 377.
24. Rhee, K. C.; Kuo, C. K.; Lusas, E. W. Texturization, In: "Protein Functionality in Foods"; Cherry, J. P., Ed.; ACS Symposium Series No. 147, American Chemical Society: Washington, D.C., 1981; p. 51.
25. Kinsella, J. E. CRC Critical Rev. Food Sci. Nutr., 1978, 10, 147.
26. Mazurs, E. G.; Schoch, T. J.; Kite, F. E. Cereal Chem., 1957, 34, 141.
27. Friedman, H. H.; Whitney, J. E.; Szczesniak, A. S. J. Food Sci., 1963, 28, 390.
28. Smith, K.; Venne, L.; Balentine, E. E. Oil Mill Gazett., 1970, 75(5), 8.
29. Martinez, W. H.; Berardi, L. C.; Pfeifer, V. F.; Crovetto, A. J. J. Am. Oil Chemists' Soc., 1967, 44, 139 A.
30. Berardi, L. C.; Martinez, W. H.; Fernandez, C. J. Food Technol., 1969, 23(10), 75.
31. A.O.C.S. "Official and Tentative Methods," 3rd ed.; American Oil Chemists' Society: Champaign, Ill., 1976.
32. A.O.A.C. "Official Methods of Analysis," 11th ed.; Association of Official Analytical Chemists: Washington, D.C., 1970.
33. Kaiser, F. E.; Gehrke, C. W.; Zumwalt, R. W.; Kuo, K. C. J. Chromatogr., 1974, 94, 113.
34. Beuchat, L. R. J. Agric. Food Chem., 1977, 25, 258.
35. Cherry, J. P., Ed. "Protein Functionality in Foods." ACS Symposium Series No. 147, American Chemical Society: Washington, D.C., 1981; 332 pp.
36. Catsimpoolas, N.; Funk, S. K.; Meyer, E. W. Cereal Chem., 1970, 47, 331.

RECEIVED September 24, 1982.

The Maillard Reaction and Its Prevention

ROBERT E. FEENEY and JOHN R. WHITAKER

University of California, Department of Food Science and Technology, Davis, CA 95616

Of all the deteriorative reactions occurring in foods, the Maillard reaction may well prove to be one of the two or three that has received, and will continue to receive, the most study. Interactions of sugars to produce insoluble substances have been known for many years. The name of the reaction comes from the first discernment of the chemical nature of the reaction by Maillard in 1912 (1). He described the production of dark colored compounds when solutions of sugars and amino acids were heated. He then extended these studies to various sugars and different amino acids and found that the reaction required a reducing group in the sugar and an amino group of the amino acid, and that different sugars and different amino acids reacted at very different rates (2, 3). Only later was it shown that proteins also reacted (4), and another one and one-half decades passed before the free amino groups of the proteins were found to be the main reacting groups in proteins (5).

The Maillard reaction is, of course, a series of reactions, and the actual brown products are the result of different processes, depending on the conditions. However, the key to the whole process is the initial reaction, which is a carbonyl-amine reaction. Carbonyl-amine reactions are very common in biological systems and, in fact, are a key process in a large number of systems. A food chemist, therefore, cannot claim that all of the subject is unique to food chemistry. Many articles and reviews have been written on the subject in the past few decades. At the present time there is a published symposium of an extensive meeting held in Uddevalla, Sweden (6). For this reason, if for no other, this present chapter can no more than skim the surfaces of all the subjects that are currently being discussed. There is also a matter of the relevant studies of the authors. These have mainly concerned the carbonyl-amine reaction itself, certain of its products on chemical treatments, and chemical modifications to prevent the reaction (7-13).

0097-6156/82/0206-0201$08.75/0

The burgeoning interest in the area of the carbonyl-amine reaction and certain of its subsequent reactions has prompted us to prepare this chapter with particular emphasis on the initial carbonyl-amine reaction and a few of the subsequent reactions.

General Steps of the Maillard Reaction

Hodge (14) presented what is probably the first review that tied together the chemical steps in the Maillard reaction. He attempted to devise reaction schemes as follows:
1) The initial stage--
   a) Sugar and amine condensation.
   b) Amadori rearrangement.
2) The intermediate stage--
   c) Sugar dehydration.
   d) Sugar fragmentation.
   e) Amino acid degradation.
3) The final stage--
   f) Aldol condensation.
   g) Aldehyde-amine polymerization with formation of heterocyclic nitrogen compounds.

Reynolds (15) suggested the following simpler classification:
1) A reversible formation of glycosylamine.
2) The rearrangement of the glycosylamine to ketosamine, 1-amino-1-deoxyketose (Amadori rearrangement), or to the aldosamine, 2-amino-2-deoxyaldose (Heyns rearrangement).
3) Formation of a diketosamine or a diamino sugar.
4) Degradation of the amino sugar, usually started by the loss of one or more molecules of water to form amino or carbonyl intermediates.
5) Reaction of amino groups from the intermediates formed in Step 4 and the subsequent polymerization of these products to brown pigments and other substances.

Amino Groups, Carbonyl Groups, and the Carbonyl-Amine Reaction

The carbonyl-amine reaction depends primarily on the properties of the carbonyl group and the amino group, which fortunately have received extensive study due to their general importance in chemistry (16-20).

Properties of amino groups. Amines have several properties that are responsible for their extensive reactivities. Paramount is their ability to act as nucleophiles by possessing a lone pair of electrons on the nitrogen atom. Another is their action as bases by accepting protons from a variety of acids, including water. Saturated aliphatic amines of widely different structures have pK_a values in aqueous solutions generally between 9 and 10. Aromatic amines, less commonly found in biological systems, are much weaker bases with pK_a values around 5.

The relative reactivity of amines is primarily a factor of the pK_a values, the steric effects of substitution, and the presence or absence of a proton on the nitrogen, as found in tertiary amines. For example, the well known Hinsberg method is based on the inability of a tertiary amine to form a stable derivative on reaction with an acid chloride (20).

Properties of carbonyl groups. The properties of the carbonyl group are primarily those of the carbon–oxygen double bond, which is both strong and reactive. It has a higher bond energy than that of two carbon–oxygen single bonds (197 kcal/mol vs two times 85.5 kcal/mol). This is the opposite of that for the carbon–carbon double bond, which is weaker than two carbon–carbon single bonds (145.8 kcal/mol vs two times 82.6 kcal/mol). The difference in electronegativity between carbon and oxygen is the primary cause of the reactivity of the carbonyl bond, and this gives a significant contribution to the dipolar resonance form, oxygen being negative and carbon being positive. Additions to the carbon–oxygen double bond constitute the more important reaction of carbonyl groups. The reaction may stop as soon as groups have been added, or subsequent reactions may take place. They can usually be classified into two types. In Type A the initial adduct loses water and the net result of the reaction is the replacement of C=O by C=Y (Equation 1).

$$\text{Type A:} \quad \underset{\substack{\| \\ O}}{A-C-B} + YH_2 \longrightarrow \underset{\substack{| \\ OH}}{\overset{\substack{YH \\ |}}{A-C-B}} \xrightarrow{-H_2O} \overset{\substack{Y \\ \|}}{A-C-B} \qquad (1)$$

In the second type, Type B, there is a rapid reaction in which the hydroxyl group of the tetrahedral intermediate is replaced by another group, given as Z. The Z is very often another YH moiety, that is, the same as in Equation 1 (Equation 2).

$$\text{Type B:} \quad \underset{\substack{\| \\ O}}{A-C-B} + YH_2 \longrightarrow \underset{\substack{| \\ OH}}{\overset{\substack{YH \\ |}}{A-C-B}} \xrightarrow{Z} \underset{\substack{| \\ Z}}{\overset{\substack{YH \\ |}}{A-C-B}} \qquad (2)$$

The strong polar character of the carbon–oxygen double bond determines the orientation of the asymmetric additions; negative species add at the carbon and positive ones add at the oxygen.

Additions to the carbon–oxygen double bond are usually subject to both acid and base catalysis. Catalysis by base can occur by converting YH_2 to the more powerful nucleophile YH^- by specific base catalysis. Since other bases may also catalyze a reaction by removing a proton from YH as it reacts, the reaction can also be

general base catalyzed. The reaction may be acid catalyzed by
protonation of the oxygen atom of the carbonyl compound, making
its carbon a much stronger electrophile. Both specific acid
catalysis and general acid catalysis may occur.

The reactivity of the addition reaction is greatly affected
by substituents. Rates are decreased if A or B is an electron-
donating group, and increased if they are electron-withdrawing
groups (see Equation 1). Steric factors may also be important.

The carbonyl-amine reaction. Schiff (21) discovered that the
condensation of primary amines with aldehydes and ketones gives
imines. Many studies have since shown that numerous substances
with amino groups condense with carbonyl compounds, according to
Equations 1 and 2. Strongly basic amines react as in Equation 1
(Type A) according to Equation 3.

$$A-\underset{\underset{O}{\|}}{C}-B + R-NH_2 \rightleftharpoons A-\underset{\underset{OH}{|}}{\overset{\overset{NHR}{|}}{C}}-B \underset{\rightleftharpoons}{\overset{-H_2O}{}} A-\underset{}{\overset{\overset{N-R}{\|}}{C}}-B \qquad (3)$$

Polymerizations may occur with aliphatic aldehydes that
possess a hydrogen atom on the carbon adjacent to the carbonyl
group. Weak bases such as amide and urea can undergo a Type B
reaction (as in Equation 2) according to Equation 4, forming
alkylidine diamines or gemdiamines.

$$A-\underset{\underset{O}{\|}}{C}-B + R-NH_2 \rightleftharpoons A-\underset{\underset{OH}{|}}{\overset{\overset{NHR}{|}}{C}}-B \underset{\rightleftharpoons}{\overset{RNH_2}{}} A-\underset{\underset{NHR}{|}}{\overset{\overset{NHR}{|}}{C}}-B + H_2O \qquad (4)$$

Plots of the rates of the carbonyl-amine reaction usually give
bell-shaped curves (Figure 1). This is because the general factors affecting the carbonyl-adduct reaction apply, as well as the
effects of protonation and deprotonation of the amino groups.
These effects are discussed by Jencks (19) and are summarized
briefly by Feeney et al. (20). At neutral pH, the loss of water
from the tetrahedral intermediate is rate determining. At lower
pH the rate of acid-catalyzed dehydration decreases and the free
amine becomes protonated, resulting in the equilibrium concentration of the addition compound decreasing. At pH values well below
the pK of the amine, these two effects offset each other and the
calculated rate becomes independent of pH. At the low pH the
attack of free amine on the carbonyl group becomes rate determining. A change in the rate determining step usually occurs
between pH 2 and pH 5 for aliphatic amines. The attack and loss
of water is fast, and the attack and loss of free amine is rate

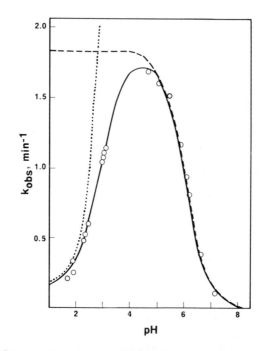

Figure 1.   Effect of pH on the pseudo first-order rate constant for the reaction of 5 × 10⁻⁴ M acetone with 0.0167 M total hydroxylamine, showing the change in rate-determining step with changing pH. Key: · · ·, rate of attack of free hydroxyl-amine on acetone; – – –, rate under conditions in which acid-catalyzed dehydration is rate determining; and ———, calculated, from the steady state. (Reproduced, with permission, from Ref. 19. Copyright 1969, McGraw–Hill Book Company.)

determining below this pH, while above this pH the opposite occurs. At high pH the attack and loss of free amine is fast and
the attack and loss of water or hydroxide ion is rate determining.
     Schiff bases are generally considered to be the important
product of carbonyl-amine reactions. Although the chemical properties of Schiff bases are similar to those of the carbonyl group,
the pK_a of the imino group is around 7, and therefore it is protonated to a much larger extent at neutral pH, and the carbon atom
is more electrophilic. One of the frequently overlooked properties of Schiff bases is that they react readily with amines by
attack on the protonated imine. Transimination is the name
usually given to the conversion of one imine into another (Equation 5).

$$
R-NH_2 \quad \overset{+}{C=N} \rightleftharpoons R-\overset{\overset{H}{|}}{\underset{\underset{H}{|}}{N}}{}^{+}-\overset{|}{\underset{|}{C}}-N \rightleftharpoons R-\overset{\overset{H}{|}}{\underset{\underset{H}{|}}{N}}-\overset{|}{\underset{|}{C}}-\overset{+}{\underset{|}{N}} \rightleftharpoons \overset{R}{\underset{H}{\diagdown}}N=C\overset{\overset{H}{|}}{\underset{|}{N}}- \qquad (5)
$$

     Aldol condensations are important reactions of carbonyl com-
pounds. These reactions are catalyzed by amines, apparently due
to the formation of an intermediate imine or Schiff base, probably
an important reaction for the Maillard sequence. Aldol conden-
sations are also very important in various biological processes,
such as enzyme reactions and biosynthetic processes.

General Reactions of Carbonyls with Amino Acids and Proteins

     Although formaldehyde is the simplest of all aldehydes and
might be a good model for carbonyl-amine reactions, it has unique
properties and reactivities which set it off from most other carbonyl compounds. There is an immense amount of literature on its
reaction with amino acids and proteins (22) and on its general
chemistry (23), and it is an extensively used reagent of protein
chemistry. In addition, some of the fragments reported in the
Maillard reaction may have many of the properties of formaldehyde.
Schiff (21, 24) suggested that the methylene derivative was formed
(Equation 6), but it is now more generally accepted that it proceeds to the dihydroxymethyl derivative, a reaction that is the
basis for the formerly much-used formol titration of amino groups
(25).

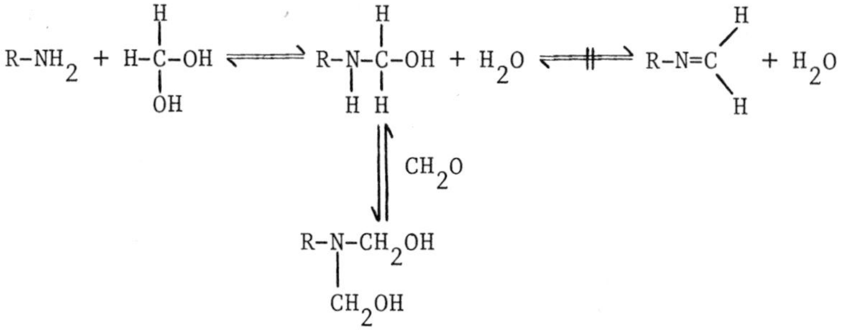

$$R\text{-}NH_2 + H\text{-}\underset{\underset{OH}{|}}{\overset{\overset{H}{|}}{C}}\text{-}OH \rightleftharpoons R\text{-}\underset{\underset{H}{|}}{\overset{\overset{H}{|}}{N}}\text{-}\underset{\underset{H}{|}}{\overset{}{C}}\text{-}OH + H_2O \rightleftharpoons R\text{-}N{=}C\overset{\diagup H}{\diagdown H} + H_2O \qquad (6)$$

$$\Big\Updownarrow CH_2O$$

$$R\text{-}\underset{\underset{CH_2OH}{|}}{N}\text{-}CH_2OH$$

The long-time uses of formaldehyde in the pharmaceutical and
tanning industries (20) prompted many studies on the reactions of
formaldehyde with a variety of protein side chain groups. It is
now accepted that formaldehyde reacts with groups containing an
active hydrogen atom with the formation of the hydroxymethyl compound, which is usually reactive and may condense with another
nucleophilic site to give a methylene bridge (23). This condensation may take place intermolecularly or intramolecularly in the
individual amino acid side chains themselves, e.g., the cyclic
compound from asparagine. Tome and Naulet (26) have recently reported on the addition and condensation reactions of formaldehyde
with polyfunctional amino acids in neutral or mildly acidic solutions using ^{13}C NMR. In addition to the amino group, many of the
other side chains, such as guanidyl, hydroxyl, indole, and imidazole groups, react with different rates and different equilibrium
constants. In the case of asparagine, threonine, histidine, and
tryptophan, the NH_2-hydroxymethyl condenses with the basic group
of the side chain to give a cyclic compound. With cysteine the
S-hydroxymethyl condenses with the alpha amino group. The following are the kinds of products that have been reported as a result
of long and extensive treatment of proteins with formaldehyde:

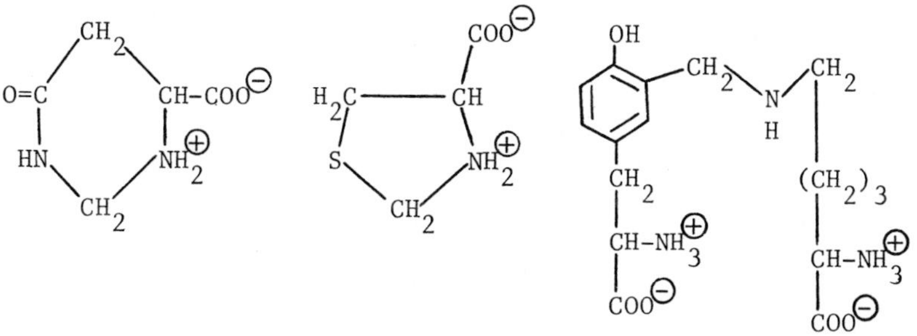

The reduction of Schiff bases, or the hydroxymethyl adducts
with formaldehyde, has become a popular work horse of chemistry
and biochemistry. The process, termed reductive alkylation, was
first used by Fischer (27) as a special interaction between
certain enzymes and their substrates or coenzymes, such as pyri-

doxal 5'-phosphate. Means and Feeney (7), using different sized substituents, introduced the method for reductive alkylation of proteins as a general procedure for chemical modification. Probably the most important modification has been the introduction of the small methyl group, usually as a dimethyl derivative of the epsilon amino groups of lysine. The key to the successful modification with formaldehyde was the prior addition of the reducing agent (then sodium borohydride) to the protein, and then the addition of the formaldehyde, thereby preventing any side reactions with the very reactive formaldehyde (8). With formaldehyde, the reaction is rapid, giving ε-N-N-dimethyl lysine residues as the principal products (Equation 7).

$$
\begin{array}{c}
\text{RNH}_2 \\
\updownarrow \text{H}^{\oplus} \\
\text{RNH}_3^{\oplus}
\end{array}
\quad + \quad \text{R'CHO} \rightleftharpoons \text{RN=CHR'} \xrightarrow{[H]}
$$

$$
\begin{array}{c}
\text{RNHCH}_2\text{R'} \\
\updownarrow \text{H}^{\oplus} \\
\text{RNH}_2\text{CH}_2\text{R'}
\end{array}
\quad \xrightleftharpoons{\text{R'CHO}} \quad
\begin{array}{c}
\overset{\oplus}{\text{RN=CHR'}} \\
| \\
\text{CH}_2 \\
| \\
\text{R'}
\end{array}
\quad \xrightarrow{[H]} \text{RN(CH}_2\text{R')}_2
$$

(7)

Because of the relatively small changes in basicity and the relatively small space occupied by the methyl groups, ε-N-N-dimethyl lysine residues appear to give minimal changes in the properties of proteins. With other aldehydes and ketones, it is much more difficult to put in a second substituent; at low levels of modification, it may be entirely monosubstitution that occurs. For a general review, see Means (28).

The methylation procedure has received considerable impetus because of the simplicity of labeling. Double labeling can be done by the use of a tritiated reducing agent and ^{14}C formaldehyde. Many applications are now being made by labeling with ^{13}C formaldehyde for NMR experiments (29).

The more conventionally used reducing agents have been sodium borohydride and sodium cyanoborohydride, but more recently amine boranes have been suggested as having certain superior qualities (12). Dimethylamine borane (12) proved quite satisfactory, and pyridine borane may have particularly superior qualities (30).

The alkylated derivatives of lysine produced by reductive alkylation survive acid hydrolysis in most cases and can be conveniently determined on amino acid analyzers. The high stability, however, of the alkylated derivatives has not previously allowed

them to be useful as reversibly modified products, but recently
Geoghegan et al. (11) described a reversible form of reductive
alkylation. Amino groups are modified to a 2-hydroxyalkyl form by
mono-alkylation and restored to their original form by treatment
of the proteins with low concentrations of periodate (Equation 8;
11).

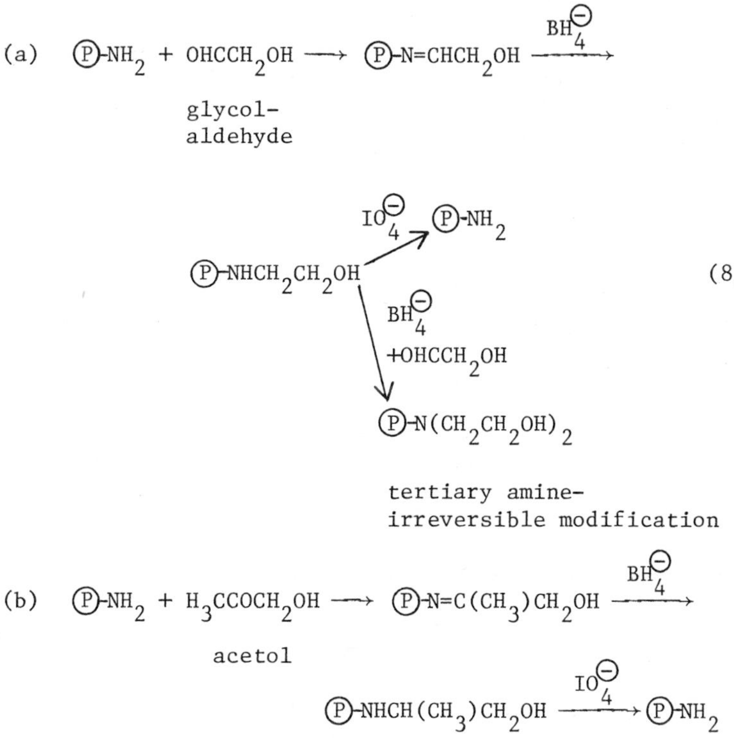

(a) $\text{P}-NH_2 + OHCCH_2OH \longrightarrow \text{P}-N=CHCH_2OH \xrightarrow{BH_4^{\ominus}}$

glycol-
aldehyde

$$\text{P}-NHCH_2CH_2OH \begin{array}{c} \nearrow^{IO_4^{\ominus}} \text{P}-NH_2 \\ \searrow_{BH_4^{\ominus}} \\ +OHCCH_2OH \\ \downarrow \\ \text{P}-N(CH_2CH_2OH)_2 \end{array}$$ (8)

tertiary amine-
irreversible modification

(b) $\text{P}-NH_2 + H_3CCOCH_2OH \longrightarrow \text{P}-N=C(CH_3)CH_2OH \xrightarrow{BH_4^{\ominus}}$

acetol

$\text{P}-NHCH(CH_3)CH_2OH \xrightarrow{IO_4^{\ominus}} \text{P}-NH_2$

The procedure works well for many proteins, but some proteins are
easily inactivated by periodate (31) and the method would therefore not be useful with such proteins.

Chemistry of the Maillard Reaction

The Maillard reaction is far from being understood today.
The first step, the carbonyl-amine reaction, is comparatively
simple when compared to many of the subsequent reactions, and yet
there is still no good, acceptable, satisfactory procedure for
controlling this first step in foods and biological materials.
The rearrangements immediately following the carbonyl-amine reaction have been studied extensively, and considerable information
is available, but this is the point of departure from more conventional chemistry into a wide variety of types of reactions and

products. These in turn are involved in numerous reactions involving chain splittings, oxidations, dehydrations, aldol con-
densations, and finally, polymerizations and formation of pigments.

The initial steps - formation of the N-substituted glycosylamines. The first step is the carbonyl-amine reaction between
the carbonyl group, sugar, and the amino group of an amine, an
amino acid, peptide, or protein, to produce an N-substituted
glycosylamine in equilibrium with a Schiff base. The reaction
has a broader pH optimum than for simple carbonyl amine reactions
(Figure 1) because the presence of other substances permits acid
or base catalysis, or the interactions supply their own catalysis.
The overall Maillard reaction increases approximately linearly
with increasing alkalinity from pH 3 up to as high as 8, and perhaps higher (32). Such data, however, also include a multitude of
reactions from which it is difficult to separate individual
effects.

Although pentoses are generally considered to be more reactive than hexoses, and monosaccharides more reactive than di-
saccharides (33), these data again are usually taken from the
overall Maillard reaction and are not based on the initial formation of glycosylamines. Glucose exists nearly 100% in the alpha
and beta pyranose forms, with only a trace in the carbonyl form,
and yet it is one of the most reactive sugars. Fructose shares
with most other ketones the relatively low rate of reaction as
compared to aldehydes, and yet it is considered one of the most
reactive sugars in the overall Maillard reaction. The relative
reactivities of glucose, fructose, and lactose are also reflected
in the coupling of these sugars to the lysine groups of casein on
reduction with cyanoborohydride (Figure 2).

Amadori and Heyns rearrangements. The next step in the
Maillard sequence is not commonly encountered in carbonyl-amine
reactions but during the last decade has been found in several
important biological reactions. This is either the Amadori or
Heyns rearrangement. The aldosamines are converted to a 1-1
deoxyketose (in the case of glucose, to 1-amino-1-deoxyfructose)
(Equation 9), while the ketosamines are transformed to 2-amino-
2-deoxyaldoses (in the case of fructose, to 2-amino-2-deoxyglucose
(Equation 10).

$$(9)$$

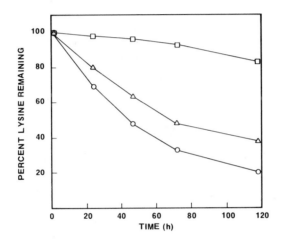

Figure 2.   The rate of coupling glucose (○), fructose (△), and lactose (□), to the ε-amino group of lysyl residues of casein in the presence of cyanoborohydride at 37°C in 0.2 M potassium phosphate buffer, pH 9.0. Losses in lysine were determined by amino acid analysis. (Reproduced from Ref. 9. Copyright 1979, American Chemical Society.)

$$\text{(Heyns)} \qquad\qquad (10)$$

The Amadori reaction and Amadori compounds have been extensively studied, but the Heyns reaction and products have not received much attention. In fact, products from some Heyns rearrangements may sometimes be called Amadori products. Amadori first demonstrated the condensation of glucose with an aromatic amine to yield different isomers, but it remained for Kuhn and Weygand (34) to show that an isomerization had occurred from the aldosylamine structure to a substituted deoxyketose structure. In the Amadori rearrangement (Equation 11; 35) a critical step for ketosamine formation is the initial formation of protonated Schiff base with subsequent prototropic shifts.

$$
\begin{array}{ccc}
\text{HC--NH}\cdot\text{R} & \text{HC--NH}\cdot\text{R} & \text{HC}=\overset{\oplus}{\text{NHR}}\\
| & | & \\
\text{HC--OH} \; \underset{-\text{HA}}{\overset{\text{HA}}{\rightleftharpoons}} \; & \text{HC--OH} \; \longrightarrow & A^{\ominus}\; \text{H--C--OH}\\
| & | & |\\
-\text{C}- & -\text{C}- & \\
| & |\; A^{\ominus} & \text{Imonium ion}\\
& & \text{(transition state)}
\end{array}
$$

$$(11)$$

$$
\begin{array}{ccc}
\text{H}_2\text{C--NHR} & \text{H}_2\text{C--NHR} & \text{HC--NHR}\\
| & | & \|\\
\text{HOC} \rightleftharpoons & \text{C}=\text{O} \longleftarrow & \text{C--OH}\\
| & | & |\\
-\text{C}- & & \\
| & & \text{Eneaminol}\\
& & +\ \text{HA}
\end{array}
$$

Once formed, a ketosamine can react with another molecule of an aldose. Such a product can undergo another Amadori rearrangement to give diketosamine. One such compound, a difructose glycine, has been crystallized. While aldosylamines are unstable compounds, some Amadori compounds can be readily crystallized, preferentially as the β-anomer of the D-pyranose form (35).

Amadori compounds are obviously at a key juncture in the Maillard reaction sequence. They have therefore been studied ex-

tensively. Comprehensive older reviews are by Hodge (36) and
Gottschalk (35).

These initial reactions, both the initial carbonyl-amine reactions and the formation of Amadori and Heyns products, are still
the subject of intensive research and are probably more complex
than originally visualized. One of the difficulties has been the
distinction between glycosylamines and Amadori products, and this
has recently been studied by Funcke et al. (37) by mass spec-
trometry of the trimethylsilyl derivatives. Hayashi and Namiki
(38), studying the reaction of glucose with t-butylamine, have
postulated the formation of reactive 2-carbon sugar fragments at a
stage even prior to an Amadori rearrangement, and the subsequent
reaction of these with the amine, followed by an Amadori rearrangement to form glyoxal-dialkylamine (Figure 3). They have
further suggested that a free radical may be formed by the reduction of N,N'-dialkylpyrazinium, possibly formed by the condensation of two carbon enaminol followed by oxidation (39).

Later processes in the Maillard sequence of reactions. As
newer techniques are used to study the myriad of chemical reactions and products occurring in the later stages of the Maillard
reaction, researchers should find more and more products, equilibrium forms, and different pathways. Suggested pathways include
those listed by Hodge (14), such as sugar dehydration, sugar
fragmentation, amino acid degradation, aldol condensation, and
many types of polymerization in the formation of numerous heterocyclic nitrogen compounds. One example is the breakdown of the
Amadori compound, di-D-fructose-glycine, to a variety of compounds
(Figure 4).

Among the numerous products that are produced near the end of
the Maillard sequence are low molecular weight substances, which
in many cases have color and taste, as well as sufficient volatil-
ity for odor, and the higher molecular weight ones which give the
dark brown color to the products. Included in these are numerous
polymers and heterocyclics. For some of the reactions leading to
them, and for a description of their properties and, in a few
cases, structures, current reviews should be consulted (6). A few
will be mentioned under the next section.

Deleterious and Beneficial Aspects

Effects on flavor in foods. The products of Maillard reactions are so generally distributed in food products that the
elimination of the reaction in foods would cause widespread and
general unacceptability of many food products. This is because
the Maillard reactions are critical for the acceptance of many
processed foods as well as many home-cooked foods. Amadori compounds have been reported to be the key intermediates whose
thermal breakdown gives many of the particular flavors and aromas
of browned food products (40, 41). No exact classifications are

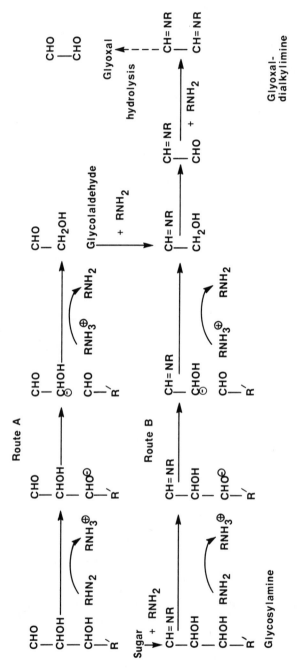

Figure 3. Possible pathways for the formation of a two-carbon compound by the reaction of sugar with amine. (Reproduced, with permission, from Ref. 38. Copyright 1980, Agricultural Chemical Society of Japan.)

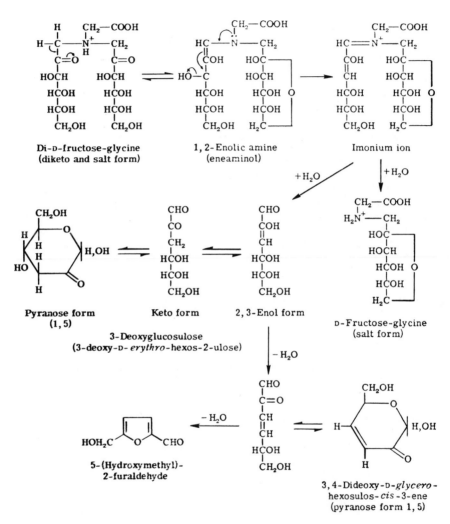

Figure 4. Reaction mechanism of the degradation of di-D-fructose–glycine to 5-(hydroxymethyl)-2-furaldehyde and D-fructose–glycine. (Reproduced, with permission, from Ref. 35. Copyright 1972, Elsevier/North-Holland Biomedical Press.)

possible of the many compounds with distinct aromas, but a partial
one is their separation into nitrogen-heterocyclic compounds with
responsibility for nutty, corny, and bread aroma flavors, and
oxygen-heterocyclic compounds with responsibility for caramel
aroma (Figure 5). Some Amadori products have been added to foods
to give characteristic flavors, such as the one derived from
6-deoxy-D-galactose and L-proline in order to increase the buttery
flavor of margarine (42).

   Antioxidant effects. Maillard products are considered beneficial for their antioxidant properties. These are thoroughly
discussed elsewhere (6).

   Effects on physical properties. Sugar-protein interactions
in foods and in model systems have long been known to result in
changes in physical properties such as lowered solubilities and
losses in functional capacities. Merely recalling the many years
of the natural fermentation of egg white prior to its drying shows
how an industry was once based on an art rather than science.
Without the prior removal of the sugars (mainly glucose) by fermentation, the dried egg white quickly lost its functional capacities for baking and candy making (43). The Maillard reaction
between egg-white proteins and glucose can cause many changes in
physical properties (44, 45).
   Changes in egg white may occur even on incubation of shell
eggs or separated egg white for three or four days or storage at
5°C for two weeks. The rapidity of the reactions is apparently
due to the relatively high pH (approximately 9.5) that an in-
fertile egg and a separated white will quickly assume due to
losses of CO_2. Within four days of incubation of bird eggs at
37°C, the gel electrophoretic patterns of egg-white proteins
changed significantly (46) (Figure 6). The ovotransferrin, a
protein containing large numbers of lysyl ε-amino groups, showed
extensive heterogeneity. Since gel electrophoretic patterns have
been extensively used to detect genetic differences among birds
and to establish taxonomic lines (47), and since bird eggs may be
collected at widely different geographical areas and transported
elsewhere for analyses, these rapid glucose-protein deteriorations
can possibly cause confusion and misinterpretations. Of course,
the same concern applies to work with many other materials that
have reactive amino groups.

   Nutritional aspects. Although experts may disagree as to
when the nutritional and health aspects of Maillard products began
to receive extensive attention, there are many who say that this
did not occur until during, and perhaps even after, World War II.
First publications began to appear on the nutritional value,
particularly related to lysine, and then later on toxic and mutagenic compounds possibly arising from the Maillard reaction. Lea
and Hannan (48, 49) showed that when casein is treated with glu-

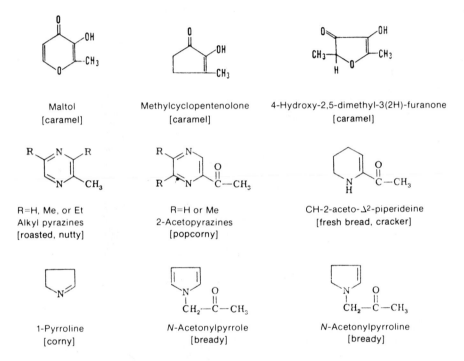

Maltol
[caramel]

Methylcyclopentenolone
[caramel]

4-Hydroxy-2,5-dimethyl-3(2H)-furanone
[caramel]

R=H, Me, or Et
Alkyl pyrazines
[roasted, nutty]

R=H or Me
2-Acetopyrazines
[popcorny]

CH-2-aceto-Δ2-piperideine
[fresh bread, cracker]

1-Pyrroline
[corny]

N-Acetonylpyrrole
[bready]

N-Acetonylpyrroline
[bready]

Figure 5.   Alicyclic and o-heterocyclic caramel-aroma compounds and N-heterocyclic nutty-, corny-, and bready-aroma compounds from the Maillard reaction. (Reproduced, with permission, from Ref. 40. Copyright 1972, American Association of Cereal Chemists, Inc.)

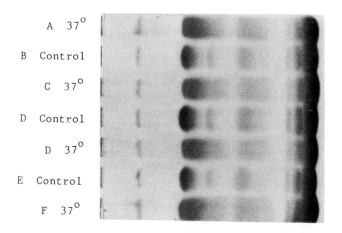

Figure 6.   Starch-gel electrophoretic patterns of incubated fertile eggs. Egg whites were all from white Leghorn hens and all contained globulin A_1. Eggs were incubated at 37°C for 6 d or stored at 2°C for 6 d (controls). Letters identify the hen. (Reproduced, with permission, from Ref. 46.)

cose, most of the lysyl residues were modified, but only a small
percentage of other amino acids reacted. It would still appear
that lysine is involved in the main deteriorative reaction affecting the nutritional quality, whereas reactions involving other
amino acids, such as arginine, histidine, tyrosine and methionine,
become important with more extensive deteriorations.

One of the difficulties in determining the losses in the
nutritional availability of lysine has been the lack of suitable
chemical or physical methods for this determination (50). The
nutritional value of lysine is lowered during the early stages of
the Maillard reaction, but on amino acid analysis after acid
hydrolysis the lysine is found to be unaffected, being liberated
during the hydrolytic procedure. This has resulted in many workers attempting to develop chemical methods not requiring acid
hydrolysis, such as the reaction of lysine residues with fluoro-
dinitrobenzene (50, 51). Hurrell and Carpenter (52) have found
some discrepancies with longer heating of experimental samples and
would rather term the results of the chemical tests as "reactive
lysine content". The problem is still unsolved, although more
refined methods continue to appear (53). In addition to losses in
specific amino acids, particularly lysine, there are reports of
decreased availability of other amino acids, perhaps as a result
of lower digestibility of the proteins.

Possible toxicities or mutagenicities. During the last decade there have been numerous studies on the toxicities and possible mutagenicities of Maillard products. Nagao et al. (54), using
the Ames assay, reported strong mutagenic activity with smoke con-
densates of various heated proteins. Shibamoto et al. (55) have
reported on the formation of substances that are mutagenic in the
Ames test by treating maltol and ammonia in sodium hydroxide
solution for 5 hr at 100°C. Other reports have indicated mutagenic activities of a variety of browning products in the Ames
test (56-59). Although these findings in no way prove that these
substances would be carcinogenic in humans even at levels much
higher than found in foods, it does indicate the general importance of the subject and the need for further studies. The large
number of compounds that have been identified as being produced by
the Maillard reaction is probably only a small fraction of the
many that are actually formed, so the need for study is obvious.

Unfortunately, extrapolation of some of these results to
human usage at this time is very difficult or impossible. Furthermore, removal of all Maillard products from the human diet
would necessitate cultural changes and possibly result in bland
and boring diets. On the opposite side of the fence, Pintauro
et al. (60) have reported that extracts of Maillard browned egg
albumin did not show any mutagenic effects in the Salmonella-
mammalian microsome plate assay.

Inhibitory effects of Maillard products on nutritional utilization of protein have been noted by feeding Maillard browned

egg albumin to rats over a period of 1 to 12 months (61). After a few months the rats fed the browned diet exhibited a lag in weight gain relative to a control group (Figure 7), and there were changes in the composition of blood serum as well as enlargement of some organs. It was concluded that the poor nutritional quality of the browned protein is not restricted to the loss of amino acids and that inhibitory substances might have been formed during the browning process.

An indication of a number of compounds formed during the Maillard reaction is seen in Table I, where some of the fructose-containing Amadori compounds are listed, a listing which is only the tip of the iceberg of Maillard products.

Amadori and Other Possible Maillard Products Formed in Biological Systems Other Than Foods

Although Schiff bases have been known for a number of decades to occur in biological processes (20), deteriorative products such as Amadori compounds were not usually considered important until perhaps two decades ago. Since then at least three systems in which Amadori compounds, and perhaps others, are important have been reported. Amadori products have been reported in hemoglobin (63–66), erythrocyte membrane protein (67), lens crystallins (68, 69), and collagen (70), as well as in some other proteins (71).

Hemoglobin A_{Ic}. Probably the first well documented covalent attachment of a sugar-like compound to an important protein in a biological system was the report by Holmquist and Schroeder (65) that there was a heretofore unrecognized N-terminal blocking group involving a Schiff base in hemoglobin A_{Ic}. This created general interest because this form of hemoglobin was first thought to be a genetic variant in humans. Identification of the group in protein as a glucose derivative and a possible Amadori product soon followed (63, 66). Earlier, Rahbar (72) had reported an abnormal hemoglobin in red cells of diabetics. Dixon (64) characterized the reaction of glucose with valylhistidine (the amino terminal residues of the beta chain of hemoglobin A) and showed that this product was identical to that obtained by Holmquist and Schroeder (65). It was Dixon (64) who suggested that the levels of hemoglobin A_{Ic} usually observed (that is, 3 to 6%) could be due purely to nonenzymatic glucosylation (i.e., there would be no translational enzymatic modification in the biosynthetic process). High levels of hemoglobin A_{Ic} in the blood of diabetics could therefore be due to the higher level of glucose in diabetics' blood serum. Higgins and Bunn (73) have calculated rate constants for the initial formation of the aldimine ($k_1 = 0.3 \times 10^{-3}$ $mM^{-1}hr^{-1}$; $k_{-1} = 0.33 \ hr^{-1}$) and for the Amadori rearrangement to the ketoamine ($k_2 = 0.0055 \ hr^{-1}$). An estimate of the amounts of pre-hemoglobin A_{Ic} (the aldimine) in normal red blood cells was

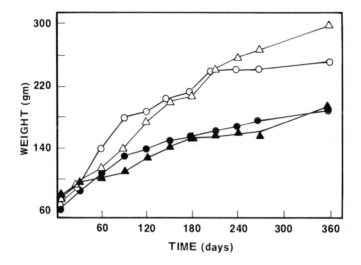

Figure 7.   Growth curve for rats on brown and control diets. Key: brown, male,
▲; *brown, female,* ●; *control, male,* △; *and control, female,* ○. *(Average of five*
or six rats in each group; the weight gains shown here are for the rats which were
fed for 12 months). (Reproduced from Ref. 61. Copyright 1980, American Chem-
ical Society.)

Table I. Occurrence of fructose-containing Amadori compounds (62)

Source	Amino Acid Moiety
Browned, dried apricots and peaches	Aspartic acid, threonine, serine, proline, valine, alanine, glycine, leucine, asparagine, β-alanine, γ-aminobutyric acid
Sugar beet molasses and high-sugar end liquors	Asparagine, serine, glycine, alanine, valine, leucine, tyrosine, glutamic acid, γ-aminobutyric acid, aspartic acid
Licorice root	Proline, alanine, aspartic acid, asparagine
Soy sauce, miso, sake	Glycine, alanine, valine, isoleucine, leucine
Ultra-heat-treated (UHT) milk, sterilized and condensed milk, dried skim milk, dried whey and whey concentrates, baby food, chocolate	Lysine

Tomato powder	Aspartic acid, glutamic acid, glycine, alanine, γ-amino-butyric acid
Black tea	Threonine
Roasted meat	Glycine
Fresh and dried hog liver	Glutamic acid, alanine, glycine
Commercial calf liver extract	Glutamic acid, alanine, glycine, valine, serine, threonine, aspartic acid, leucine
White wine	Leucine, isoleucine, valine, alanine, proline
Beer malt	Glycine, alanine, valine
Dehydrated cabbage	Lysine, glycine
Flue- and sun-cured tobaccos	Proline, alanine, γ-aminobutyric acid, valine, threonine, phenylalanine, tyrosine, asparagine
Human hemoglobin (HbA_{Ic})	Valine

0.5% for the total hemoglobin, or approximately 10% of the total
hemoglobin A_{Ic} (aldimine and ketoamine).

Lens crystallins. Lens crystallins of the eyes are long-
lived proteins, comprising more than 90% of the dry weight of lens
fiber cells (74). Many changes occur in the lens proteins during
the lifetime of an individual, and since lens proteins are "lifetime proteins", many types of slow processes that would ordinarily
not be of much significance in proteins with a life of less than a
month, can accumulate to cause extensive effects. Forty percent
of the protein is reported to be insoluble in the normal lenses of
80 to 90 year old individuals. In cataracts, insoluble protein
increases. High molecular weight aggregates are found among the
crystallins in older eyes. Cross-links are formed; pigments
develop, and other deteriorative reactions such as racemizations
and formation of fluorescent products occur. A recent report
states that the incubation of lens proteins with reducing sugars
leads to the formation of fluorescent yellow pigments and cross-
links similar to those reported in human lenses from individuals
with cataracts (75). They also cite work showing that Amadori
products were found in lens crystallins and that they accumulated
in aging lenses (76). They conclude that in diabetes the Maillard
reaction may accelerate aging in certain tissues and contribute to
the earlier onset of cataracts and even atherosclerosis (75). We
are left with the possibility that the Maillard reaction, in at
least its earlier stages, such as the formation of Amadori products, may prove to be one of the more important deteriorative
reactions in living systems. Unfortunately, methods for the study
of deteriorated proteins in general are not well developed (77).

Model systems. Amadori compounds have recently been reported
from treatments of hemoglobin with a three-carbon hydroxyaldehyde,
glyceraldehyde (78) or the two-carbon hydroxyaldehyde, glycolaldehyde (79). The Amadori products apparently can cause crosslinking between molecules, resulting in the formation of polymers.
The findings of such polymers suggest that similar products may be
formed naturally in biological systems as in the case of hemoglobin A_{Ic}. In comparing the facility of the Amadori rearrangement with glycolaldehyde, one can see the value of adding the
reducing agent first in reversible reductive alkylation with
glycolaldehyde (11) (see Equation 8).

Suggested evolutionary significance. Bunn and Higgs (80)
have recently proposed an evolutionary significance to the
Maillard reaction. From studies showing that glucose was the
least reactive of a series of aldohexoses in the formation of
Schiff base linkages with hemoglobin, they suggested that the
emergence of glucose as the primary metabolic fuel may be due in
part to the relatively high stability of its ring structure. This
would allow high concentrations of glucose and proteins to exist
together with the least interactions.

Possible Methods to Control or Prevent Maillard Reactions

A long list of procedures have been suggested for control or prevention of the Maillard reaction. Among those which have received particular attention are the following:

1)   Removal of the offending carbohydrates by dialysis or by enzymatic action (81).
2)   Addition of chemical substances to block the carbonyl group on the sugars, such as sulfite (20).
3)   Addition of a chemical·substance to compete with the amino groups of the product and to form covalent products with the carbonyls of the sugar (82), such as cysteine to form the thiazolidine.
4)   The blocking of amino groups, usually by covalent modifications such as acylation (83, 84) or methylation (85), and by compartmentalization (such as by separating the carbonyl compounds and the nitrogen compounds with starch) (86).
5)   Lowering the water activity to produce a very dry protein.
6)   Lowering the pH (43).
7)   Changing the recipe.

All of these methods may present problems, and the best that can be said at this time is that one or more of them, either alone or in combination, has been used to control, or to decrease the speed of, the formation of Maillard products in foods.

Sometimes a misunderstanding of the mechanisms can cause the use of a less desirable method for controlling or retarding the Maillard reaction. An example of this was the method developed to retard the formation of the repulsive odors and flavors that developed in whole dried eggs which were manufactured during the time of World War II and consumed by many hundreds of thousands of nauseated members of the armed forces in the South Pacific theater. The reactions were first considered to be due to a Maillard-type reaction between the amino groups of phospholipids and carbonyl compounds formed on oxidation of egg lipids. Measures were therefore instituted to inhibit the oxidations and to retard the reactions by acidification. However, it was nearly eight years later before it was discovered that it was actually the glucose in the egg reacting with the amino groups of the phospholipids (81, 87).

Conclusions

The Maillard reaction is one of the most important and widely occurring reactions in foods. Much is known about its chemistry, but it is so complex that only a fraction of the total reaction is understood. The primary reaction is the carbonyl-amine addition, between sugars and amines (usually the ε-amino groups of the side chain of lysines in proteins). This is followed by rearrangments, scissions, hydrolyses, and more carbonyl-amine reactions, as well as numerous other reactions. The products are both beneficial and deleterious to the acceptability of foods. The Amadori compounds, one of the earlier products, have also been reported to occur in other important biological systems, such as hemoglobin A_{Ic} and in aging lens proteins from the eye.

As is necessary for the prevention of all deteriorative chemical reactions, definitive knowledge of the successive steps and bifurcations in the sequences of the Maillard reaction is necessary for its control and prevention. Because of the complexities of the chemistry involved, the necessary information can be supplied only by the continued applications of newer techniques, such as the current ones of high performance liquid chromatography and nuclear magnetic resonance spectroscopy.

Acknowledgments

The authors would like to thank Chris Howland for editorial assistance and Clara Robison for typing the manuscript. Background researches for this article were supported in part by NIH Grants AM 26031 and GM 23817.

Literature Cited

1. Maillard, L. C. Compt. Rend., 1912, 154, 66.
2. Maillard, L. C. Ann. Chim. (Paris), 1916, 5, 258.
3. Maillard, L. C. Ann. Chim. (Paris), 1917, 7, 113.
4. Ramsey, R. J.; Tracy, P. H.; Ruehe, H. A. J. Dairy Ind., 1933, 16, 17.
5. Mohammed, A.; Olcott, H. S.; Fraenkel-Conrat, H. Arch. Biochem., 1949, 24, 270.
6. Eriksson, C., Ed.; "Maillard Reactions in Food. Proceedings Int. Symp., Uddevalla, Sweden"; Pergamon Press: Oxford, 1981; pp. 500.
7. Means, G. E.; Feeney, R. E. Biochemistry, 1968, 7, 2192.
8. Galembeck, F.; Ryan, D. S.; Whitaker, J. R.; Feeney, R. E. J. Agric. Food Chem., 1977, 25, 238.
9. Lee, H. S.; Sen, L. C.; Clifford, A. J.; Whitaker, J. R.; Feeney, R. E. J. Agric. Food Chem., 1979, 27, 1094.
10. Fretheim, K.; Iwai, S.; Feeney, R. E. Int. J. Peptide Protein Res., 1979, 14, 451.
11. Geoghegan, K. F.; Ybarra, D. M.; Feeney, R. E. Biochemistry, 1979, 18, 5392.
12. Geoghegan, K. F.; Cabacungan, J. C.; Dixon, H. B. F.; Feeney, R. E. Int. J. Peptide Protein Res., 1981, 17, 345.
13. Sen, L. C.; Lee, H. S.; Feeney, R. E.; Whitaker, J. R. J. Agric. Food Chem., 1981, 29, 348.
14. Hodge, J. E. J. Agric. Food Chem., 1953, 1, 928.

15. Reynolds, T. M. In "Symposium on Foods: Carbohydrates and Their Roles"; Schultz, H. W.; Cain, R. F.; Wrolstad, R. W., Eds.; Avi Publ. Co.: Westport, Conn., 1969; p. 219.
16. Patai, S., Ed.; "The Chemistry of the Carbonyl Group"; Vol. I; Wiley Interscience: New York, 1966; pp. 1027.
17. Patai, S., Ed.; "The Chemistry of the Amino Group"; Wiley Interscience: New York, 1968; pp. 813.
18. Patai, S., Ed.; "The Chemistry of the Carbonyl Group"; Vol. II; Wiley Interscience: New York, 1970; pp. 428.
19. Jencks, W. P. "Catalysis in Chemistry and Enzymology"; McGraw-Hill: New York, 1969; pp. 644.
20. Feeney, R. E.; Blankenhorn, G.; Dixon, H. B. F. Adv. Protein Chem., 1975, 29, 135.
21. Schiff, H. Justus Liebig's Ann. Chem., 1900, 310, 25.
22. French, D.; Edsall, J. T. Adv. Protein Chem., 1945, 2, 277.
23. Walker, J. F. "Formaldehyde"; 3rd ed.; Van Nostrand-Reinhold: Princeton, New Jersey, 1964; pp. 701.
24. Schiff, H. Justus Liebig's Ann. Chem., 1901, 319, 59.
25. Sörensen, S. P. L. Biochem. Z., 1908, 7, 45.
26. Tome, D.; Naulet, N. Int. J. Peptide Protein Res., 1981, 17, 501.
27. Fischer, E. H. In "Structure and Activity of Enzymes"; Goodwin, T. W.; Harris, J. I.; Hartley, B. S., Eds.; Academic Press: New York, 1964; p. 111.
28. Means, G. E. Methods Enzymol., 1977, 47, 469.
29. Jentoft, J. E.; Jentoft, N.; Gerken, T. A.; Dearborn, D. G. J. Biol. Chem., 1979, 254, 4366.
30. Cabacungan, J. C.; Ahmed, A. I.; Feeney, R. E. Abstr. Pacific Slope Biochem. Conf., Univ. Calif., Davis, June 28-July 1, 1981; p. 31.
31. Geoghegan, K. F.; Dallas, J. L.; Feeney, R. E. J. Biol. Chem., 1980, 255, 11429.
32. Lea, C. H. Chem. Ind. (London), 1950, 155.
33. Ellis, G. P. Adv. Carbohydr. Chem., 1959, 14, 63.
34. Kuhn, R.; Weygand, F. Chem. Ber., 1937, 70, 769.
35. Gottschalk, A. In "Glycoproteins. Their Composition, Structure and Function, Part A"; Gottschalk, A., Ed.; Elsevier Publ. Co.: New York, 1972; p. 141.
36. Hodge, J. E. Adv. Carbohydr. Chem., 1955, 10, 169.
37. Funcke, W.; Henneberg, D.; von Sonntag, C.; Klemer, A. Organic Mass Spectro., 1979, 14, 200.
38. Hayashi, T.; Namiki, M. Agric. Biol. Chem., 1980, 44, 2575.
39. Hayashi, T.; Namiki, M. Agric. Biol. Chem., 1981, 45, 933.
40. Hodge, J. E.; Mills, F. D.; Fisher, B. E. Cereal Sci. Today, 1972, 17, 34.
41. Mills, F. D.; Hodge, J. E. Carbohydr. Res., 1976, 51, 9.
42. Doornbos, T.; van den Ouweland, G. A. M. Ger. Offen. 2,529,320, 22 Jan., 1976; Chem. Abstr., 1976, 84, 149557d.

43. Lineweaver, H.; Feeney, R. E. In "Crops in Peace and War";
    Stefferud, A., Ed.; Yearbook of Agriculture, 1950-1951; U.S.
    Dept. Agric., Govmt. Printing Office: Washington, D.C.; p.
    642.
44. Kato, Y.; Watanabe, K.; Sato, Y. Agric. Biol. Chem., 1978,
    42, 2233.
45. Watanabe, K.; Sato, Y.; Kato, Y. J. Food Process. Preserv.,
    1980, 3, 263.
46. Feeney, R. E.; Abplanalp, H.; Clary, J. J.; Edwards, D. L.;
    Clark, J. R. J. Biol. Chem., 1963, 238, 1732.
47. Feeney, R. E.; Allison, R. G. "Evolutionary Biochemistry of
    Proteins: Homologous and Analogous Proteins from Avian Egg
    Whites, Blood Sera, Milk, and Other Substances"; Wiley and
    Sons: New York, 1969; pp. 290.
48. Lea, C. H.; Hannan, R. S. Biochim. Biophys. Acta, 1950, 4,
    518.
49. Lea, C. H.; Hannan, R. S. Biochim. Biophys. Acta, 1950, 5,
    433.
50. Carpenter, K. J. Nutr. Abstr. Rev., 1973, 43, 423.
51. Carpenter, K. J.; Booth, V. H. Biochem. J., 1960, 77, 604.
52. Hurrell, R. F.; Carpenter, K. J. Brit. J. Nutr., 1974, 32,
    589.
53. Bujard, E.; Finot, P. A. Ann. Nutr. Alim., 1978, 32, 291.
54. Nagao, M.; Honda, M.; Seino, Y.; Yahagi, T.; Kawachi, T.;
    Sugimura, T. Cancer Lett., 1977, 2, 335.
55. Shibamoto, T.; Nishimura, O.; Mihara, S. J. Agric. Food
    Chem., 1981, 29, 643.
56. Mihara, S.; Shibamoto, T. J. Agric. Food Chem., 1980, 28,
    62.
57. Spingarn, N. E.; Garvie, C. T. J. Agric. Food Chem., 1979,
    27, 1319.
58. Shibamoto, T. J. Agric. Food Chem., 1980, 28, 883.
59. Toda, H.; Sekizawa, J.; Shibamoto, T. J. Agric. Food Chem.,
    1981, 29, 381.
60. Pintauro, S. J.; Page, G. V.; Solberg, M.; Lee, T.-C.;
    Chichester, C. O. J. Food Sci., 1980, 45, 1442.
61. Kimiagar, M.; Lee, T.-C.; Chichester, C. O. J. Agric. Food
    Chem., 1980, 28, 150.
62. Coughlin, J. R. "Formation of N-Nitrosamines from Maillard
    Browning Reaction Products in the Presence of Nitrite"; Ph.D.
    Thesis, University of California, Davis, 1979; pp. 329.
63. Bookchin, R. M.; Gallop, P. M. Biochem. Biophys. Res.
    Commun., 1968, 32, 86.
64. Dixon, H. B. F. Biochem. J., 1972, 129, 203.
65. Holmquist, W. R.; Schroeder, W. A. Biochemistry, 1966, 5,
    2504.
66. Bunn, H. F.; Haney, D. N.; Gabbay, K. H.; Gallop, P. M.
    Biochem. Biophys. Res. Commun., 1975, 67, 103.
67. Miller, J. A.; Gravalles, E.; Bunn, H. F. J. Clin. Invest.,
    1980, 65, 896.

68. Stevens, V. J.; Rouzer, C. A.; Monnier, V. M.; Cerami, A. Proc. Nat. Acad. Sci. U.S.A., 1978, 75, 2918.
69. Pande, A.; Garner, W. H.; Spector, A. Biochem. Biophys. Res. Commun., 1979, 89, 1260.
70. Bailey, A. J.; Robins, S. P. Front. Matrix Biol., 1973, 1, 130.
71. Day, J. F.; Thorpe, S. R.; Baynes, J. W. J. Biol. Chem., 1979, 254, 595.
72. Rahbar, S. Clin. Chim. Acta, 1968, 22, 296.
73. Higgins, P. J.; Bunn, H. F. J. Biol. Chem., 1981, 256, 5204.
74. Zigler, J. S., Jr.; Goosey, J. Trends Biochem. Sci., 1981, 6, 133.
75. Monnier, V. M.; Cerami, A. Science, 1981, 211, 491.
76. Chiou, S.-H.; Chylack, L. T.; Tung, W. H.; Bunn, H. F., J. Biol. Chem., 1981, 256, 5176.
77. Feeney, R. E. In "Chemical Deterioration of Proteins"; Whitaker, J. R.; Fujimaki, M., Eds.; American Chemical Society: Washington, D. C., ACS Symp. Series, 123, 1980; p. 1.
78. Acharya, A. S.; Manning, J. M. J. Biol. Chem., 1980, 255, 7218.
79. Acharya, A. S.; Manning, J. M. Fed. Proc., Fed. Am. Soc. Exper. Biol., 1981, 40 (6), Abstr. 423, 1613.
80. Bunn, H.; Higgins, P. J. Science, 1981, 213, 222.
81. Kline, L.; Hanson, H. L.; Sonoda, T. T.; Gegg, J. E.; Feeney, R. E.; Lineweaver, H. Food Technol., 1951, 5, 323.
82. Kline, R. W.; Fox, S. W. Iowa State Coll. J. Sci., 1946, 20, 265.
83. Bjarnason, J.; Carpenter, K. J. Brit. J. Nutr., 1969, 23, 859.
84. Bjarnason, J.; Carpenter, K. J. Brit. J. Nutr., 1970, 24, 313.
85. Lee, H. S.; Sen, L. C.; Clifford, A. J.; Whitaker, J. R.; Feeney, R. E. J. Nutr., 1978, 108, 687.
86. Mohammed, K. U.S. Patent 4,144,357, 1979.
87. Kline, L.; Gegg, J. E.; Sonoda, T. T. Food Technol., 1951, 5, 181.

RECEIVED June 1, 1982.

Lysinoalanine Formation in Soybean Proteins: Kinetics and Mechanisms

MENDEL FRIEDMAN

United States Department of Agriculture, Western Regional Research Center,
Agricultural Research Service, Berkeley, CA 94710

Commercial and home processing of foods employ heat, sunlight, ultraviolet radiation, acids, alkali, and combinations of these. The purpose of such treatments is to make food edible (cooking, broiling), to alter texture and flavor, to permit storage, to destroy toxins, to prepare protein concentrates, to pasteurize milk, kill bacteria, etc. On the other hand, such treatments cause undesirable changes. These include formation of cross-linked amino acids such as lysinoalanine (LAL). LAL has been indentified in hydrolysates of alkali-treated and heat-treated proteins (1, 2).

Feeding alkali-treated soy proteins to rats induces changes in kidney cells characterized by enlargement of the nucleus and cytoplasm, increased nucleoprotein content, and disturbances of DNA synthesis and mitosis (3,4). The lesions affect epithelial cells of the straight portion (pars recta) of the proximal renal tubules. Foods containing the crosslinked amino acid lysinoalanine and D-amino acids are widely consumed (5,6).

These observations cause concern about the nutritional quality and safety of alkali-treated foods. Chemical changes that govern formation of unnatural amino acids during alkali treatment of proteins need to be studied and explained. The nutritional and toxicological importance of these changes need to be defined. Appropriate strategies to minimize or prevent these reactions need to be developed.

Previous papers covered the following aspects of protein-alkali reactions: 1) elimination reactions of disulfide bonds in amino acids, peptides, and proteins (7); 2) alkali-induced amino acid crosslinking (8); 3) inhibitory effects of certain amino acids and inorganic anions on lysinoalanine formation (9-11); 4) prevention of lysinoalanine formation by protein acylation (10, 11); 5) transformation of lysine to lysinoalanine, and of cysteine to lanthionine residues in proteins and polyamino acids (12); 6) effects of lysine modification on chemical, nutritional, and functional properties of proteins (13); 7) alkali-induced racemization of casein, lactalbumin, soy protein, and wheat gluten (14); 8) inverse relationship between lysinoalanine and D-amino acid content of casein and in vitro digestibility

by trypsin and chymotrypsin (6); 9) chemical, nutritional, and
toxicological consequences of alkali-induced racemization of
proteins (15); 10) alkali-induced formation of dehydroalanine
residues in casein and acetylated casein (16); 11) biological
utilization of D-isomers of nine essential amino acids by mice
(17); and 12) kinetics of racemization of amino acid residues in
casein (18).

This Chapter reports and discusses conditions that favor or
prevent lysinoalanine formation in soy protein and the consequent
mechanistic implications. Related results for other proteins are
compared. Although important research in this area is being done
in other laboratories, no attempt is made to critically examine
the literature of this subject (1-137), but rather to interpret
the factors found to govern lysinoalanine formation in soy protein. The findings have theoretical interest, but also have
profound practical implications for nutrition and food safety.
Hopefully, they will be useful to food processors and consumers
who wish to minimize the formation of unnatural, deleterious
compounds, during commercial and home food processing.

Factors Affecting Lysinoalanine Formation

pH. Alkali treatments and amino acid analyses were carried
out as previously described (6, 10, 11). The effect of pH on
the amino acid composition of soy protein subjected to 0.1N NaOH
for 3 hours at 75°C is given in Table I in terms of mole per cent
of amino acids accounted for. The mole per cent or ratio methods
are more consistent measures of amino acid composition of processed, commercial, or modified proteins because they avoid errors
due to moisture content (acylated proteins are highly hygro-
scopic), protein content (the commercial wheat gluten, soy
protein, and lactalbumin contain carbohydrates, lipids, etc.),
molecular weight changes, and errors occurring during amino acid
analysis.

Table II records the same information in terms of weight per
cent or grams of amino acid per 100 grams of protein. The data in
both tables show similar trends in alkali-induced destruction of
labile amino acids.

The following amino acids are degraded, destroyed, or modified during the alkaline treatment of soy protein: threonine,
serine, cystine, lysine, and arginine. Thus, treating 1% (w/v)
soy protein in 0.1N NaOH at 75°C decreases the threonine content
from 4.23 to 2.92%. The corresponding values for serine are 6.41
and 3.97%; for cystine, 0.47 and 0.14%; for lysine, 5.55 and 4.33%;
and for arginine, 5.84 and 4.33%, respectively (Table I). The
treatment also induces the formation of lysinoalanine, which
begins to appear at pH 9.0 and continuously increases with pH up
to pH 12.5 and then decreases at pH 13.9 (Tables I and II).

Table I. Effect of pH on amino acid composition of soy protein (Promine-D).[1]

Amino Acid	Untreated Control	Water-Dialyzed Control	pH 8.0	9.0	9.5	10.0	10.5	10.9	12.0	12.5	13.9
ASP	11.73	11.74	11.66	11.71	11.59	11.62	11.75	11.70	11.63	12.47	9.64
THRE	4.23	4.12	4.09	4.03	4.22	4.16	4.17	4.04	3.99	2.92	3.08
SER	6.41	6.57	6.29	6.25	6.35	6.26	6.08	5.98	5.53	3.97	3.59
GLU	17.29	18.25	17.69	18.22	17.55	17.36	16.62	17.91	15.29	17.82	12.28
PRO	5.84	5.96	5.46	5.43	5.76	5.49	5.43	5.70	5.37	5.27	5.39
GLY	7.43	7.28	7.31	7.34	6.97	7.44	7.29	7.50	7.82	8.05	9.13
ALA	6.24	6.14	6.15	6.09	6.08	6.36	6.57	6.51	7.31	6.98	9.21
CYS	0.47	0.37	0.66	0.34	0.16	0.18	0.18	0.00	0.15	0.14	0.00
VAL	5.43	5.33	5.77	5.74	6.39	6.01	6.41	5.43	6.71	6.68	8.12
MET	1.08	1.00	0.92	0.89	0.81	0.75	0.85	0.93	0.88	1.02	0.00
ILEU	4.91	4.80	4.95	4.98	5.04	5.07	5.30	5.07	5.56	5.23	6.52
LEU	8.32	8.22	8.32	8.29	8.41	8.55	8.82	8.50	9.45	9.39	13.94
TYR	2.77	2.70	2.69	2.61	2.79	2.70	2.80	2.78	3.12	3.05	2.48
PHE	4.32	4.28	4.52	4.51	4.58	4.64	4.68	4.52	4.89	4.64	5.96
HIS	2.15	2.11	2.16	2.14	2.10	2.12	2.09	2.14	2.01	2.18	2.23
LYS	5.55	5.48	5.86	5.83	5.82	5.80	5.70	5.69	5.22	4.33	2.85
LAL	0.00	0.00	0.00	0.28	0.40	0.61	0.68	0.95	1.14	1.86	1.47
ARG	5.84	5.77	5.86	5.83	5.82	5.80	5.70	5.69	5.22	4.33	2.85

[1]Conditions: 0.5 g protein; 50 cc 0.1 N NaOH; 75°C; 3 hours. Numbers in mole % are averages from two separate experiments.

Table II. Effect of pH on amino acid composition of soy protein (Promine-D).[1]

Amino Acid	Untreated Control	Water-Dialyzed Control	pH 8.0	9.0	9.5	10.0	10.5	10.9	12.0	12.5	13.9
ASP	10.53	10.82	10.55	10.40	10.33	10.93	10.68	10.61	10.72	10.70	7.17
THRE	3.40	3.40	3.32	3.20	3.36	3.50	3.39	3.28	3.29	2.24	2.05
SER	4.55	4.78	4.50	4.83	4.47	4.65	4.36	4.28	4.02	2.69	2.12
GLU	17.16	18.60	17.71	17.89	17.29	18.04	16.70	17.95	15.60	16.90	10.11
PRO	4.53	4.75	4.28	4.17	4.44	4.47	4.27	4.47	4.28	3.90	3.45
GLY	3.76	3.75	3.73	3.68	3.50	3.94	3.74	3.84	4.06	3.89	3.82
ALA	3.75	3.79	3.73	3.62	3.63	4.00	3.98	3.95	4.50	4.01	4.56
CYS	0.38	0.26	0.54	0.28	0.13	0.15	0.15	0.00	0.12	0.10	0.00
VAL	4.29	4.32	4.60	4.49	5.02	4.97	5.12	4.34	5.44	5.05	5.27
MET	1.09	1.03	0.94	0.89	0.81	0.79	0.86	0.94	0.90	0.99	0.00
ILEU	4.35	4.36	4.41	4.36	4.42	4.69	4.75	4.52	5.05	4.42	4.73
LEU	7.37	7.47	7.42	7.25	7.38	7.92	7.90	7.59	8.58	7.94	10.11
TYR	3.39	3.39	3.31	3.15	3.38	3.46	3.48	3.43	3.91	3.55	2.51
PHE	4.81	4.90	5.90	4.97	5.07	5.42	5.28	5.09	5.60	4.93	5.47
HIS	2.25	2.27	2.28	2.22	2.18	2.32	2.21	2.27	2.17	2.17	1.92
LYS	5.47	5.54	5.42	5.14	4.84	5.00	4.64	4.59	4.00	3.46	3.34
ARG	6.87	6.96	6.95	6.78	6.78	7.13	6.78	6.75	6.30	4.86	2.79
LAL	0.00	0.00	0.00	0.57	0.81	1.28	1.42	1.79	2.42	3.66	2.49

[1]Conditions: 0.5 g protein; 50 cc of 0.1 M borate buffer; 75°C; 3 hours. Numbers show average values for two experiments in weight % (g/100 g protein).

Evidently, lysinoalanine is destroyed as well as formed during alkaline treatment. To obtain evidence for this hypothesis, the hydrolytic stability of pure lysinoalanine under acidic and basic conditions was examined. The results summarized in Tables III and IV show that 1) commercial lysinoalanine has about 2% lysine; 2) lysinoalanine appears stable to acid hydrolysis conditions, although the small decrease in lysinoalanine and accompanying increase in lysine may be outside of experimental error; 3) hydrolysis in 1 $\underline{N}$ NaOH destroys a large fraction of lysinoalanine, as it does leucine (added as an internal standard). (Note that the lysinoalanine/leucine ratio is essentially unchanged during both acid and base hydrolysis); and 4) about 10% of lysinoalanine is destroyed during co-hydrolysis with soy flakes or starch in acid and 32 to 52% in base.

Time. Data in Table V show that heating a 1% (w/v) solution of soy protein in 0.1 $\underline{N}$ NaOH at 75°C for various time periods induces a progressive increase in lysinoalanine formation which levels off after about three hours and then starts to decrease.

These results and those mentioned above on the pH-dependence of lysinoalanine formation indicate that for each protein, conditions may exist where the rate of lysinoalanine formation may be as fast as its rate of destruction. Although lysinoalanine could, in principle, regenerate lysine by a base-catalyzed nucleophilic displacement, examination of the last-column in Table I shows that combined values of lysinoalanine and lysine do not add up to the original lysine content of untreated soy protein, suggesting that lysinoalanine may act as a precursor for other compounds in addition to lysine.

Temperature. Exposure of a 1% (w/v) solution of soy protein in 0.1 $\underline{N}$ NaOH (pH 12.5) for three hours at temperatures ranging from 25°C to 85°C reveals a progressive degradation of the following labile amino acids: threonine, serine, lysine, and arginine (Table VI). Serine and threonine destruction starts at about 45°C and lysine and arginine at about 25°C. Thus, the value of threonine decreases from 3.40 mole % for the untreated soy protein to 1.84 mole % for the sample treated at 85°C. The corresponding values for serine are 6.41 and 2.73 mole %; for lysine, 5.55 and 3.24 mole %; and for arginine, 5.84 and 3.15 mole %; respectively. The treatment also induces the progressive disappearance of cystine (not shown) and the appearance of lysinoalanine residues. These start to appear at 25°C (0.29 mole %) and continuously increase up to 2.24 mole % at 85°C. Although imidazole NH group of histidine could, in principle, also combine with dehydroalanine to form histidinoalanine (8), the data in Table VI show that histidine appears not to have been affected by the treatment.

Table III.   Stability of free lysinoalanine to acid and
base hydrolysis.[1]

	6 N HCl[1]				1 N NaOH[2]			
Time (hrs)	LAL	LYS	LAL/LEU		LAL	LYS	LAL/LEU	SER
0	4.69	0.09	1.124		4.22	0.0	1.124	0.0
4	4.36	0.11	1.107		2.90	0.13	1.123	0.0
8	4.39	0.13	1.171		---	---	---	---
24	4.60	0.17	1.081		3.01	0.20	1.122	0.11
48	4.40	0.23	1.113		---	---	---	
72	4.57	0.29	1.073		2.11	0.29	1.039	0.0

[1]Conditions: 4.69 μmoles LAL and 5.27 μmoles LEU; 10 ml 6N HCl;
110°C; 24 hours.
[2]4.22 μmoles LAL and 4.74 μmoles LEU; 10 ml 1N NaOH; 110°C;
24 hours.

Table IV.   Stability of added lysinoalanine to acid and
base hydrolysis.[1]

Solvent	% Recovery lysinoalanine in the presence of:	
	Soy flakes	Starch
6N HCl	90.8	87.7
1N NaOH	68.2	47.6

[1]Conditions: 5.08 uM lysinoalanine
5 mg soy flakes or starch in 10 ml
6 N HCl or 1 N NaOH were hydrolyzed
in evacuated, sealed tubes for 24
hours at 110°C.

Table V. Effect of time of treatment on LAL content of alkali-treated soy protein.[1]

Time (min)	Lysinoalanine	
	Weight %	Mole %
10	0.76	0.46
20	0.96	0.63
40	1.64	1.01
60	1.94	1.10
120	3.14	1.75
180	4.04	2.39
300	4.24	2.98
480	3.66	2.05

[1]Conditions: average of two experiments; 0.5 g protein in 50 ml 0.1 $\underline{N}$ NaOH; 75°.

Table VI. Effect of temperature on amino acid composition of alkali-treated soy protein.[1]

Sample Number	THRE Wt. %	THRE Mole %	SER Wt. %	SER Mole %	HIS Wt. %	HIS Mole %	LYS Wt. %	LYS Mole %	LAL Wt. %	LAL Mole %	ARG Wt. %	ARG Mole %
1. Untreated control	3.40	4.23	4.55	6.41	2.25	2.15	5.47	5.55	0.00	0.00	6.87	5.84
2. H2O-dialyzed control	3.40	4.12	4.78	6.57	2.27	2.11	5.54	5.48	0.00	0.00	6.96	5.77
3. 25°C	3.39	4.49	4.47	6.33	2.28	2.27	4.91	5.32	0.69	0.29	6.49	5.86
4. 35°C	3.35	4.32	4.48	6.24	2.16	2.14	4.88	5.08	0.81	0.41	6.52	5.81
5. 45°C	3.25	4.29	4.07	6.08	2.17	2.19	4.63	4.97	1.04	0.79	6.29	5.67
6. 55°C	3.00	4.16	4.08	6.13	2.12	2.16	4.29	4.64	1.64	0.85	6.11	5.54
7. 65°C	2.83	3.87	3.32	5.14	2.06	2.15	3.72	4.13	2.64	1.40	5.48	5.11
8. 75°C	2.24	2.92	2.69	4.97	2.17	2.18	3.46	3.68	3.66	1.86	4.86	4.33
9. 85°C	1.84	2.56	1.73	2.73	2.29	2.45	2.86	3.24	4.13	2.24	3.30	3.15

[1]Conditions: 0.5 g protein; 50 cc 0.1 N NaOH; 3 hours. Numbers are averages from two experiments.

Although conditions described for the formation of lysino-alanine are more severe than generally used in food processing, the results of this temperature study and those cited earlier strikingly demonstrate that significant amounts of lysinoalanine appears to be formed in soy protein under relative mild conditions of pH, time, and temperature. Consequently, preventative measures against lysinoalanine formation may be justified even if proteins are subjected to mild alkaline conditions during food processing.

Concentration. Data in Table VII show that the concentration of soy protein in the range 0.5 to 5.0% (w/v) does not appear to signficantly affect the extent of lysinoalanine formation. This observation and the previous finding (10) that the extent of lysinoalanine formation in wheat gluten in 1 N NaOH at 65°C also does not appear to vary significantly in 1% or 10% gluten suspensions suggests that this may be a general phenomenon. However, protein concentration does seem to influence the extent of lysino-alanine formation at lower pH (5), where ε-NH$_2$ addition to dehydroalanine may be rate-controlling.

Table VII. Effect of concentration on lysinoalanine content of alkali-treated soy protein (Promine-D).[1]

Protein (%)	Lysinoalanine Wt. %	Mole %
0.5	2.98	1.59
2.0	2.71	1.43
5.0	2.72	1.31

[1]Conditions: 0.1 N NaOH; 65°C, 3 hours.

The observed absence of a concentration effect can probably be explained as follows. Since elimination reactions of serine, threonine, and cystine to form dehydroalanine side chains are second-order reactions that depend on the concentration of both hydroxide ion and susceptible amino acid side chains, the extent of crosslinking should be a function of hydroxide ion concentration. This is indeed the case (19). However, because the reaction of the ε-amino group of lysine with a vinyl-type compound such as dehydroalanine is also a second-order reaction (20-22), it would have been expected that an increase of protein concentration should lead to a corresponding increase in the rate of formation of lysinoalanine. Since this apparently does not occur, the major factor controlling formation of lysinoalanine, once the rate-determining dehydroalanine precursors are formed, may be the location of required partners for crosslink formation. Only dehydroalanine and lysine residues situated on the same or closely adjacent protein chains are favorably placed to form crosslinks. Once the convenient sites are used up, additional lysinoalanine or other crosslinks form less readily. Each protein, therefore, may have a limited fraction of potential sites for "productive" formation of crosslinked residues. The number of such sites may vary and is presumably dictated by the protein's size, composition, conformation, chain mobility, and steric factors.

Carbohydrates. The ε-amino group of lysine reacts under the influence of heat with the potential carbonyl groups of sugars such as glucose to form Schiff's bases that are further transformed to incompletely characterized Maillard browning products (23). If such reactions were to occur in strongly alkaline media, they would be expected to reduce the formation of lysinoalanine because the blocked lysine amino groups would be unable to combine with dehydroalanine. To test this hypothesis, the influence of several reducing and nonreducing sugars on the extent of lysinoalanine formation was evaluated.

Exposure of a 1% (w/v) solution of soy protein to 0.1 N NaOH for 1 hour at 45 or 75°C in the absence and presence of 0.5 to 2.0% (w/v) glucose did not affect the amount of lysinoalanine produced (Tables VIII, IX; Figures 1, 2). Glucose, however, did produce two noticeable effects: 1) a sharp decrease in the arginine content which varied directly with the amount of glucose present; and 2) the formation of a new compound, X, (Figure 2; Tables VIII, IX) eluting just before lysinoalanine.

The concentration of the unknown compound X increased sharply when the temperature was raised to 75°C ranging from 0.645 mole % in the presence of 0.5% glucose to 1.87 mole % with 2% glucose. The arginine content dropped from 5.84 mole % for untreated soy protein to 0.805% for soy protein treated with base in the presence of 2% glucose (Table IX).

Table VIII. Effect of glucose on amino acid composition of alkali-treated soy protein (Promine-D).[1]

Sample Number	THRE	SER	MET	HIS	LYS	ARG	X	LAL	N %
1. Untreated control	4.23	6.41	1.08	2.15	5.55	5.84	0.00	0.00	13.47
2. H₂0-dialyzed control	4.12	6.57	1.00	2.11	5.48	5.77	0.00	0.00	14.33
3. NaOH only	4.02	6.34	0.555	2.16	5.05	5.48	0.00	0.41	13.44
4. NaOH plus 25 mg glucose	4.03	6.27	0.815	2.24	4.90	4.83	0.51	0.50	13.37
5. NaOH plus 50 mg glucose	4.03	6.28	0.755	2.22	5.05	4.50	0.45	0.44	13.17
6. NaOH plus 100 mg glucose	4.12	6.31	0.655	2.16	5.07	3.63	0.33	0.40	13.15

[1]Conditions: 0.5 g protein; 50 cc 0.1 N NaOH; 1 hour; 45°C. Numbers are averages from two experimental treatments in mole per cent.

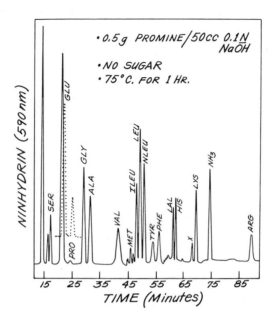

Figure 1. *Amino acid analysis of alkali-treated soy protein. Note that lysinoalanine (LAL) peak and unknown peak appear before lysine (LYS). Conditions: 0.5 g Promine/50 cm³ 0.1 N NaOH; no sugar; and 75°C for 1 h.*

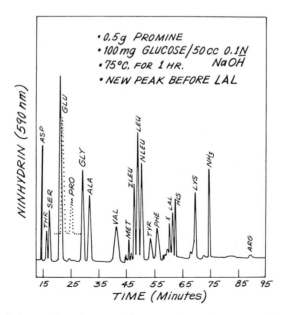

Figure 2. *Amino acid analysis of alkali-treated soy protein in the presence of glucose. Note the new peak before lysinoalanine (LAL). Conditions: 0.5 g Promine; 100 mg glucose/50 cm³ 0.1 N NaOH; and 75°C for 1 h.*

Table IX. Effect of glucose on amino acid composition of alkali-treated soy protein (Promine-D).[1]

Sample Number	THRE	SER	MET	HIS	LYS	ARG	X	LAL	N %
1. Untreated control	4.23	6.41	1.08	2.15	5.55	5.84	0.00	0.00	13.47
2. H₂0-dialyzed control	4.12	6.57	1.00	2.11	5.48	5.77	0.00	0.00	14.33
3. NaOH only	1.99	2.77	0.78	2.53	4.21	4.22	0.00	1.72	12.18
4. NaOH plus 25 mg glucose	2.02	2.79	0.76	2.33	4.14	2.92	0.645	1.50	11.84
5. NaOH plus 50 mg glucose	2.09	2.92	0.70	2.29	4.88	1.88	1.15	1.42	10.89
6. NaOH plus 100 mg glucose	2.14	3.08	1.05	2.27	5.12	0.805	1.87	1.38	11.54

[1]Conditions: 0.5 g protein; 50 cc 0.1N NaOH; 75°C; 1 hour. Numbers are mole per cent values (average from two experimental treatments).

Because the lysine content of the soy protein treated in the presence of glucose was not systematically altered (Table X), these results suggest that glucose does not appear to modify lysine to derivative(s) stable to protein acid hydrolysis conditions. Nevertheless, glucose, under alkaline conditions, seems to modify arginine irreversibly to derivatives stable to acid hydrolysis.

To assess the generality of these reactions, the effect of xylose on lysine, lysinoalanine, and arginine content of soy protein was evaluated. Table X shows that the effect of xylose is similar to that of glucose, i.e. lysinoalanine did not change whereas about 85% of the arginine residues disappeared. In addition, a new ninhydrin-positive compound (X_3) was formed eluting just prior to lysinoalanine.

Metal salts. Several metal ions can interact with basic groups of amino acid residues including the guanidino group of arginine, the imidazole group of histidine, and the ε-amino group of lysine (24). Complexation with lysine may minimize or prevent lysinoalanine formation if the complex is stable in alkali. To test this hypothesis, the influence of several metal salts on

Table X. Effect of xylose on lysine, lysinoalanine, and arginine content of alkali-treated soy protein. Values listed are mole %.[1]

Amino Acid	Elution time (min)	Untreated control	NaOH only (No xylose)	NaOH plus xylose
Lysine	69.24	5.55	4.07	3.81
Lysinoalanine	60.42	0.00	1.24	1.13
Arginine	88.23	5.84	4.52	0.83
X_1	13.29	0.00	0.58	0.34
X_2	46.01	0.00	0.50	0.37
X_3	60.17	0.00	0.00	1.44
X_4	67.18	0.00	0.92	1.37

[1]Conditions: 0.5 g protein plus 200 mg xylose; 50 cc 0.1 $\underline{N}$ NaOH; 1 hour; 75°C
Note: X_3 peak apparently due to presence of xylose.

the amino acid composition of alkali-treated soy protein was examined. Table XI shows that, within experimental error, only $CuSO_4$ and possibly $FeCl_2$ caused a significant decrease in lysinoalanine content. The complete amino acid profiles also revealed that histidine, arginine, and lysine were not significantly affected by the presence of any of the listed salts.

Dimethyl sulfoxide. Dimethyl sulfoxide (DMSO) and other dipolar aprotic solvents strongly affect the reactivity of ionizable reactive sites in proteins. Several effects contribute to this result. In particular, such solvents, because of differences in dielectric constant and hydrogen bonding properties, influence ionization equilibria, hydrogen-bonding and hydrophobic interactions, and the nucleophilic strengths of protein functional groups. The stabilities of ground and transition states are therefore modified as are the chemical reactivities of protein functional groups, protein conformations, and, for instance, enzyme reactivities (13, 25-27).

Friedman (25) measured reaction rates of amino groups in structurally different amino acids, peptides, and proteins with conjugated vinyl compounds in a medium consisting of borate buffer and DMSO and compared them with analogous rates in an aqueous buffer.

Table XI.   Effect of metal salts on lysinoalanine content of alkali-treated soy protein[1]

Metal salt treatment	Lysinoalanine	
	Weight %	Mole %
Untreated control	0.00	0.00
Alkali only	1.94; 1.57 [2]	1.10; 1.24
$MgSO_4 \cdot 7H_2O$	1.69	0.958
$ZnSO_4 \cdot 7H_2O$	1.59; 1.60	0.930; 0.899
$CaCl_2 \cdot 2H_2O$	1.88; 1.94	1.08; 1.13
$FeCl_2 \cdot 4H_2O$	1.26; 1.39	0.888; 0.901
$FeSO_4 \cdot 7H_2O$	1.54; 1.82	1.00; 1.05
$CuSO_4 \cdot 5H_2O$	0.00; 0.457	0.00; 0.270

[1]Conditions: 0.5 g soy protein plus 0.5 millimoles metal salt in 50 ml 0.1N NaOH; 75°C; 1 hour.
[2]Results from two separate experiments.

The results were described by the following equation:

$$\log \frac{k_2 \text{ (buffer - DMSO, 50:50 v/v)}}{k_2 \text{ (buffer)}} = R = 0.855 \text{ pK}_2 - 6.505$$

In the mixed solvent, the rate increase (R), which ranged from 1.4 for tetraglycine to 209 for ε-aminocaproic acid, varied directly with the pK_a value of the amino group. This rate increase can be rationalized as follows. Water strongly solvates an amino acid anion of structure $NH_2CH(R)COO^-$ as illustrated in Formula I. In this case, solvation decreases the electron density of nitrogen and, consequently, decreases the nucleophilicity of the amino group. Water can also hydrogen bond to the amino groups as shown in Formula II. This solvation process increases the electron density on nitrogen and therefore increases the nucleophilicity of the amino group. Evidently, because water can act as both a hydrogen donor and acceptor, its effect results from a balance of these opposing actions. Because DMSO acts only as an acceptor, only the hydrogen-bonding interactions shown in Formula III and IV are possible in this medium. These interactions strengthen both the basicity and nucleophilicity of the amino group. In addition, by hydrogen-bonding interactions with protonated amino groups shown in Formulas V-VI, DMSO can lower the basicity of amino groups as measured by pK.

Friedman (25) also examined the reaction of acrylonitrile with amino groups in three proteins. The rate enhancement factor in the mixed solvent ranged from 20 for bovine serum albumin to 60 for polylysine. Although such rate studies suggest that proteins can be chemically modified more rapidly in mixed aqueous-nonaqueous media than in dilute aqueous solvents, it is difficult to ascertain how much the rate enhancement by DMSO is due to the described direct influences of the nonaqueous compound on electrostatic and hydrogen-bonding environments of protein amino groups and how much to effects on protein conformation.

Since the nucleophilic addition of the ε-NH$_2$ group of lysine to the double bond of dehydroalanine is mechanistically analogous to the addition of NH$_2$ groups to the double bond of acrylonitrile, DMSO should profoundly influence the extent of LAL formation if the reaction with the double bond is the rate-determining step. Limited studies summarized in Table XII show that the extent of lysinoalanine formation in a pH 10 buffer is essentially the same as in a solvent medium consisting of 50 : 50 (v/v) pH 10 buffer : DMSO. This result implies that the rate-determining step is probably the β-elimination reaction of serine or cystine to form dehydroalanine rather than the subsequent addition of the amino group to the former. However, the situation in DMSO may be complicated since the data in Table XII also show that the amount of lysinoalanine produced in 0.1N NaOH is nearly twice that in a medium consisting of 50:50 (v/v) 0.1N NaOH : DMSO. This result suggests that the negative dipole of DMSO probably suppresses dehydroalanine formation by destabilizing the incipient negative carbanion by charge repulsion.

Table XII. Effect of dimethyl sulfoxide on lysinoalanine content of alkali-treated soy protein.[1]

Solvent	Lysinoalanine	
	Wt. %	Mole %
pH 10 buffer	0.53 + 0.04	0.30 + 0.0
pH 10 : DMSO (50 : 50)	0.70 + 0.20	0.44 + 0.06
0.1 N NaOH	1.57 + 0.07	1.24 + 0.01
0.1 N NaOH : DMSO (50 : 50)	0.89 + 0.02	0.58 + 0.01

[1]Conditions: 0.5 g protein in 50 cc solvent; 1 hour; 75°C. Values are averages from two separate experiments.

Mechanism of Lysinoalanine Formation

    A postulated mechanism of lysinoalanine formation (Figures 3-5) is a two-step process at least. First, hydroxide ion-catalyzed elimination reactions of serine, threonine, and cystine (and to a lesser extent probably also cysteine) give rise to a dehydroalanine intermediate. Since such elimination reactions are second-order reactions that depend directly on the concentration of both hydroxide ion and susceptible amino acid, the extent of lysinoalanine formation should vary directly with hydroxide ion concentration. This is indeed the case within certain pH ranges. The dehydroalanine residue, which contains a conjugated carbon-carbon double bond, then reacts with the ε-amino group of lysine in a second, second-order step, to form a lysinoalanine crosslink. This step is governed not only by the number of available amino groups but also by the location of the amino group and dehydroalanine potential partners in the protein chain. When convenient sites have reacted, additional lysinoalanine (or other) crosslinks form less readily or not at all. Each protein, therefore, may have a limited fraction of potential sites for forming crosslinked residues. The number of such sites is presumably dictated by the protein's size, composition, conformation, chain mobility, steric factors, extent of ionization of reactive amino (or other) nucleophilic centers, etc.

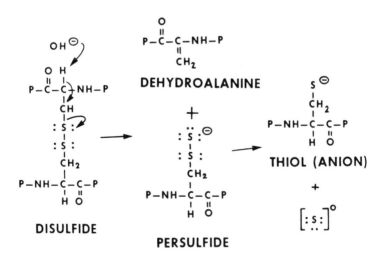

Figure 3.   Postulated mechanism for base-catalyzed transformation of a protein-disulfide bond to two dehydroalanine side chains, a sulfide ion, and sulfur.

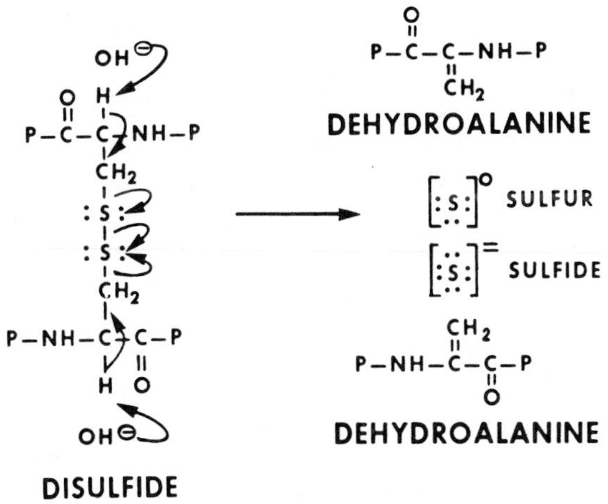

Figure 4. Postulated mechanism for base-catalyzed formation of one dehydro-alanine side chain and one persulfide anion from a protein–disulfide bond. The persulfide can decompose to a thiol anion (cysteine) and elemental sulfur.

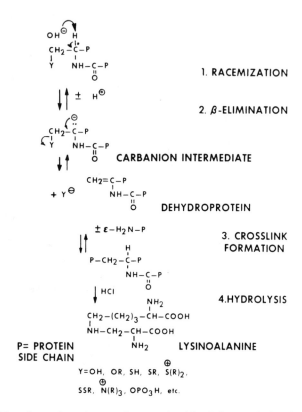

Figure 5.   Transformation of a reactive protein side chain to a lysinoalanine side chain via elimination and cross-link formation. Note that the intermediate carbanion has lost the original asymmetry of the reactive amino acid side chain. The carbanion can combine with a proton to regenerate the original amino acid side chain, which is now racemic, or undergo an elimination reaction to form dehydroalanine.

These considerations suggest that a cascade of reactions occurs leading to lysinoalanine residues. Thus, dehydroalanine formation is governed not only by the absolute concentration of serine and cystine residues but by their relative susceptibilities to base-catalyzed eliminations. On the other hand, reaction of the ε-amino groups with dehydroalanine to form lysinoalanine depends not only on the cited steric and conformational factors but also on the pH of the medium, which governs the concentration of reactive nonprotonated amine. Since the pK of the ε-amino groups of lysine residues is near 10 for most proteins, complete ionization of all amino groups does not occur until pH 12. At pH 9, only about 10% of the amino groups are ionized, and thus available for reaction (all of the amino groups can eventually react, however, since additional amino groups are formed by ionization of the protonated ammonium ions as the nonprotonated amino groups are used up).

To gain insight into the reactivity of the double bond of dehydroalanine, the factors that influence the reaction rates of SH and NH_2 groups in amino acids, peptides, proteins, and related compounds with dehydroalanine and dehydroalanine methyl ester were investigated ($\underline{12}$, $\underline{21}$, $\underline{22}$). Kinetic studies revealed that the double bond of dehydroalanine readily combines with the sulhydryl group of cysteine and other thiols to form lanthionine and related compounds. Related studies showed that the amino group of ε-N-acetyl-L-lysine interacts and combines at a slower rate with dehydroalanine than do sulfhydryl groups (Table XIII). Since comparison of relative reactivities of ionized nucleophiles such as R-S$^-$ and R-NH$_2$ with the double bond of dehydroalanine are valid only in terms of pH-independent rate constants associated with the ionized forms and not with observed pH-dependent rate constants, second-order anion rate constants (k_A- anion) in Table XIII were calculated from the observed rate constants k_2 by means of equation 1 ($\underline{20}$, $\underline{28}$, $\underline{29}$), where K is the dissociation constant of the protonated amino group.

$$k_{A^-} = k_2 \left[1 + \frac{(H^+)}{(K)} \right] \tag{1}$$

Results ($\underline{20}$, $\underline{28}$) showed that the second order anion (inherent) rate constants (k_A-) are essentially invariant over the pH range studied, and demonstrate that equation 1 agreed with experimental observations.

To calculate predicted rate ratios of two functional groups as a function of hydrogen ion concentration, as was done for the rate ratio for reaction of the sulfhydryl group in cysteine and the ε-amino group of α-N-acetyl-L-lysine with dehydroalanine methyl ester (Tables XIII and XIV), the equation 2 version of

Table XIII. Reaction rates of N-α-acetyl-L-lysine and
mercaptoamino acids with N-acetyl dehyroalanine
and its methyl ester (9, 20, 22).

Reactant	Dehydroalanine form	pH	pK	k_2 obs $(M^{-1} min^{-1})$	k_{A^-} (anion) (calculated) $(M^{-1} min^{-1})$
N-α-acetyl-L-lysine	Ester	9.5	10.53	0.71	11.6
Cysteine	Ester	8.5	8.15	---	358.7
DL-Penicillamine	Ester	8.0	7.90	5.08	9.1
Thiosalicylic acid	Ester	6.4	7.79	1.76	45.7
Cysteine	Acid	8.5	8.15	0.0217	0.0314
DL-Penicillamine	Acid	8.5	7.90	0.0006	0.0008

Table XIV. Calculated second-order rate constants for reaction
of N-acetyldehydroalanine methyl ester with the ε-NH_2
group of N-α-acetyl-L-lysine and the SH group of
cysteine at various pH's (9, 20, 22).

pH	α-N-acetyl-L-lysine	L-cysteine	Rate Ratio: SH Cysteine / ε-NH_2 Lysine
7.0	5.14×10^{-3}	25.62	5000
8.0	6.48×10^{-2}	148.84	2300
9.0	0.763	314.4	410
10.0	2.58	344.6	133
11.0	8.66	357.2	43
12.0	11.22	358.6	34

equation 1 may be used. In this, k_2, k_{A^-}, and K are associated with one functional group (e.g. sulfhydryl), and k_2', k_{A^-}', and K' with the second (e.g. amino):

$$\frac{k_2}{k_2'} = \left(\frac{k_{A^-}}{k_{A^-}'}\right)\left(\frac{K}{K'}\right)\left[\frac{K' + (H^+)}{K + (H^+)}\right] \tag{2}$$

Thus, provided the cited approximations hold for both functional groups, all that is necessary to calculate the ratio of second-order rate constants at any hydrogen-ion concentration is to determine the k_2 values for the two functional groups at one hydrogen-ion concentration, calculate the k_{A^-} values by means of equation 1 and then use equation 2 to calculate the rate ratio at any other hydrogen-ion concentration.

The results in Table XIV show that the rate ratio $R-S^-/R-NH_2$ is strongly pH-dependent ranging from 34 at pH 12.0 to 5000 at pH 7.0. These results suggest that it is possible to chose conditions to maximize or minimize relative reactivities of sulfhydryl and amino groups with dehydroalanine.

These results and those of a previous study (12) on reactions of amino and sulfhydryl groups in proteins with added dehydro-alanine show that lysinoalanine and lanthionine residues can be introduced into proteins under conditions that avoid strong alkaline treatment and that dehydroalanine is an intermediate in forming lysinoalanine and lanthionine crosslinks under the influence of alkali. The last conclusion is reinforced by a recent study (16) in which we demonstrated that alkali-treated casein and acetylated casein contain significant amounts of dehydroalanine residues.

These findings and the reported observations that acylation prevents or minimizes lysinoalanine formation but not racemization (15, 18) suggest that it should be possible to differentiate chemical, metabolic, nutritional, and toxicological effects of dehydroalanine, lysinoalanine, and racemization.

Inhibiting Lysinoalanine Formation

Added nucleophiles. Since sulfhydryl groups react more rapidly than ε-amino groups with dehydroalanine, addition of thiols such as cysteine or reduced glutathione should trap the residues of dehydroalanine as S-alkyl derivatives and/or eliminate potential precursors for dehydroalanine as described below. These competitive reactions should minimize lysinoalanine formation. This is indeed the case (Tables XV, XVI). Inhibition of lysino-alanine formation by added thiols or other nucleophiles such as sulfite ions can occur by at least three distinct mechanisms.

Table XV. Effect of thiols on lysinoalanine
content of alkali-treated soy protein.[1]

| Additive | Lysinoalanine | |
(1 mmole)	Weight %	Mole %
None	1.67; 1.58	1.08; 1.09
2-Aminoethanethiol	0.73	0.58
L-Cysteine	0.56	0.36
N-Acetyl-L-Cysteine	0.55	0.38
Reduced Glutathione	0.59	0.38
DL-Penicillamine	0.97	0.60

[1]Conditions: 1% protein in 0.1 N NaOH; 65°C; 3 hours.

First, by direct competition the added nucleophile can trap dehydroalanine residues derived from protein amino acid side chains as mentioned earlier (equations 3, 4). Second, the added nucleophile can cleave protein disulfide bonds (equations 5, 6) and thus generate free sulfhydryl groups, which may, in turn, combine with dehydroalanine residues as illustrated in equation 7 (secondary competition). Third, by cleaving disulfide bonds the added nucleophile can diminish a potential source for dehydro-alanine, inasmuch as cystine residues would be expected to undergo β-elimination reactions more readily than negatively charged cysteine (P-S$^-$) or sulfo-cysteine (P-S-SO$_3^-$) protein side chains (suppression of dehydroalanine formation mechanism).

It may be possible to distinguish among these mechanisms. With radioactive-labeled cysteine as the added nucleophile, for example, direct competition will give rise to labeled lanthionine, secondary competition to unlabeled lanthionine, while the suppression mechanism, to the extent it is successful, to no lanthionine. More than one of the cited pathways can, of course, operate simultaneously (58).

Table XVI. Lysine (LYS), lysinoalanine (LAL), and
lanthionine (LAN) content (μmoles/g)
of alkali-treated proteins[1,2] (8).

Protein & Treatment	LYS	LAL	LAN	$\frac{LAL}{LAL + LYS} \times 100$
Casein: untreated	482.7	0.0	0.0	
pH 11.6	294.7	145.2	0.0	33
pH 11.6 + Na$_2$SO$_3$	373.0	65.7	0.0	15
Lysozyme: untreated	397.3	0.0	0.0	
pH 11.6	222.2	135.2	0.0	38
pH 11.6 + Na$_2$SO$_3$	317.2	35.6	0.0	10
Soybean trypsin inhibitor: untreated	319.3	0.0	0.0	
pH 11.6	129.3	235.1	0.0	35
pH 11.6 + Na$_2$SO$_3$	239.0	36.8	0.0	13
Wool: untreated	222.5	0.0	0.0	
pH 11.6	139.0	58.7	119.5	30
pH 11.6 + Na$_2$SO$_3$	176.9	19.1	0.0	10

[1]Alkali treatments were carried out as follows: to 100 mg of
wool fibers was added 50 ml of pH 11.6 borate buffer and, where
appropriate, 1.26 g (0.01M) Na$_2$SO$_3$. The flasks were placed in a
60°C water bath for 3 hours. The soluble proteins were treated
similarly, except that they were also dialyzed and lyophilized.

[2]The data show that 1) the amount of LAL formed under similar
conditions varied greatly among the four proteins tested; and
2) the presence of Na$_2$SO$_3$ results in a decrease in LAL content.

Direct Competition:

$$R-S^- + CH_2=C(NHCOP)-P + H^+ \rightleftharpoons R-S-CH_2-CH(NHCOP)-P \qquad (3)$$

$$SO_3^- + CH_2=C(NHCOP)-P + H^+ \rightleftharpoons {}^-O_3S-CH_2-CH(NHCOP)-P \qquad (4)$$

Suppression and First Step of Indirect Competition:

$$2R-S^- + P-S-S-P \rightleftharpoons 2P-S^- + R-S-R-R \qquad (5)$$

R = thiol side chain

$$SO_3^- + P-S-S-P \rightleftharpoons P-S^- + P-S-SO_3^- \qquad (6)$$

P = protein side chain

Indirect Competition:

$$P-S^- + CH_2=C(NHCOP)-P + H^+ \rightleftharpoons P-S-CH_2-CH_2(NHCOP)-P \qquad (7)$$

Acylation of protein ε-NH$_2$ groups. Since lysinoalanine formation from lysine requires participation of an ε-amino group, protection of the amino groups by acylation (acetylation, succinylation, etc.) should minimize or prevent lysinoalanine formation under alkaline conditions if the protective group survives the treatment. This expection was, indeed, realized (11, 18). Results in Table XVII demonstrate that acylations of soy protein, casein, or wheat gluten prevented or minimized lysinoalanine formation.

Preliminary findings summarized in Table XVIII show that the protein efficiency ratios (PER's) in rats of acetylated and unmodified soybean proteins did not differ significantly. In contrast, succinylated soybean proteins did not support the growth of rats (30).

Labeled Lysinoalanine and Total Number of Crosslinks

Friedman and Boyd (31) reported on the synthesis of a deuterium-labeled phenylalanine derivative by adding its amino group to the double bond of a vinyl compound. This result is of mechanistic interest because the method should be applicable for synthesizing deuterium and tritium labeled lysinoalanine, ornithinoalanine, and lanthionine, as follows:

$$HOOCCH(NHCOCH_3)(CH_2)_3CH_2NH_2 + CH_2=C(NHCOCH_3)COOCH_3$$

α-N-acetyllysine   N-Acetyldehydroalanine methyl ester

1. X_2/Et$_3$N; 2. Hydrolysis

$$HOOCCH(NH_2)(CH_2)_3CH_2NH-CH_2CX(NH_2)COOH$$

X = D or T   Deuterium or tritium labeled lysinoalanine

TABLE XVII.   Effect of acetylation on lysinoalanine content of alkali-treated protein (10, 11, 15).

Protein	Lysinoalanine (mole %)
Soy protein + alkali[1]	1.46
Acetylated soy protein + alkali[1]	0.412
Casein + alkali[2]	2.35
Acetylated casein + alkali[2]	0.00
Wheat gluten + alkali[3]	0.654
Acetylated wheat gluten + alkali[3]	0.00

[1] 1% protein in 0.1 N NaOH; 75°C; 3 hours.
[2] 1% protein in 0.1 N NaOH; 65°C; 3 hours.
[3] 1% protein in 1 N NaOH; 65°C; 1 hour.

The deuterium and tritium labeling method results in introducing one label for each crosslink formed, and could, in principle, also be used to determine the total number of crosslinks formed during alkali-treatment of a protein and to characterize α-amino acid-crosslinked amino acids. The total number of crosslinks includes not only those derived from reaction of dehydroalanine side chains with free protein functional groups such as ε-amino groups of lysine, imidazole groups of histidine, and sulfhydryl groups of cysteine side chains as discussed in detail elsewhere (8), but also those formed by adding α-amino groups to dehydroalanine. These α-amino groups are present in N-terminal amino acids and can arise also from peptide bond cleavage that may occur during the alkaline treatment.

This approach should also make it possible to synthesize double-labeled lysinoalanine for biological studies in which the lysine part is labeled with C[14] and the alanine part with tritium or deuterium.

Table XVIII. Per and digestibility of soy protein and acylated soy proteins.

Dietary Source of Protein[1]	Final Body Weight[2,3]	Total Feed Consumption[2,3]	PER[4,5] Actual[3]	Adjusted	% Digestibility[6] Diet	Nitrogen
1. Casein	163 ± 5^A_a	309 ± 14^a	3.54 ± 0.04^A_a	2.50	96	95
2. Soy protein	94 ± 3^B_b	257 ± 11^{ab}	1.58 ± 0.06^B_b	1.12	94	88
3. Soy protein, acetylated[7]	90 ± 4^B_b	230 ± 10^b	1.60 ± 0.09^B_b	1.13	94	91
4. Soy protein, succinylated[7]	46 ± 2^B_b	129 ± 7^B_b	-0.82 ± 0.09^B_b	-0.66	94	9

[1]Diets contained 10% protein. Nitrogen factor: 6.25
[2]All weights in grams.
[3]Mean ± S.E. Duncan's Multiple Range Test: Means without a superscript letter in common are significantly different: lower case, P 0.05; upper case, P 0.01. Five rats per group. Means without a letter are not significantly different.
[4]Male, Sprague-Dawley rats, initial age = 21 days; initial weight = 53 grams.
[5]PER (protein efficiency ratio) = weight gain/protein intake.
[6]Digestibility: Diet = (feed intake - fecal weight)/ feed intake X 100. Nitrogen = (N intake - fecal N)/ N intake X 100. Pooled data from 6th through 13th test days.
[7]The extent of acylation of amino groups was about 90% as determined by the manual ninhydrin reaction described in 23.

Mathematical Analysis of Kinetics of Lysinoalanine Formation

As mentioned earlier, lysinoalanine is formed by a multi-step process involving initially the generation of dehydroalanine residues from serine, cystine, and cysteine side chains followed by addition of ε-amino groups of lysine to the reactive double bond of dehydroalanine. Since the ionization of the protonated amino ε-amino groups to the reactive ionized form is governed by their pK values and since the rate of dehydroalanine formation from cystine precursors is more rapid than from serine residues (11), kinetic studies are needed to define the rate-controlling step during lysinoalanine formation in structurally different proteins.

The following mathematical analysis should be useful for interpreting the reactivity of amino acid side chains in proteins to form lysinoalanine. The analysis will hopefully stimulate studies designed to assess the applicability of the derived equations to a variety of proteins and glycoproteins. (For related analyses of reaction kinetics of lysine amino groups with vinyl compounds which may also be applicable to some aspects of the present case, see 28, 29)

Definitions

T = $Pr-CH(NH-CO-Or)-CH_2-Y$

Pr = protein side chain

Y = OH; $S-S-Pr$; SH; etc.

T_0 = serine; cystine; cysteine at t = zero time

A_T = $R-NH_2$ (unprotonate lysine) + $R-NH_3^+$ (protonated lysine) + lysinoalanine

AH^+ = $R-NH_3^+$ (protonated lysine); A = $R-NH_2$ (unprotonated or ionized lysine)

V = $Pr-\underset{NH-CO-Pr}{C}=CH_2$ = dehydroalanine

AV = lysinoalanine

$$T + OH^- \underset{k_{-2}}{\overset{k_2}{\rightleftharpoons}} V + Y^- + H_2O$$

$$AH^+ \overset{K}{\rightleftharpoons} A + H^+$$

$$A + V \underset{k_{-1}}{\overset{k_1}{\rightleftharpoons}} AV$$

Ignore k_{-2}, then:

$$dV/dt = k_2(T)(OH^-) + k_{-1}(AV) - k_1(A)(V) \qquad (8)$$

$$(AV)/(A)(V) = k_1/k_{-1} = K_{AV} \text{ or}$$
$$k_1(A)(V) = k_{-1}(AV) \qquad (9)$$

$$V_{total} = \text{CONSTANT} = T_0 = (T) + (V) + (AV) \qquad (10)$$

$$A_T = (AH^+) + (A) + (AV) \qquad (11)$$

$$= \frac{(A)(H^+)}{K} + (A) + (AV)$$

$$AV = A_T - A(1 + H^+/K); \text{ Therefore, when} \qquad (12)$$

$$AV = 0, \text{ then:}$$

$$A = \frac{A_T}{(1 + H^+/K)} \quad ; \quad K = \frac{(A)(H^+)}{(AH)}$$

CASE I. Pseudo-first-order

Ignore (AV), i.e. (AV) not significant:

Equations (9) and (10) into (8) gives the following key differential equation:

$$dV/dt = k_2(T_0)(OH^-) - k_2(V)(OH^-) - \underbrace{k_2(AV)(OH^-) + k_{-1}(AV)}_{\text{ignore}}$$
$$- k_1(A)(V) \qquad (13)$$

Assume (OH^-) is constant and in excess (pseudo-first-order):

$$(OH^-) = K_W / H^+, \text{ where } K_W = (H^+)(OH^-) / H_2O$$

$$dV/dt = \underbrace{k_2 T_0 K_W/H^+ -}_{\text{a(constant)}} - k_2(OH^-) + \left[\underbrace{k_1 A_T / (1 + H^+/K)}_{\text{b(constant)}}\right](V)$$

$$dV/dt = a - bV = +abe^{-bt}; \quad \text{Integrate:}$$

$$V_{(t)} = a(-e^{-bt} + 1/b) = \text{DEHYDROALANINE concentration at time t}$$

$$dAV/dt = k_1(A)(V) = \left[k_1 A_T / (1 + H^+/K) \right] V_{(t)}$$

Integrate:

$$(AV)_t = k_1 A_T a/(1 + H^+/K) \quad (1/b)(t) + be^{-bt}$$

Since $(AV)_0 = 0$, therefore:

$$0 = c \left\{ 0 + b \right\} - x; \quad x = bc$$

$$(AV)_t = c \left\{ (1/b)(t) + be^{-bt} \right\} - bc$$

$$= \text{LYSINOALANINE concentration at time t}$$

$$(dAV/dt)_t = c \left\{ 1/b - b^2 e^{-bt} \right\}$$

$$(dAV/dt)_0 = \text{INITIAL RATE} = c \left\{ 1/b - b^2 \right\}$$

Make plots at different pH values and calculate k_1 (two equations with two unknowns) = rate of lysinoalanine formation. Then calculate k_2 = rate for dehydroalanine formation.

CASE II. Do not ignore (AV), i. e. (AV) is significant:

$(AV) = K_{AV}(A)(V)$ in equation (13) gives equation (14):

$$dV/dt = k_2 T_0(OH^-) - k_2 V(OH^-) - k_2 K_{AV}(A)(V)(OH^-)$$
$$+ k_{-1} K_{AV}(A)(V) - k_1(A)(V) \quad (14)$$

$$A = \frac{A_T}{(1 + H^+/K) + K_{AV}(V)}$$

$$dV/dt \ = \ \underbrace{k_2 T_0 (OH^-)}_{a} \ - \ \underbrace{k_2 (OH^-)}_{b} \ \underbrace{(V)}_{x}$$

$$+ \ \overbrace{\frac{A_T \ k_{-1} K_{AV} \ - \ k_2 K_{AV} (OH^-) \ - \ k_1 (A_T)}{\underbrace{1 \ + \ (H^+/K) \ + \ K_{AV}(V)}_{e \quad x}}}^{c} \ \underset{x}{(V)}$$

$$dV/dt \ = \ a \ + \ bx \ + \ \frac{cx}{d \ + \ ex} \tag{15}$$

Equation (15) is too complicated to integrate. Might be solved by numerical approximations.

CASE III. <u>Steady-state assumption for DEHYDROALANINE formation and disappearance</u>

A. <u>Ignore (AV), i.e. (AV) not significant:</u>

$$0 \ = \ dV/dt \ = \ k_2 T_0 (OH^-) \ - \ k_2 V_{ss}(OH^-) \ - \ k_1(A)V_{ss}, \text{ where}$$

V_{ss} = the steady-state concentration of dehydroalanine, which implies that it reacts as fast as it is formed.

Therefore,

$$V_{ss} \ = \ \frac{k_2 T_0 (K_W/H^+)}{k_1(A) \ + \ k_2(K_W/H^+)} \tag{16}; \qquad A \ = \ \frac{A_T}{(1 \ + \ H^+/K)} \tag{17}$$

$$V_{ss} \ = \ \frac{k_2 T_0 (K_W/H^+)}{\dfrac{k_1 A_T}{(1 + H^+/K)} \ + \ k_2(K_W/H^+)} \ = \ \text{CONSTANT} \tag{18}$$

$$dAV/dt \ = \ k_1(A)(V) \ = \ \left\{ k_1 A_T \ / \ (1 \ + \ H^+/K) \right\} \cdot (V_{ss}); \text{ or}$$

$$(AV) \ = \ Ct \text{ form}; \ C \text{ depends on } (H^+)$$

B. Do not ignore (AV), i. e. (AV) is significant:

$$0 = k_2 T_0 (K_W/H^+) - k_2 V_{ss} (K_W/H^+) - k_2 (AV)(K_W/H^+) \tag{19}$$
$$+ k_{-1}(AV) - k_1(A)V_{ss}$$

$$V_{ss} = \frac{k_2 T_0 (K_W/H^+) - k_2 (AV)(K_W/H^+) + k_{-1}(AV)}{k_1(A) + k_2(K_W/H^+)} \tag{20}$$

but,

$$K_{AV} = \frac{(AV)}{(A)\,(V)} \tag{21}$$

$$(AV) = A_T - A(1 + H^+/K) \tag{22}$$

With equations (20), (21), and (22) can eliminate (A), (AV) and obtain equation for V_{ss} in terms of constants.

Equation (21) into (20):

$$V_{ss} = \frac{k_2 T_0 (K_W/H^+) - k_2 K_{AV} V_{ss}(K_W/H^+) + k_{-1} K_{AV}(A)V_{ss}}{k_1(A) + k_2(K_W/H^+)} \tag{23}$$

Equation (21) into (22):

$$K_{AV}(A)V_{ss} = A_T - A(1 + H^+/K) \tag{24}$$

$$(A) = \frac{A_T}{K_{AV}V_{ss} + [(1 + H^+/K)]} \tag{25}$$

Equation (25) into (23):

$$V_{ss} = \frac{k_2 T_0 (K_W/H^+) - k_2 K_{AV} V_{ss}(K_W/H^+) + \dfrac{k_{-1} K_{AV} V_{ss} A_T}{K_{AV}V_{ss} + (1 + H^+/K)}}{k_2(K_W/H^+) + \dfrac{k_1 A_T}{K_{AV}V_{ss} + (1 + H^+/K)}}$$

It may be possible to find a solution for V_{ss} in the form of a quadratic equation.

Alternate approach

$$A = A_T / [(1 + H^+/K) + K_{(AV)}(V_{ss})]$$

$$dV/dt = k_2(T)(OH^-) - k_1(A)(V) + k_{-1}(AV)$$
$$\text{very small, ignore}$$

At steady state:

$$dV/dt = [k_2 K_W / H^+](T); \quad (T) = T_0 - (V_{ss}) - (AV)$$

$$(V_{ss}) = T_0 - (AV) = T_0 - K_{AV}(A)(AV)$$

$$dV/dt = 0 = [k_2 K_W / H^+](T_0 - V_{ss} - AV)$$

$$= k_2 T_0 K_W / H^+ - k_2 V_{ss} K_W / H^+ - k_2(AV)K_W / H^+$$

$$= k_2 T_0 K_W / H^+ - k_2 V_{ss} K_W / H^+ - k_2(A)(V)K_W / H^+$$

$$= k_2 T_0 K_W / H^+ - k_2 V_{ss} K_W / H^+$$

$$- k_2 AV[(A_T / 1 + H^+/K + K_{AV}V_{ss})] \cdot (V_{ss}K_W/H^+)$$

Multiplying by $H^+/k_2 K_W$ gives:

$$0 = T_0 - V_{ss} - K_{AV}A_T V_{ss} / \left[(1 + H^+/K) + K_{AV}V_{ss} \right]$$

Multiplying by $(1 + H^+/K + K_{AV}V_{ss})$ gives:

$$0 = T_0(1 + H^+/K + K_{AV}V_{ss}) - V_{ss}\left[(1 + H^+/K + K_{AV}V_{ss}) - K_{AV}A_T V_{ss} \right]$$

$$= T_0 + T_0 H^+/K + K_{AV}T_0 V_{ss} - V_{ss}H^+/K - K_{AV}V_{ss}^2 - K_{AV}A_T V_{ss}$$

$$= T_0 + T_0 H^+/K + \left[K_{AV}T_0 - K_{AV}A_T - 1 - H^+/K \right] V_{ss} - K_{AV}V_{ss}^2$$

$$= T_0(1 + H^+/K) + \left[K_{AV}(T_0 - A_T) - (1 + H^+/K) \right] V_{ss} - K_{AV}V_{ss}^2$$

The last quadratic equation can be written as:

$$\underset{r}{K_{AV}(V_{ss})^2} + \underset{s}{\left\{ (1 + H^+/K) + K_{AV}A_T - K_{AV}T_0 \right\}} V_{ss} - \underset{t}{(1 + H^+/K) V_T} = 0$$

$$r(V_{ss})^2 = s(V_{ss}) - t = 0; \quad \text{Solving the quadratic equation:}$$

$$(V_{ss}) = \frac{-s \pm \sqrt{s^2 - 4rt}}{2r} = \begin{array}{l}\text{steady-state}\\\text{DEHYDROALANINE}\\\text{concentration}\end{array}$$

LYSINOALANINE concentration (AV):

$$dAV/dt = k_1(A)(V_{ss}) - k_{-1}(AV)$$

Equation (21) into (22):

$$K_{AV}(A)V_{ss} = A_T - A(1 + H^+/K)$$

$$A = \frac{A_T}{\left[K_{AV}V_{ss} + (1 + H^+/K)\right]}$$

Define:

$$a' = k_1(A)(V_{ss}); \quad b' = k_{-1} \text{ gives the following form:}$$

$$dAV/dt = a' - b'(AV) \text{ with solution:}$$

$$(AV) = \frac{a'}{b'}\left[1 - e^{-b't}\right]$$

Substituting for a' and b':

$$(AV)_t = \frac{k_1(A)(V_{ss})}{k_{-1}}\left[1 - e^{-k_{-1} \cdot t}\right]$$

$$= K_{AV}(A)(V_{ss})\left[1 - e^{-k_{-1} \cdot t}\right] = \text{LYSINOALANINE concentration at time t.}$$

We are challenged to establish whether the derived mathematical relationships can be used to describe kinetic pathways for dehydroalanine and lysinoalanine formation in proteins and related model compounds.

Sample calculations (numbers are from Table 1):

$$A = A_T - AH^+ - AV; \text{ At pH 12.5, } AH^+ \text{ (protonated lysine) is small and can be neglected.}$$

$$K_{AV} = \frac{(AV)}{(A)(V)} = \frac{(AV)}{(A_T - AV)(T_o - T - AV)} = \frac{1.86}{(5.55-1.86)(6.88-4.11-1.86)}$$

$$= 0.554$$

$$V_{ss} = T_o - T - AV$$

total   actual   actual
CYS + SER   CYS + SER   lysinoalanine

$$= 6.88 - 4.11 - 1.86 = 0.91 \text{ mole \%}$$

These calculations do not depend on the above derivations.

Chemical Basis for Biological Action of Lysinoalanine

Although the mechanism of the observed cellular action of lysinoalanine is still not well understood (3, 4, 15), several largely speculative points, based on the chemical properties of dehydroalanine and lysinoalanine, could include one or more of the following events. First, dehydroalanine may react with essential sulfhydryl and amino groups of proteins and nucleic acids in vivo. Such reactions have been postulated for peptide antibiotics that contain dehydroalanine and methyldehydroalanine (32). Second, lysinoalanine may exert its pharmacological effect on kidneys by undergoing a reverse Michael reaction in situ to reform dehydroalanine as the reactive alkylating agent. Lysinoalanine in effect may act as a carrier of dehydroalanine or similar alkylating agent to kidney target sites. Third, transition metal ions also show strong affinity for sulfhydryl and amino groups (33). Therefore, they may compete with dehydroalanine for such biological nucleophiles, thus minimizing or preventing the formation of lanthionine and lysinoalanine. These possibilities and the possible role of lysinoalanine in metal binding in vivo deserve careful study. Fourth, lysinoalanine may compete with lysine or other amino acid(s) during protein biosynthesis, i.e., it may act as a competitive inhibitor of protein biosynthesis. Fifth, lysinoalanine may competitively inhibit lysine catabolism by reacting with α-ketoglutarate to prevent the analogous transformation of lysine to saccharopine, a key intermediate of lysine metabolism (34). Sixth, the secondary amino group of lysinoalanine may interact with nitrites both in vitro and in vivo to form a toxic N-nitroso derivative. Seventh, lysinoalanine may act as a local irritant or allergen of kidney tissue. Finally, lysinoalanine may exert its effect by inhibiting or affecting normal reabsorption of amino acids by kidney tubules.

The possible interaction of lysinoalanine in vivo with metal ions deserves special comment. Since lysinoalanine contains three amino and two carboxyl groups and structurally resembles ethylenediamine tetraacetic acid (EDTA), it could, in principle, act as a metal-chelating agent. In fact, the long residence time of lysinoalanine in the kidney (35), may be related to possible chelating action of lysinoalanine to metal ions of kidney metallothioneins. This hypothesis raises several important questions which include: 1) Can lysinoalanine toxicity be prevented by giving equivalent (stoichiometric) amounts of lysine, histidine, arginine, or cysteine? 2) Does lysinoalanine in fact bind particular metal ions in vitro and in vivo? 3) Does co-administration of lysinoalanine with metals affect toxicity? 4) Does lysinoalanine compete with physiologically active proteins for essential metal ions? 5) Does lysinoalanine bind to hemoglobin, metallothioneines or other specific metalloproteins in vitro and in vivo? 6) Does administering lysinoalanine have any observable (transient) effect on excretion of either metal ions, amino acids, or proteins in urine?

Finally, alkali-treatment of proteins also catalyzes concurrent racemization of optically active amino acids. Factors that favor racemization have been examined in detail elsewhere (6, 14, 15, 18). These include high pH-temperature, time of exposure to alkali, and the inductive or steric properties of the various amino acid side chains. The presence of D-amino acid residues along a partially-racemized protein chain decreases its digestibility and nutritional quality and introduces specific toxicity of D-serine and possibly other D-isomers. The evidence suggests that D-amino acid residues may reinforce the toxic action of lysinoalanine at the cellular level. Additional studies are needed to define the ability of various D-amino acids to potentiate the biological action of lysinoalanine.

In summary, understanding chemical events governing protein-alkali reactions provides insight into events occurring during food processing and can be decisive for producing better and safer foods. This area of food science, with many unsolved problems, affords a fascinating interplay among chemistry, nutrition, and toxicology. The results discussed in this essay and in the evergrowing worldwide literature (36-137) attest to the importance of these studies to human welfare.

Conclusions

Factors favoring lysinoalanine formation in soybean proteins include high pH, time and temperature of exposure; protein concentration had no effect. Those factors minimizing lysinoalanine formation include additives such as N-acetyl-L-cysteine, sodium sulfite and acylation of basic ε-NH$_2$ groups to neutral amides. Alkali treatment of soy protein in the presence of carbohydrates such as glucose and xylose did not influence the extent of lysinoalanine production but induced a drastic decrease in arginine content. These and related observations of effects of diemthyl sulfoxide (DMSO) and metal ions on lysinoalanine formation can be rationalized by a two-step mechanism in which generation of dehydroalanine residues from serine and cysteine is followed by nucleophilic addition of ε-NH$_2$ groups to the electrophilic double bond of dehydroalanine residues. Based on steady-state assumptions about the rate of dehydroalanine formation, mathematical relationships were derived that relate pH of the medium and pK of ε-NH$_3^+$ groups to kinetics of lysinoalanine production. Understanding conditions and mechanisms that govern lysinoalanine formation should help in designing food processes that minimize both loss of essential amino acids and formation of unnatural derivatives such as dehydroalanine and lysinoalanine.

Acknowledgements

Special appreciations are extended to: Carol E. Jones for excellent technical assistance; Amy T. Noma for her help with the amino acid analyses; Alan D. Friedman for his help with the mathematical derivations for lysinoalanine formation; Michael R. Gumbmann for the PER data; James C. Zahnley and my colleagues, including those listed in the cited references, for helpful comments and fruitful scientific collaboration.

Presented at the symposium on "Mechanisms of Food Protein Deterioration," Food Biochemistry Subdivision, Division of Agricultural and Food Chemistry, American Chemical Society Meeting, New York, August 23-28, 1981.

Reference to a brand or firm name does not constitute endorsement by the U. S. Department of Agriculture over others not mentioned.

Literature Cited

1. Friedman, M., Ed. "Protein Crosslinking: Nutritional and Medical Consequences," Plenum Press: New York, 1977; 704 pp.
2. Friedman, M., Ed. "Protein Crosslinking: Biochemical and Molecular Aspects," Plenum Press: New York, 1977; 760 pp.
3. Woodard, C. J.; Short, D. D.; Alvarez, M. R.; Reyniers, J. "Protein Nutritional Quality of Foods and Feeds," Part 1, Friedman, M., Ed.; Marcel Dekker: New York, 1975; p. 595.
4. Gould, D. H.; MacGregor, J. T. Reference 1; p. 29.
5. Sternberg, M.; Kim, C. Y. Reference 1; p. 73.
6. Friedman, M.; Zahnley, J. C.; Masters, P. M. J. Food Sci., 1981, 46, 127.
7. Friedman M. "Chemistry and Biochemistry of the Sulfhydryl Group in Amino Acids, Peptides, and Proteins," Pergamon Press: Oxford, England, and Elmsford, New York; p. 135.
8. Friedman, M. Reference 1; p. 1.
9. Finley, J. W.; Snow, J. T.; Johnston, P. H.; Friedman, M. Reference 1; p. 85; J. Food Sci., 1978, 43, 619.
10. Friedman, M. "Proceedings of the 10th National Conference on Wheat Utilization Research," U. S. Department of Agriculture, Science and Education Administration, Western Regional Research Center, Berkeley, CA 94710, 1978, ARM-W-4; p. 81.
11. Friedman, M., Ed. "Nutritional Improvement of Food and Feed Proteins," Plenum Press: New York, 1978; p. 613.
12. Friedman, M.; Finley, J. W.; Yeh, Lai-Sue. Reference 1; p. 213.
13. Friedman, M. "Food Proteins," Whitaker, J. R. and Tannenbaum, S. R., Eds.; Avi Press: Westport, CT, 1977; p. 446.

14. Masters, P. M.; Friedman, M. J. Agric. Food Chem., 1979, 27, 165.
15. Masters, P. M.; Friedman, M. "Chemical Deterioration of Proteins," Whitaker, J. R. and Fujimaki, M., Eds.; ACS Symp. Series 123; Washington, D. C., 1980; p. 165.
16. Masri, M. S.; Friedman, M. Biochem. Biophys. Res. Commun., 1982, 104, 321.
17. Friedman, M.; Gumbmann, M. R. Fed. Proc., 1982, 41, 392.
18. Friedman, M.; Masters, P. M. J. Food Sci., 1982, 47, 760.
19. Whitaker, J. R.; Feeney, R. E. Reference 1; p. 155.
20. Friedman, M.; Wall, J. S. J. Amer. Chem. Soc., 1964, 86, 3735.
21. Snow, J. T.; Finley, J. W.; Friedman, M. Int. J. Peptide Protein Res., 1976, 5, 177.
22. Snow, J. T.; Finley, J. W.; Friedman, M. Int. J. Peptide Protein Res., 1975, 7, 461.
23. Friedman, M. Diabetes, 1982, 31, Supplement 3, 5.
24. Henkin, R. I. "Protein-Metal Interactions," Friedman, M., Ed.; Plenum Press: New York, 1974; p. 299.
25. Friedman, M. J. Amer. Chem. Soc., 1967, 89, 4709.
26. Friedman, M. Quarterly Repts. Sulfur Chem., 1968, 3, 125.
27. Snow, J. T.; Finley, J. W.; Friedman, M. Biochem. Biophys. Res. Commun., 1975, 64, 441.
28. Friedman, M.; Cavins, J. F.; Wall, J. S. J. Amer. Chem. Soc. 1965, 87, 3572.
29. Friedman, M.; Williams, L. D. Reference 1; p. 299.
30. Friedman, M.; Gumbmann, M. R. J. Nutrition, 1981, 111, 1362.
31. Friedman, M.; Boyd, W. A. Reference 2; p. 727.
32. Gross, E. Reference 1; p. 131.
33. Friedman, M. Reference 7; p. 25.
34. Hutzler, J.; Dancis, J. Biochem. Biophys. Acta., 1975, 377, 42.
35. Finot, P. Reference 1; p. 51.
36. Abe, K.; Ozawa, M.; Homma, S.; Fujimaki, M. Kaseigaku Zasshi, 1981, 32, 367; Chem. Abst. 95, No. 167335.
37. Almquist, H. J.; Chistensen, H. L.; Maurer, S. Proc. Soc. Expt. Biol. Med., 1961, 122, 913.
38. Achon, I. M.; Richardson, T.; Draper, N. R. J. Agric. Food Chem., 1981, 29, 27.
39. Asquith, R. S.; Booth, A. K.; Skinner, J. D. Biochim. Biophys. Acta, 1969, 181, 164.
40. Asquith, R. S.; Otterburn, M. S. Reference 1; p. 93.
41. Asquith, R. S.; Carthew, P. Biochim. Biophys. Acta, 1972, 278, 8.
42. Aymard, C.; Cuq, J. L.; Cheftel, J. C. Food Chem. 1978, 3, 1.
43. Bohak, Z. J. Biol. Chem., 1964, 239, 2878.
44. Cheng, C. C.; Huang, P. Chung-Kuo Nung Yeh Hua Hsueh Hui Chih, 1977, 15, 12.

45. Chu, N. T.; Pellet, P. L.; Nawar, W. W. J. Agric. Food Chem., 1976, 24, 1084.
46. Colvin, B. M.; Ramsery, H. A. J. Dairy Sci., 1969, 52, 270.
47. Creamer, L. K.; Matheson, A. R. New Zealand J. Dairy Sci. Technol., 1977, 12, 253.
48. Dastidar, P. G.; Nickerson, J. W. FEMS Microbiol. Lett., 1978, 4, 331.
49. De Groot, A. P.; Slump, P. J. Nutr., 1969, 98, 45.
50. De Groot, A. P.; Slump, P.; Feron, V. J.; Van Beek, L. J. Nutr., 1967, 106, 1527.
51. DeRham, O.; Van de Revaart, P.; Bujard, E.; Mottu, F.; Hidago, J. Cereal Chem., 1977, 54, 238.
52. Dworschak, E.; Orsi, F.; Zsigmond, A.; Trezl, L.; Rusznak, I. Nahrung, 1981, 25, 441.
53. Ebert, C.; Ebert, G.; Rossmeissl, G. Reference 1; p. 205.
54. Ellison, M. S.; Lundgren, H. P. Text. Res. J., 1978, 48, 269.
55. Engelsma, J. W.; Van der Meulen, J. D.; Slump, P.; Haagsma, N. Lebensm.-Wiss. Technol., 1979, 203.
56. Ebersdobler, H. F.; Holstein, B.; Lainer, E. Lebensm. Unters. Forsch., 1979, 168, 8.
57. Ebersdobler, H. F.; Holstein, A. B. Milchwiss., 1980, 35, 374.
58. Feairheller, S. H.; Taylor, M. M.; Bailey, D. G. Reference 1; p. 177.
59. Feeney, R. E. "Chemical Deterioration of Proteins," Whitaker, J. R. and Fujimaki, M., Eds.; ACS Symp. Series 123; Washington, D. C., 1980; p. 1.
60. Feron, V. J.; Van Beek, L.; Slump, P.; Burns, R. B. "Biochemical Aspects of New Protein Foods," Adler-Nissen, J., Ed.; Pergamon Press: Oxford, 1978.
61. Finley, J. W.; Friedman, M. Reference 1; p. 123.
62. Finely, J. W.; Kohler, G. O. Cereal Chem. 1979, 56, 130.
63. Florence, T. M. Biochem. J., 1980, 189, 507.
64. Freimuth, U.; Krause, W.; Doss, A. Nahrung, 1978, 22, 557.
65. Freimuth, U.; Krause, W. Nahrung, 1980, 24, 317.
66. Freimuth, U.; Notzold, H.; Krause, W. Nahrung, 1980, 24, 351.
67. Freimuth, U.; Notzold, H. Nahrung, 1980, 24, 627.
68. Friedman, M. Proc. Royal Soc. Med., 1977, 70(3), 50.
69. Friedman, M. "Functionality and Protein Structure," Pour-El., A., Ed.; ACS Symp. Series 92, Washington, D. C., 1979; p. 225.
70. Friedman, M. U. S. Patent 4,212,800, July 15, 1980.
71. Friedman, M.; Krull, L. H. Biochim. Biophys. Acta, 1970, 207, 301.
72. Fritsch, R. J.; Klostermeyer, H. Z. Lebensm.-Unters. Forsch., 1981, 172, 435.
73. Fritsch, R. J.; Klostermeyer, H. Z. Lebensm.-Unters. Forsch., 1981, 173, 101.

74. Fujimaki, M.; Haraguchi, T.; Abe, K.; Homma, S.; Arai, S. Agric. Biol. Chem., 1980, 44, 1911.
75. Gowri, C.; Joseph, K. Leather Sci. (Madras), 1970, 17, 177.
76. Gross, E. Adv. Chem. Ser., 1977, 160, 37.
77. Haagsma, N.; Gortemaker, B. G. M. Z. Lebensm.-Unters. Forsch., 1979, 168, 550.
78. Haagsma, N.; Slump, P. Z. Lebensm.-Unters. Forsch., 1978, 167, 238.
79. Haraguchi, T.; Abe, K.; Arai, S.; Homma, S.; Fujimaki, M. Agric. Biol. Chem., 1980, 44, 1951.
80. Hasegawa, K.; Okamoto, N. Agric. Biol. Chem., 1980, 44, 649.
81. Hasegawa, K,; Okamoto, N.; Ozawa, H.; Kitajima, S.; Takado, Y. Agric. Biol. Chem., 1981, 45, 1645.
82. Hayakawa, S.; Katsuta, K. Nippon Shokuhin Kogyo Gakkaishi, 1981, 28, 347; Chem. Abst. 95, No. 167336n.
83. Hayashi, R.; Kameda, I. Agric. Biol. Chem., 1980, 44, 175; 1980, 44, 891.
84. Jones, G. P.; Rivett, D. E.; Tucker, D. J. J. Sci. Food Agric., 1981, 32, 805.
85. Karayiannis, N. I.; MacGregor, J. T.; Bjeldanes, L. F. Food Cosmet. Toxicol., 1979, 17, 585; 1979, 17, 591.
86. Kearns, J. E.; MacLaren, J. A. J. Text. Inst., 1979, 70, 534.
87. Kemp, B. E. FEBS Lett., 1980, 110, 308.
88. Kleyn, D. H.; Klostermeyer, H. Z. Lebensm.-Unters. Forsch., 1980, 170, 11.
89. Kondo, K. Memoirs Faculty Agric. Hokkaido Univ., 1979, 11, 265.
90. Krull, L. H.; Friedman, M. J. Polym. Sci. A-1, 1967, 5, 2535.
91. Krull, L. H.; Friedman, M. Biochem. Biophys. Res. Commun., 1967, 29, 273.
92. Leegwater, D. C. Food Cosmet. Toxicol., 1978, 16, 405.
93. Leegwater, D. C.; Tas, A. C. Lebensm.-Unters. Forsch., 1980, 13, 87.
94. Manson, W.; Carolan, T. J. Dairy Res., 1980, 47, 193.
95. Meyer, M.; Klostermeyer, H.; Kleyn, D. H. Z. Lebensm.-Unters. Forsch., 1981, 172, 446.
96. Muindi, P.; Thomke, S. J. Sci. Food Agric., 1981, 32, 139.
97. Murase, M. J. Agric. Chem. Soc. Japan, 1980, 54, 13.
98. Nashef, A. S.; Osuga, D. T.; Lee, H. S.; Ahmed, A. I.; Whitaker, J. R.; Feeney, R. E. J. Agric. Food Chem., 1977, 25, 245.
99. O'Donovan, C. J. Food Cosmet. Toxicol., 1976, 14, 483.
100. Okuda, T.; Zahn, H. Chem. Ber., 1964, 98, 1164.
101. Otterburn, M. S. Text. Res. J., 1975, 95, 88.
102. Provansal, M. M. P.; Cuq, J. L. A.; Cheftel, J. C. J. Agric. Food Chem., 1975, 23, 938.

103. Patchornik, A.; Sokolowski, M. J. Amer. Chem. Soc., 1964, 86, 1206; 1964, 86, 1860.
104. Pickering, B. T.; Li, C. H. Archiv. Biochem. Biophys., 1964, 104, 119.
105. Raymond, M. L. J. Food Sci., 1980, 45, 56.
106. Robbins, K. R.; Baker, D. H.; Finley, J. W. J. Nutr., 1980, 110, 907.
107. Robson, A.; Williams, M. J.; Woodhouse, J. M. J. Chromatogr., 1967, 31, 284.
108. Robson, A.; Zaidi, A. H. J. Text. Inst. Trans., 1967, 58, 267.
109. Sakamoto, M.; Nakayama, F.; Kajiyama, K. Reference 2; p. 687.
110. Sanderson, J.; Wall, J. S.; Donaldson, G. L.; Cavins, J. F. Cereal Chem., 1978, 55, 204.
111. Schade, J. E.; Schultz, W. G.; Neumann, H. H.; Morgan, J. P.; Finley, J. W. J. Agric. Food Chem., 1980, 28, 512.
112. Serassi-Kyriakou, K.; Hadjichristidis, N.; Touloupis, C. Chem. Chron., 1978, 7, 161.
113. Shetty, J. K.; Kinsella, J. E. J. Agric. Food Chem., 1980, 28, 798.
114. Slump, P.; Jongerius, G C.; Kraaikamp, W. C.; Schreuder, H. A. W. Ann. Nutr. Alim., 1978, 32, 271.
115. Sternberg, M.; Kim, C. Y.; Plunkett, R. A. J. Food Sci., 1975, 40, 1168.
116. Sternberg, M.; Kim, C. Y.; Schwende, F. J. Science, 1975, 190, 992.
117. Sternberg, M.; Kim, C. Y., J. Agric. Food Chem. 1979, 27, 1130.
118. Struthers, B. J. J. Amer. Oil Chem. Soc., 1981, 58, 501.
119. Struthers, B. J.; Brielmaier, J. R.; Raymond, M. L.; Dahlgren, R. R.; Hopkins, D. T. J. Nutr., 1980, 110, 2065.
120. Struthers, B. J.; Hopkins, D. T.; Dahlgren, R. R. J. Food Sci., 1978, 43, 616.
121. Tas, A. C.; Kleipool, R. J. C. Z. Lebensm. Wiss. Technol., 1976, 9, 360.
122. Touloupis, C.; Vassiliadis, A. Reference 1; p. 187.
123. Van Beek, L.; Feron, V. J.; DeGroot, A. P. J. Nutr., 1974, 104, 1630.
124. Venkatasubramanian, K.; Gowri, C.; Joseph, K. T. Leather Sci. (Madras), 1972, 19, 248.
125. Walsh, R. G.; Nashef, A. S.; Feeney, R. E. Int. J. Pept. Prot. Res., 1979, 14, 290.
126. Whitaker, J. R. Reference 59; p. 145.
127. Whitaker, J. R.; Feeney, R. E. Reference 1; p. 155.
128. Whiting, A. H. Biochim. Biophys. Acta, 1971, 243, 332.
129. Wood-Rethewill, J. C.; Warthesen, J. J. J. Food Sci., 1980, 45, 1637.
130. Woodard, J. C.; Alvarez, M. R. Archiv. Pathol., 1969, 84, 153.

131.   Woodard, J. C.; Short, D. D.   Food Cosmet. Toxicol., 1977, 15, 117.
132.   Woodard, J. C.; Short, D. D.; Strattan, C. E.; Duncan, J. H. Food Cosmet. Toxicol., 1977, 15, 109.
133.   Zahn, H.; Lumper, L.   Z. Physiol. Chem., 1968, 349, 77.
134.   Ziegler, K.   J. Biol. Chem. 1964, 239, 2713.
135.   Ziegler, K.   Appl. Polym. Symp., 1971, 18, 257.
136.   Ziegler, K.; Melchert, I.   Nature, 1967, 214, 404.
137.   Zsigmond, A.; Orsi, F.; Kalman, J.; Dworschak, E.   Proc. Hungarian Annual Meeting Biochem., 1979, 294.

RECEIVED July 14, 1982.

Chemical Modification of Food Proteins

TRIVENI P. SHUKLA

Krause Milling Company, Milwaukee, WI 53215

During the past decade, a number of non-conventional proteins have been identified as human food ingredients, e.g., single-cell proteins, fish, leaf, and cereal concentrates, and soybeans, cottonseeds, and rapeseeds. Successful utilization of these new proteinaceous materials depends on their digestibility, nutritive value, and overall functional and organoleptic properties as related to processed food formulations. Many of them, although to varying degrees, fail to meet one or more such utilization criteria.

Although the science, much less the art and technology, of chemical modification of food proteins is in its infancy, there exists a potential for the use of existing fundamental knowledge on protein derivatization - modification, replacement, and removal of amino acid residues, and protein-lipid, protein-carbohydrate, and protein-ligand interactions for commercially viable food products and process research. Such research programs may lead to the development of modified food-grade proteins of improved functionality, nutritive value, and organoleptic quality and of reduced propensity to deterioration by physicochemical agents. This chapter is a review of 1) theory and known mechanisms of modification, 2) applied and basic research on chemical modification of food proteins, 3) interaction of modified proteins with other food ingredients and 4) legal, regulatory, public health, and consumer acceptance implications of modified food protein products. In general, future research must address three major objectives: 1) selecting a structure-function specific modification scheme; 2) developing a process and product control strategy from commercial standpoints; and 3) testing the efficacy and safety of the product for human consumption.

Chemistry of Food Protein Modification

Food protein structure. The knowledge of food protein structure, its capacity for covalent and non-covalent interactions, and structure-function relationships in the native state must be

known before the selection of a modifying reagent and the prediction of the properties of the modified protein can be made. Table
I contains a list of details a food protein should be investigated
for before any attempt of chemical modification is made.

Table I. Proteins: Structure and Reactivity.
                 PRIMARY STRUCTURE
                     Amino Acid Composition
                     Amino Acid Sequence
                     Overall Hydrophobicity

               MICROENVIRONMENT
                     Isoelectric Point
                     Charge Distribution
                     Solvation
                     Hydrogen Bonding
                     Steric Factors

               MOTILITY OF PROTEIN STRUCTURE
                     pH Dependence
                     Ionic Strength Dependence
                     Dielectric Constant

               PROTEIN CONFORMATION
                     Alpha-Helix
                     Pleated Beta-Sheets
                     Random Coils

    Proteins are composed of one or more linear copolymers of
alpha-L-amino acids. Most food proteins are often heterogeneous
and consist of two or more polypeptide chains crosslinked by the
disulfide (-S-S-) bridge of cystine. The polypeptide chains are
folded or coiled into a highly specific conformation. To describe a food protein for the purpose of chemical reactivity, the
following must be known: 1) the number, length, and amino acid
composition of polypeptide chains; 2) the linear arrangement (or
sequence) of their amino acids; and 3) the number and the
position of crosslinkages between chains. In other words, the
primary structure of a target protein for chemical derivatization
should be determined, a priori. Furthermore, the protein in
question should be examined for phosphate crosslinkages and other
unusual bonds; e.g., ester bonds of collagen. To complicate the
issue further, food proteins may exist in conjunction with nucleic
acids, carbohydrates, and lipids. These details must be worked
out fully in order to understand proteins as reagents and modified proteins as functional food ingredients.
    Once the primary structure is known, the chemical reactivity
of a protein can be more accurately defined in terms of molecular
weight, end groups, residues per unit, sulfhydryls per unit, disulfides per unit, average methyls per segment, large non-polar

side chains per unit, average methyls per non-polar side chain,
maximum positive side chain charge, maximum negative side chain
charge, and number and distribution of various electron-rich
(nucleophilic) amino acid residues per unit. The preceding details more or less fix the configuration of a protein in a given
environment (solvents plus other solutes) which in turn determines
its ionization behavior, the number of "exposed" and "buried"
groups in it, its overall topology, and the motility of its structure (21, 36). None of the food proteins subjected to chemical
modification heretofore are well defined. Some that have been
defined to some degree (47), unfortunately, have not been
included in recent studies.

The literature in molecular biology and biochemistry contains
in-depth accounts of chemical modifications of well-defined
proteins (10, 32, 36, 44, 48), and this information has been useful to the fundamental protein chemist. On the contrary, the
food protein chemist has yet to define a target protein before
successfully subjecting it to covalent chemical modifications to
effect a predictable change in its primary structure. The resultant modified protein should exhibit equally predictable changes
in its structure and its behavior in solution, etc. Even if a
process of chemical modification was perfected, the product there-
from should be tested from the viewpoint of in vivo proteolysis,
biological value, nutritive value, and toxicity. Furthermore,
the economics of production and utilization of modified food
proteins is probably the most important consideration for judging
the state of the art of the chemical modification of food
proteins.

Tables II and III present data on side chain hydrophobicities (2), non-polar side chains (NPS), polarity ratios
(P), and net charge for a number of proteins (47). The physical
properties of proteins correlate with the type and density of nonpolar side chain amino acids. In other words, hydrophobic bonds
by and large determine the globular structure of proteins (3).
The polarity ratio, $P = V_e/V_i$, which is an indicator of hydrophobicity, therefore, relates well with the molecular structures
of globular proteins. V_e and V_i stand for the external (polar)
and internal (apolar) shell volumes of a protein molecule in water
and are calculated as the sum of the volume of polar (arginine,
histidine, lysine, aspartic acid, glutamic acid, tyrosine, serine,
and threonine) and apolar (tryptophan, isoleucine, tyrosine,
phenylalanine, proline, leucine, and valine) amino acids. Table
IV illustrates the amino acid composition of economical and
readily available proteins that have been subjected to chemical
modification. Research programs directed to developing data-
bases on parameters NPS, P, and net charge of Table III and
hydrophobicity (transfer free energy) of Table II for various
unconventional food proteins should be defined and undertaken.
The data so obtained will be very useful for predicting the
functional properties of proteins. Given the understanding of

Table II. Amino acid side chain hydrophobicities (transfer free energies, H_ϕ [1]/, and residue volumes). [2]

Amino Acid	H_ϕ (Kcal/Residue)	Volume (A^{o3})
Tryptophan	3.00	135.4
Isoleucine	2.95	102.0
Tyrosine	2.85	116.2
Phenylalanine	2.65	113.9
Proline	2.60	73.6
Leucine	2.40	102.0
Valine	1.70	85.1
Lysine	1.50	105.1
Methionine	1.30	97.7
Half Cystine	1.00	68.3
Alanine	0.75	52.6
Arginine	0.75	109.1
Threonine	0.45	72.2
Glycine	-	36.3
Serine	-	54.9
Histidine	-	91.9
Aspartate	-	68.4
Glutamate	-	84.7
Amide	-	4.0

[1]/$H_\phi = F_t = -RT \ln \dfrac{S_{EtOH}}{S_{H_2O}}$

Table III. Hydrophobicity and polarity parameters for various food proteins. ([2], [47])

Protein	NPS	$P = V_e/V_i$	Charges
Edestin (hemp)	0.30	1.41	0.23
Avidin (hen egg)	0.30	1.41	0.25
Alpha-Lipovitelin (hen egg)	0.34	1.09	0.23
Ovalbumin (hen egg)	0.34	0.92	0.24
Alpha-Lactalbumin	0.34	1.11	0.38
Alpha-Casein	0.38	1.27	0.24
Beta-Lactoglobulius	0.37	0.96	0.28
Zein (Maize)	-	-	0.03

the chemical basis for the expression of a physical property, it would be easier to define the type and degree of chemical modification.

The introduction of new side chain residues in a native protein by way of chemical modification may interfere with other interactions, e.g., formation of disulfide, hydrophobic and hydrogen bonds, and thus affect its non-covalent interactions.

Table IV. Summary of the amino acid composition of food protein targeted for chemical modification.

	Hen's Egg (49)	Baker's Yeast (52)	Yeast Protein Isolate (50)	Soybean Protein 7S Globulin (51)	Fish Protein (26)
Alanine	5.6	6.1	7.2	1.8	5.0
Arginine	6.4	5.9	4.7	7.1	6.3
Aspartic Acid	9.1	11.4	12.7	10.3	10.1
Half Cystine	2.4	1.1	1.3	-	0.8
Glutamic Acid	12.4	11.5	12.4	15.3	16.9
Glycine	3.2	4.5	5.0	1.4	3.0
Histidine	2.6	2.8	0.9	1.7	2.0
Isoleucine	6.0	5.6	4.0	3.7	4.8
Leucine	8.7	8.9	9.4	5.9	4.2
Lysine	6.7	9.3	9.8	4.8	10.6
Methionine	3.1	1.9	1.5	0.2	2.6
Phenylalanine	5.6	5.6	3.9	5.1	3.6
Proline	4.0	4.3	3.4	3.3	3.4
Serine	7.2	5.0	6.3	3.2	3.9
Threonine	4.8	5.0	5.0	1.4	4.7
Tryptophan	1.7	1.7	0.5	-	1.4
Tyrosine	3.8	4.5	3.4	3.2	3.8
Valine	7.0	5.8	5.6	3.1	4.8

This necessitates quantitative evaluation of the non-covalent interaction of a given protein (Table V). The experimental work on the enzyme-susceptibility, toxicology, and nutritive value of chemically modified protein would be better interpreted with a fuller knowledge of non-covalent interactions and consequent motility and topology of the protein molecule. At the present time, neither the native proteins nor their chemically modified counterparts are well understood.

Residue modification. The four types of electron-rich centers in a protein molecule are 1) the nitrogen nucleophiles or amino group of lysine, the imidazole group of histidine and the guanidine group of arginine, 2) the oxygen nucleophiles or hydroxyl groups of serine and threonine, the carboxyl groups of glutamate and aspartic acid, and the phenolic hydroxyl of tyrosine, 3) the sulfur nucleophiles or thiol group of cystine and the thioether group of methionine, and 4) the carbon nucleophiles or phenolic ring of tyrosine and the indole ring of tryptophan. Approximately 12 out of 20 amino acids that make up the backbone of proteins contain electron-rich, potentially nucleophilic centers. Since their reactivities are different and subject to variation depending on the intra- and extra-molecular

Table V. Role of covalent and non-covalent interactions in
protein functionality.

Interaction	Origin	Role	Strength (Kcal/Mole)	Attraction Distance (Å)
Non-Covalent:				
van der Waals	Induced Dipole	Packing Density Stabilization, Protein Interior	1-3	5
H-Bond	Attraction (Partial Charges)	Protein-Protein Complex, Protein - Lipid Complex, Protein Structure	3-5	3
Electro-static	Attraction or Repulsion	Structure Stability pH Sensitivity - Sensitivity to Ions and Ionic Strength	1-10	20
Hydrophobic Bonds	Non-polar Side Chains	Solubility Surfac-tancy, Protein-Lipid Complex	1-3	5
Covalent:				
Disulfide Bonds	(-S-S-)	Structure, Mechani-cal Property, Solu-tion Property	Weak	3
Others	Nucleo-philes	Change in Net Charge and Physical Properties	100	3

environment, their reaction with electrophilic reagents is non-specific (15, 32, 42). The selective modification of proteins, therefore, depends more on luck than sound judgment (32). However, the number of reactive nucleophiles can be narrowed some-what by proper selection of pH. Also, since the order of nucleophilicity is: sulfur > nitrogen > oxygen > carbon, the knowledge of amino acid composition can better define the reac-tion conditions. For discussions of nucleophilicity, reference should be made to Jencks (60). Table VI is a summary of possible reactions with various functional groups of proteins.

The conjugate acid of the basic groups such as ammonia, guanidine, and imidazole that bind metal ions or the positive end of dipolar groups (carbonyl) constitute the electron-deficient amino acid residues of proteins. Although not very reactive, the amino acid side chains which are protonated near or above their

Table VI. Summary of the chemical reactivity of proteins with
           various reagents. (15, 32)

Electrophilic Reagents:
  Acid Anhydrides -
    Acetic, Succinic, Citraconic

Miscellaneous Aryl Transfer Reagents:
    Carbon Disulfide, Cyanate,
    O-Methyl Isourea, Imidic Ester,
    Sulfonyl Halides

Arylating Reagents:
    Fluorodinitrobenzene (FDNB)
    Trinitrobenzene Sulfonic Acid (TNBS)
    Cyanuric Fluoride

Alkylating Reagents:
    Alpha-Halo Acids
    Activated Benzene Halide
    Sulfamryl Halide

Addition to Double Bond:
    N-Alkylanaleimide
    Acrylonitrile

Electrophilic Substitution:
    Tetranitromethane
    Diazonium-H-Tetrazole (DHT)
    Iodine

Condensation with Carbonyls

Oxidizing Agents:
    Hydrogen Peroxide
    N-Bromosuccinamide
    Photochemical

Reducing Agents:
    Disulfide Reduction
    Carboxyl Reduction

---

pK_a behave as electrophiles. Such electron-deficient centers in
proteins may be susceptible to reduction. The reduction of
disulfide bonds in proteins can be easily effected by either
electrolytic or chemical methods. A good reagent used to reduce
carboxyl groups is lithium aluminum hydride. The carboxyl groups
can also be reduced by diborane.
    Amino acids with bulky hydrocarbon side chains, although
neutral with respect to electrons, play crucial roles in protein
folding, protein stabilization, and ligand-binding. Those of

high hydrophobicity and a polarity ratio greater than unity are
largely responsible for dictating physical boundaries (loci) for
ligand binding. The reactions of neutral side chains are limited
to those with free-radicals, carbenes, and nitrenes.

With only minor exceptions, the published work on chemical
modification of food proteins deals mainly with acetylation and
succinylation reactions (Table VII). Although the general
chemistry of these reactions has been previously described in the
context of food protein modification (15, 31), the general
findings and their implications have not been examined
critically.

Acylation of Food Proteins

Acylation reactions are known to follow the carbonyl-
addition pathway (Figure 1). While amino- and tyrosyl-groups are
easy to acylate, histidine and cystine residues are only rarely
observed to undergo acylation. The serine and threonine hydroxyls
are rather weak nucleophiles and, therefore, are usually never
acylated in aqueous medium. Acylation reactions, in general, are
greatly influenced by pH. There is an inverse relationship
between the rates of acylation and their pK except in rather high
pH environments. Acylation of alpha and epsilon-amino groups in
proteins with electrophilic reagents depends on three parameters:
1) relative concentration of the nucleophile which is dependent
on pH and pK_a, 2) relative activity of the amino group, and 3)
steric factors. Stark (42) offers an excellent treatment of
linear free energy relationships of reactions with reagents as
follows: 1) preserve the positive charge, 2) change the positive
to negative charge, and 3) abolish net charge on the modified
proteins.

Acetic and succinic anhydrides are the most commonly used
reagents because they are simple to use. Acetic anhydride is
relatively selective for amino groups, acylates low pK_a groups at
low pH, and renders the cationic groups neutral. Succinic
anhydride, on the other hand, is more selective than acetic anhydride for amino groups. The net negative charge of the protein
increases upon succinylation causing changes in the electrostatic
environment and conformation of the protein molecule. The end
products tend to dissociate, are more soluble, possess higher
intrinsic viscosity, and are more heat stable (10, 15, 42). The
target protein may influence the reactivity of its own functional
groups and also that of the electrophilic reagent (10) as shown
in Table VIII. However, the scientist can make constructive
judgments as to the overall reactivity and the reaction mechanism
only when the protein structure and its behavior in aqueous
systems are known.

Table VII. List of chemical modification work on food proteins.

Protein	Modification Reaction	Changes in Physical and Chemical Properties	Reference
Albumin, γ-Globulin	Succinylation Acetylation	Ionization behavior	(58)
Casein, β-Casein	Succinylation, Acetylation,	Inefficient tryptic hydrolysis	(34)
	Succinylation, Acylation with acid anhydrides	Lack of aggregation, increased solubility, lack of sensitivity to calcium ions	(14,28)
Myosin	Succinylation	Increased solubility	(40)
Fish Protein	Acetylation	Lower isoelectric point - Inefficient tryptic hydrolysis	(7,26)
	Succinylation	Lower isoelectric point - Solubility at pH 4.0, improved foaming capacity, improved emulsion activity but decreased stability, improved heat stability	(24,26)
Soybean	Acetylation	Decreased water-binding increased solubility, decreased gel-strength,	(1,17)
	Citraconylation	increased net charge, ease of separation, solubility at pH 4.5	(4)
	Succinylation	Lower isoelectric point, increased emulsifying activity and stability, increased foaming capacity	(17,38)
	Oxidation Reduction (sulfite)	Reduced viscosity, resistance to aggregation, Reduced viscosity in aqueous solutions, and increased viscosity in salt solution	(38) (31,38)
	Alkali (pH 10.0)	Increased dispersibility and solubility, low viscosity and sensitivity to calcium ions, and resistance to aggregation, increased elasticity	(38)
	Calcium Binding	Lower isoelectric point change in solubility profile	(46)

Table VII (continued).

Protein	Modification Reaction	Changes in Physical and Chemical Properties	Reference
Cottonseed	Acetylation	Improve water-holding, oil-holding, emulsifying and foam capacities	(8)
	Succinylation	Improved water solubility, improved heat stability, decreased bulk density, and improved oil holding, emulsion capacity, gel-strength, water hydration, water retention and viscosity	(8,9)
Leaf	Succinylation	Increased bulk density, solubility and viscosity, improved flavor	(18)
	Calcium Binding	Enhanced emulsifying activity lower isoelectric point	(46)
Yeast	Succinylation	Reduced emulsion stability, increased viscosity, change in solubility, inefficient pancreatic and peptic hydrolysis	(41)
Peanut	Succinylation	Enhanced solubility at low pH, increased water-absorption and viscosity, dissociation into subunits, swelling	(59)
Rapeseed	Reductive alkylation (HCHO, Sodium borohydride	Loss of gelation	(20)
Egg White	Glutarination (3,3-dimethyl-glutaric anhydride)	Change in net charge, improved heat stability	(19)
Wheat	Acylation	Increased solubility and viscosity, dissociation into subunits	23)
	Calcium binding Oxidation-reduction (dough improvement)	Lower isoelectric point, increase/decrease in dough-strength	(46)

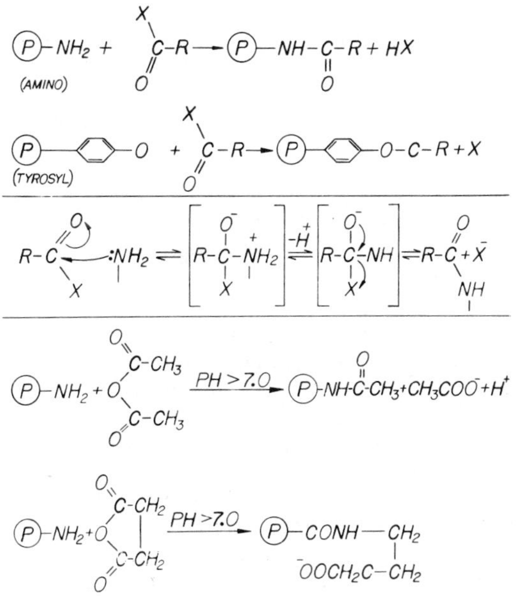

Figure 1. Acylation of protein amino groups with acetic and succinic anhydride. (Reproduced, with permission, from Ref. 36. Copyright 1971, Holden-Day, Inc.)

Table VIII. Effects of protein environment on functional groups
and the electrophilic reagents. (10)

PROTEIN EFFECTS ON:	
Functional Groups	Electrophiles
Steric protection	Selective absorption of reagent prior to reaction
H-bond: pK_a versus activity	Electrostatic interaction with charged groups
Field effect of charged groups: pk_a versus activity	Partial steric protection by the protein matrix
Polarity of local environment, repressed ionization due to hydrophobic regions	Polarity of the local environment
Reversible covalent inter-action (hemithioacetals decrease nucleophilicity)	Other functional groups in the vicinity
Charge-transfer interactions (pi electrons)	Conformational restriction against permissible reaction mechanism

Beta-casein. Native beta-casein is characterized by a disordered conformation dictated by the presence of highly charged
negative regions (serine phosphate) separated by hydrophobic
regions (14, 28, 53). Its primary structure is known (53). A
complete sequence of beta-casein A is shown in Figure 2. Given
the net negative charge of 11 per mole, the regions of pronounced
hydrophobicity, and a random coil conformation because of uniform
distribution of proline, beta-casein serves as a good ampholyte
(surfactant) both for foams and emulsions. However, emulsion
stability and solubility in aqueous medium are strongly affected
by changes in pH and calcium ion concentration.

Acetyl, succinyl, butyryl, and higher series acyl derivatives of beta-casein acquire fairly predictable properties (Table
IX). Succinylated beta-casein is completely soluble below pH 5.0
(Figure 3) and not precipitated by calcium ions (the net gain of
negative charge is far more than what can be counteracted by
calcium binding by five phosphate and six imidizole groups). The
two publications reviewed herein (14,28) and others (55-57)
describe the research aimed at studying the aggregation and the
calcium ion sensitivity behavior of beta-casein. However, since
the knowledge of casein's primary structure and ionization
behavior is available, it may serve as a model food protein for
chemical modification and final assessment of the functional and
nutritional properties of its derivatives. As a natural
extension of such research, the hydrophobicity, polarity ratio
and NPS values for beta-casein (the type of data in Tables II and
III) and its acyl derivatives should be calculated and related to
functional properties. Acylation of beta-casein with acid

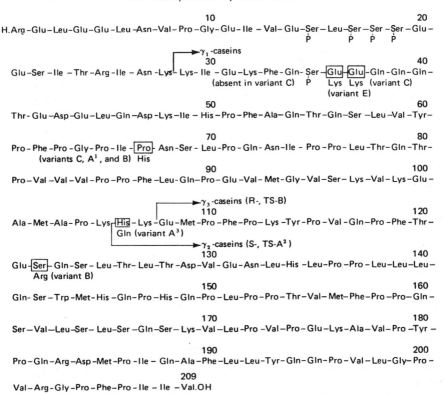

10   20
H.Arg-Glu-Leu-Glu-Glu-Leu-Asn-Val-Pro-Gly-Glu-Ile - Val- Glu-Ser—Leu-Ser—Ser -Ser -Glu–
                                                          P   P   P   P

                          ►γ₁-caseins
                          30   40
Glu-Ser -Ile -Thr-Arg-Ile - Asn -Lys⌐ Lys- Ile - Glu-Lys-Phe-Gln- Ser —Glu—Glu—Gln-Gln-Gln–
                          (absent in variant C)   P   Lys Lys (variant C)
                                                      (variant E)

                          50   60
Thr- Glu –Asp–Glu–Leu–Gln–Asp–Lys–Ile – His –Pro –Phe-Ala–Gln–Thr–Gln–Ser –Leu –Val –Tyr–

                          70   80
Pro- Phe -Pro -Gly-Pro - Ile -⌐Pro⌐ Asn-Ser - Leu-Pro - Gln- Asn-Ile - Pro -Pro- Leu -Thr-Gln –Thr–
(variants C, A¹, and B)   His

                          90   100
Pro – Val –Val –Val –Pro –Pro –Phe –Leu- Gln–Pro – Glu –Val- Met-Gly–Val–Ser– Lys –Val– Lys–Glu–

                          ►γ₃-caseins (R-, TS-B)
                          110   120
Ala - Met -Ala- Pro - Lys⌐His⌐- Lys –Glu–Met-Pro –Phe-Pro– Lys -Tyr-Pro - Val -Gln–Pro –Phe–Thr–
             Gln (variant A³)
                          ►γ₂-caseins (S-, TS-A²)
                          130   140
Glu-Ser⌐ Gln-Ser- Leu–Thr- Leu–Thr –Asp–Val- Glu–Asn –Leu–His –Leu-Pro –Pro – Leu-Leu-Leu–
Arg (variant B)
                          150   160
Gln- Ser – Trp-Met- His –Gln-Pro –His - Gln–Pro – Leu-Pro –Pro–Thr-Val– Met-Phe-Pro –Pro–Gln–

                          170   180
Ser –Val—Leu–Ser– Leu–Ser –Gln–Ser– Lys-Val –Leu -Pro –Val–Pro- Glu –Lys–Ala–Val- Pro –Tyr –

                          190   200
Pro – Gln –Arg –Asp –Met -Pro - Ile - Gln–Ala –Phe –Leu-Leu-Tyr– Gln-Gln–Pro - Val –Leu–Gly– Pro –

                          209
Val-Arg-Gly-Pro-Phe-Pro -Ile - Ile -Val.OH

Figure 2.   The primary structure of β-casein A. (Reproduced, with permission, from Ref. 89. Copyright 1976, American Dairy Science Association.)

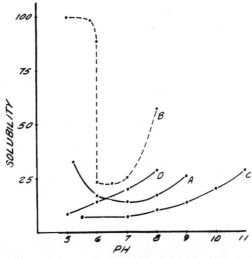

Figure 3.   The pH-solubility profile of β-casein and its acylated derivatives (1.5% in 0.25 M CaCl₂ at 30°C). Key: A, control β-casein; B, acylated β-casein; C, valerylated β-casein; and D, butyrated β-casein (28).

Table IX. Properties of beta-casein and its acyl-derivatives.
           (14, 28)

Property	Beta-Casein	Acetyl-Beta-Casein	Succinyl Beta-Casein	Higher Series of Acetylated Beta-Casein
Charge	11/mole	22/mole	33/mole	22/mole
PI	4.90	Below 4.9	Below 4.9	Below 4.9
Number of Lysyl-groups	11	-	-	-
Binding to DEAE column	Strong	Stronger	Strongest	Strongest
Electro-phoretic mobility	Normal	Higher	Highest	Higher
Solubility	-	-	High	-
Ca++ Sensitivity	High	Less	None	Less

anhydrides of varying hydrophobicity influences its solubility
predictably (Figure 3).
      Soybean protein. Soy protein isolates by definition contain
no less than 90% (dry weight basis) protein. The storage globulins constitute approximately 90% of the total protein, and the
balance is made up of various biologically active proteins
(trypsin inhibitor, 9%; hemagglutinins, 6%; and various enzymes
including lipoxygenase). In the storage globulin category, the
major constituents are 7S beta-conglycinin (27%), the 11S
glycinin dimer (34%), 15S glycinin tetramer (6%), and a small
molecular weight 2S alpha-conglycinin (18%).
      The 7S beta-glycinin is an antiparallel beta-structure with
disordered regions in the molecule. On the basis of low deuterium exchange data, it is postulated that 7S beta-conglycinin is a
rather compact molecule despite the presence of non-helical
regions (51). Most of the tyrosine residues are buried, and the
tryptophan residues are exposed. It has only 2 cystines per mole
indicating that disulfide bonds are minimal. It is composed of
nine polypeptide sub-units, is self-associating, and is characterized by an intrinsic viscosity of 0.063 dl/g (a value close to
that of most other globulins).
      Three publications (1, 4, 17) reviewed in this section deal
with acetylation (1, 17), succinylation (17), and citraconylation
(4) of soy protein. Various other chemical modifications directed
at improvement of soy protein properties for non-food industrial
applications were reviewed by Meyer and Williams (38) and Kinsella

and Shetty (31). All investigators cited above used soy protein isolate whose amino acid composition is similar to that of 7S beta-conglycinin (Table IV) except that the citraconylation study included two additional proteins isolated from the isolates - 7S beta-conglycinin and lipoxygenase. The acetylated soy protein does not exhibit substantial improvement in functional properties (i.e., a decrease in water binding activity, some increased solubility in the pH range 4.5 - 7.0, and decreased gel strength). On the other hand, succinylation of soy protein improves its emulsifying activity (30%), emulsion stability (21%), emulsifying capacity (300%), foaming capacity (20%), and foaming stability (50%). Upon succinylation, soy protein gains an additional net negative charge and, therefore, the short range attractive forces diminish and the repulsive forces predominate. Also, the succinylated soy protein attains an isoelectric pH (4.0) lower than that of the native molecule (5.4). Consequently, the derivative protein remains soluble in a broader pH range, and its overall hydration and emulsification characteristics improve.

Brinegar and Kinsella (4) found that citraconylation of soy proteins is a reversible process. They suggested a possible use of this reversible reaction in ion-exchange chromatography for separating the 7S beta-conglycinin fraction. Citraconylation is expected to magnify the net charge differences between the native and citraconylated soy proteins. The pH-solubility profile of the derivatized-soy protein does shift to a range of 3.5-5.0 from 4.0-6.0.

Since soy protein isolate is a heterogeneous mixture of proteins, the full nature and consequence of acylation reactions is hard to assess. However, succinylation, in particular, does seem to improve its hydration and surfactancy properties in acid pH food formulations.

Other vegetable proteins. A few rather preliminary studies on acetylation and succinylation of cottonseed flour (8, 9), selective reduction alkylation of 12S rapeseed glycoprotein (20), and succinylation of protein isolates from alfalfa leaves (18) were made in the last five years. In the case of cottonseed flour, all hydration and surfactancy properties improve either upon acetylation or succinylation in relation to the control flour. However, whereas acetylation yields a product of much improved surface activity, succinylation yields products that are more pH and heat tolerant. These observations are explainable in terms of differences in net charge on the two derivatives. The changes in the physicochemical properties of the alfalfa leaf protein concentrate upon succinylation manifest as alterations in color, solubility, and emulsifying and foaming properties. Acetylated leaf protein was not found to be as soluble as its succinyl counterpart, and the solubility of the latter correlated positively with its emulsifying and foaming capacities. Gill and Tung (20) used the formaldehyde and sodium borohydride treatment (16, 36) for modifying 12S rapeseed glycoprotein which has a

strong tendency to self-associate and gel at pH values higher
than 4.0, even at concentrations as low as 1.0%. The gelation
can be further augmented by thermal treatment. The authors did
not find the involvement of epsilon-amino groups of lysine in
this reaction. They hypothesized that it is possible that the
crosslinking reactions involve carbohydrate as well as amino acid
functional groups.

Other animal proteins. Myosin from the breast muscle of
chickens upon succinylation became exceedingly water soluble,
lost its calcium activated ATPase activity and ability to combine
with F-actin, and folded into a less helical conformation while
retaining its molecular weight and shape (40). About 90% of the
epsilon-amino and 50% of the sulfhydryl groups were determined to
have participated in the reaction. The chemically modified
myosin did gain net negative charge and showed exceptional
thermal stability.

Egg white proteins are very valuable for food product formulation because they form stable foams, have unique heat
denaturation properties and binding power, and high nutritional
value. However, when subjected to pasteurization temperatures,
their functional properties are partially destroyed. Chemical
modification with 3,3-dimethylglutaric anhydride results in a
derivative with excellent heat stability and only minor impair-
ment of the solubility and surfactancy properties (19). The
glutarinated egg white protein exhibits increased net negative
charge, particularly its lypozyme fraction. Egg white proteins
have been subjected to a variety of chemical modifications, e.g.,
condensation with formaldehyde (62), emulsification with synthetic
detergents (62), alkylation and reduction (64), and reaction with
acid anhydrides, ethyl acetimidate hydrochloride, iodine and
potassium cyanate (65, 66). In the glutarination reaction, the
epsilon-amino groups react faster than the sulfhydryls (steric
protection). The iron-binding capacity of glutarinated egg white
protein is significantly decreased because of the loss of free
epsilon-amino acid residues needed for binding.

Fish protein concentrate (7) and fish myofibrillar protein
(24) have also been succinylated. The succinylated fish protein
concentrate exhibits enhanced emulsifying properties at high pHs,
forms a strong cheese-like curd upon quiescent acidification, and
is resistant to pancreatin-digestion. Its isoelectric point
shifts to an acidic pH of 4.0 as opposed to 5.0 for the native
concentrate. Its pH-solubility profile shifts to the acidic
side. There is a dramatic change in the functional property of
fish myofibrillar protein upon succinylation (24); e.g., rapid
rehydration, heat stability, good dispersion in the pH range 6.0-
8.5, high emulsion capacity, and bland odor and taste occur. The
degree of change in functional property is known to depend on the
amount of succinylation.

Single-cell protein. Succinylation of yeast protein isolates after cell-disruption improves the recovery of protein from the biomass while preventing nucleic acid entrapment in the protein isolate (41). The succinylated protein can be precipitated at its isoelectric point (pH = 4.2). Succinylation improves extract-ability and thermal stability and decreases proteolysis.

Various workers have reported pH-solubility profiles of succinyl derivatives of proteins (7, 17, 18). Figure 4 shows the pH profiles of fish, leaf, and soy proteins. It is difficult to explain why three proteins of different origins acquire more or less similar solubility properties upon succinylation. The under-standing of the structure and topology of the native proteins may help explain this phenomenon.

Reductive Alkylation

Modification of epsilon-amino acid groups of proteins by dimethylation with formaldehyde and reduction, or by ethylation with acetaldehyde and reduction (15, 16, 61), leaves no traces of formaldehyde and the native protein does not suffer from changes in net charge and physical properties dictated by electrostatic interactions. The modification scheme is particularly suitable for preventing the Maillard browning reaction because the free epsilon-amino groups of lysine needed for the browning reaction are converted to dimethyl-lysino-groups. The dimethyl-lysino-proteins do occur in nature, suggesting potential safety in prac-ticing the art of reductive alkylation modification of food proteins. Ethylation might be more acceptable on an esthetic basis because ethyl alcohol is the by-product of the modification reaction. However, the fact that epsilon-ethylaminolysines do not occur in nature raises concerns as to the safety of the finished product.

The reductive alkylation procedure (16) is specific for the epsilon-amino group of lysine. The initial imine products are very rapidly reduced to alkyl-compounds by sodium borohydride as shown in Figure 5. However, the need for extensive testing with respect to nutritional adequacy and physiological effects of alkylated and reduced food proteins must be addressed prior to the development of commercial processes.

Covalent Binding of Amino Acids

Casein (54) and soybean proteins (45) have been modified by covalent attachment of various L-amino acids and methionine and tryptophan, respectively. Casein was selected as a model food protein for deliberate modification of its hydrophobic and hydro-philic groups and subsequent examination of changes in its physical and nutritional properties and digestibility. On the other hand, soy protein was selected as a candidate for incorpor-ation of methionine and tryptophan by covalent attachment. Just as in the synthesis of peptides in homogeneous solution or as in

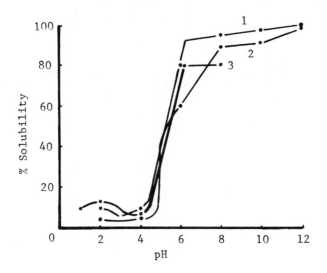

Figure 4.   The solubilities of the succinyl derivatives of leaf (1), soybean (2), and
fish proteins (3) (7, 17, 18).

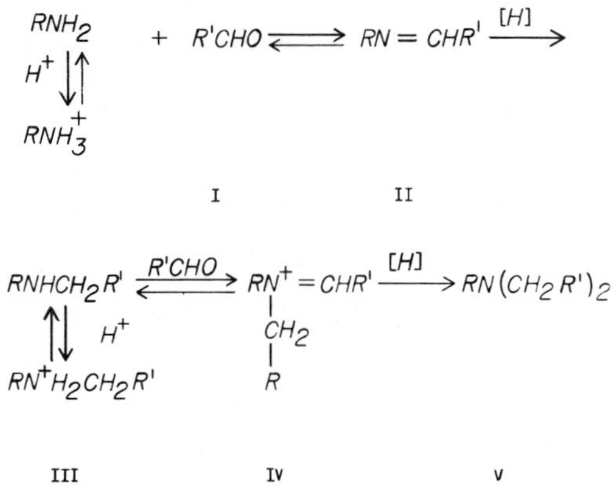

$$RNH_2 \quad + \quad R'CHO \rightleftharpoons RN = CHR' \xrightarrow{[H]}$$

$$H^+ \updownarrow$$

$$RNH_3^+$$

I   II

$$RNHCH_2R' \xrightleftharpoons{R'CHO} RN^+ = CHR' \xrightarrow{[H]} RN(CH_2R')_2$$

$$\updownarrow H^+ \qquad\qquad\quad CH_2$$

$$RN^+H_2CH_2R' \qquad\qquad\qquad R$$

III   IV   V

Figure 5.   Reductive alkylation of amino groups with carbonyl compounds (15,
36). (Reproduced, with permission, from Ref. 61. Copyright 1977, AVI Publishing Company.)

solid state peptide synthesis (67-70), the amino groups (nucleophiles) of a candidate protein are reacted with an activated
carboxyl of the amino acid to be attached. The resultant bond is
either a peptide (for alpha-amino nucleophile) or an isopeptide
(epsilon-amino nucleophile).

The N-hydroxy-succinamide esters of tert-butoxycarbonyl-L-
amino acids (boc), both hydrophobic (glycine, alanine, methionine,
N-acetyl-methionine) and hydrophilic (asparagine, asparatic acid),
have been found to be quite effective in covalent attachment to
casein (54). The chemistry employed in this case yields peptidyl-
proteins only with non poly-peptidyl-derivatives. The boc-groups
were removed by anhydrous trifluoroacetic acid. The urethane
type boc-group offers complete protection against racemization.
Since casein lacks an ordered tertiary structure or conformational complexity, changes in physical properties are neither
expected nor observed. Although the solubility and viscosity
properties of casein did not change appreciably, in vitro enzyme
digestion of modified casein decreased as the size of the
attached amino acid increased.

Methionine and tryptophan can be covalently linked to soy
protein by carbodiimide condensation reaction up to as high as
5.9% and 10.7%, respectively. The reaction involves activation
of protein carboxyls by 1-ethyl-3-(3-dimethyl-amino propyl) carbodiimide (EDC) and condensation thereof with the amino acid of
choice. Factors such as protein concentration, EDC concentration, activation time, and reaction time were found to determine
the degree of covalent binding. This offers a way to control the
fortification of proteins of low nutritive value by covalent
attachment of essential amino acids. Although there may be
potential health hazards, the excess carbodiimide and the
corresponding urea can be easily removed by working with dilute
acids or water.

Non-Covalent Interactions

Chemical modification of food proteins can also be accomplished by the manipulation of their non-covalent interactions
with themselves, carbohydrates, and lipids in formulated foods
(72, 73). For instance, lipid-protein interactions (74-76) have
been described to originate from electrostatic attraction, salt
bridging (divalent metals, i.e., calcium), and cooperative
effects of salt-bridging and electrostatic attractions. In addition to electrostatic forces and salt bridges, other protein-
protein interactions such as hydrogen bonding, disulfide bonds,
and van der Waals interactions can play a significant role in
stabilizing metal complexes. On the other hand, the glycoproteins are covalently bonded protein and carbohydrate complexes

(22, 77, 78). The Amadori-rearrangement mechanism of the food
products is a good example of the involvement of covalent bonds
(77) between proteins and carbohydrates.

If the protein, carbohydrate, and lipid constituents in a
food system were structurally defined and their environment (pH,
solutes, solvents) were accurately known, an assessment of the
contribution of various interaction forces could be made (79-81).
It is possible to safely predict the functional property of the
protein under investigation once the type and the magnitude of
such interactions are known. The determination of the secondary
structures of a protein molecule from the knowledge of amino acid
sequence (79) along with the methods of indirect evaluation of the
conformation (80), surface area, and hydrophobicity relationships
(81) are the techniques utilized for the prediction of the
physical properties.

It must be emphasized that all physical properties of matter
are a reflection of molecular structure, e.g., a protein of known
amino acid sequence would give net charge and therefore a definable standard chemical potential and solubility in an environment
of known temperature and ionic strength. This knowledge permits
estimation of both solubility and thermal stability. Likewise, a
good assessment of the surface active properties of a protein can
be determined from the knowledge of its hydrophobicity, ability
to hydrogen-bond and ionic, electrostatic, and ion-dipole interactions. Furthermore, while emulsifying capacity depends on the
shape, charge, hydrophobicity, hydration of polar groups, and
neutrality of dipoles, the emulsion stability depends on the
magnitude of these interactions (88). Once a food protein is
defined, its non-covalent interactions can be manipulated both in
kind and degree in formulated foods containing carbohydrates and
lipids.

It is an easier task, both economically and technically, to
change the environment of food proteins to induce changes in
their physical properties than to change the proteins for a given
environment. The molecular basis for the functionality of proteins resides in the structure, interactions, and their ability
to interact with other food ingredients. This is true for any
protein, modified or unmodified. In accordance with this
concept, the initial studies on proteins in coscervate systems
(86, 87) should be followed up. This author strongly believes
that an in-depth understanding of the non-covalent interactions
of proteins in food systems can be of great help in formulating
safe food products.

Protein Salts

As amphoteric molecules, proteins can bind both anions and
cations. Binding with small ions in particular is very important
in several foods. If the anions of the protein molecule can
selectively bind metal ions, it attains a net negative charge,
its pH increases, and its isoelectric point shifts to an even

more acidic pH if the binding occurs preferentially around the isoelectric point. The converse is true for cation binding.

Some ions, e.g., trichloracetic acid, form soluble salts, and some combine to form coordination complexes with proteins. The metal complexes and salts of proteins do exhibit different properties in formulated foods. A recent study compares calcium and sodium binding ability of soy protein, gluten protein, leaf protein concentrate, and blood serum albumin (46). The combined ability of various fractions of soy and leaf protein for calcium binding is approximately 55.0 and 50.0 mg/g, respectively, although their solubilities at pH 11.0 are 85% and 15%, respectively. The gluten and blood proteins bind only 25.0 and 10.0 mg/g of calcium under similar conditions, but blood protein isolate is far more soluble at pH 11.0 (92%) than gluten protein (45%).

Ion binding by proteins can be either a non-specific (pH and ligand concentration dependent) or a specific (defined stoichiometry) phenomenon which, in general, is a case of multiple equilibrium wherein the small molecules bind to an aggregating or dissociating protein system in which the monomer co-exists with the polymer and either can bind to the small molecule (82).

Other ligands such as chlorogenic acid (33), caffeic acid and isochlorogenic acid (83), flavor compounds (84) and tannins (85) are known to bind to proteins. The forces for binding may be ionic (36), covalent (53), and/or hydrophobic (84, 85). The precise knowledge of ligand-protein interactions in terms of specificity, equilibria, and reaction conditions can be employed to create more functional and stable food systems. For instance, removal of chlorogenic acid improves the biological value and protein efficiency ratio of leaf protein concentrate (33). The binding and release of flavor components to proteins improves food acceptability (84), and removal of tannins and lignins from vegetable and cereal proteins renders them biologically more available. Food protein systems should be designed wherein desired interactions are promoted in place of those that are undesired. Although production of protein salts of different bulk and solution properties is well understood (82), knowledge of ligand-protein interactions due to hydrophobic forces is poor. It is likely, however, that ligand-protein interactions result from non-covalent interactions, including metal-binding, and constitute chemistry that is not understood.

Nutritive Value and Toxicity of Chemical by Chemically Modified Food Proteins

An obvious barrier against the use of chemically modified food proteins is biological in nature. The modified protein products must have acceptable flavor, improved nutritive value, and undetectable toxicity. Most food proteins that are modified may contain other functional groups either as an integral part of their structure or as impurities. The concomitant modification

of such groups may increase the concentration of toxic substances.
These concerns have not been sufficiently answered by prior
research and, of necessity, require extensive evaluation and
experimentation.

The results from a few studies (11, 26, 34, 71) impact
rather negatively on the protein modification technology. For
example, determinations of protein efficiency ratio and toxicological trials (1) suggest that acetylated and succinylated
casein and whey proteins cannot be regarded as suitable ingredients for acid foods despite demonstratable improvement in their
solubility and heat stability around pH 4.0. Groninger and
Miller (26) investigated acetylated and succinylated myofibrillar
fish protein with respect to 1) amino acid composition, 2) protein efficiency ratio, 3) rates of enzyme digestion, and 4) the
utilization of ^{14}C-labelled derivatives by rats, and reported a
reduction in protein efficiency ratio, only partial utilization
of protein-bound acetyl-lysine, and metabolic unavailability of
both ^{14}C-labelled succinyl- and acetyl-lysine. They concluded
that acetyl- and succinyl-derivatives are only partially utilized
and that the level of utilization depends on the type of acyl-
group. Yet another succinylated protein, casein, has been
reported to resist tryptic (but not chymotryptic) hydrolysis
(34). Even the chymotrypsin activity is inhibited by N-tosyl-
lysine methyl ester. Neither carboxypeptidase A nor B release
N-succinyllysine from t-butoxy-carbonyl-N-succinyllysine.
Obviously the succinyllysyl bonds are not hydrolyzed by gastric
and pancreatic proteases.

Conclusions

A few potential edible proteins have been covalently derivatized by acylation and reductive alkylation. The modified
proteins have been found to exhibit different hydration and surfactancy properties. However, without the detailed knowledge of
the primary structure of the proteins under investigation, the
control of derivatization reactions and the physical properties
of the product are still a matter of considerable speculation.
Although some evidence as to the change in net charge, conformation, molecular weight, polydispersity, and the propensity to
associate and/or dissociate has been found, there has been very
little work done on the actual food formulations based on
covalently modified food proteins.

A number of modified food proteins including acylated-casein
and myofibrillar fish protein have been found to be resistant to
complete digestion by the pancreatic enzymes. The major process
criteria such as toxicity of reagents, cost of manufacturing, and
purification have not been identified. Considerations, such as
toxicity of the product and its in vivo digestion products,
physiological and pharmacological compatibility, flavor and odor,
and overall aesthetics and consumer acceptance, have not received

due research inputs. The questions regarding regulatory clear-
ance and test method development, although raised a few years
ago, have not been answered.

Although the benefit of fundamental knowledge for preparing
functional proteins are known, addition of amino acids to a
deficient protein, blocking free amino groups, and thereby
preventing undesirable chemical reactions in processed foods such
as formation of lysinoalanine and lanthionine, is questionable at
this time on account of toxicity, cost, and demonstratable
reduction in nutritive value.

The nature and extent of non-covalent interactions of food
protein can be quantified on the basis of primary structure, and
solution, and surface behavior. This particular concept deserves
careful study. The precise knowledge of both non-covalent and
covalent interactions between major food components (starch, protein, and lipids) along with the knowledge of the chemistry of
protein salts and protein surfactant complexes, can facilitate
food formulation. This concept needs more rigorous definition at
the molecular level in terms of interactions between major and
minor food components of a formulated food system. The anticipa-
ted research should include study of food proteins in coscervate
systems.

The covalently modified food proteins may have special significance if they are to be identified as highly functional at low
levels of incorporation. This will overcome any objections to
their food use on the grounds of lower nutritive value, toxicity,
and cost.

Acknowledgements

The author sincerely thanks the R&D staff members of Krause
Milling Company for their review and constructive criticism of
this article. A special thank you is given to the senior management of Krause Milling Company for their permission to write this
chapter.

Literature Cited

1. Barman, B.G.; Hansen, J.R.; Mossey, A.R. J. Agric. Food
    Chem., 1973, 25, 638.
2. Belikow, W.M.; Besrukow, M.G.; Walnowa, A.I. Die Nahrung.,
    1975, 19, 65.
3. Bigelo, C.C. Theoret. Biol., 1967, 16, 187.
4. Brinegar, A.C.; Kinsella, J.E. J. Agric. Food Chem., 1980,
    28, 818.
5. Bjarson, J.; Carpenter, K.J. Br. J. Nutr., 1967, 23, 859.
6. Brekke, C.J.; Eisele, T.A. Food Technol., 1981, 35, 231.
7. Chen, L.F.; Richardson, T.; Amundson, C.H. J. Milk Food
    Technol., 1975, 38, 89.
8. Child, E.A.; Parks, K.K. J. Food Sci., 1976, 41, 713.

9. Choi, Y.R.; Lusas, E.W.; Rhea, K.C. J. Food Sci., 1981, 46, 954.
10. Cohen, L.A. Ann. Rev. Biochem., 1968, 37, 695.
11. Creamer, L.K. Aust. J. Dairy Sci. and Technol., 1971, 6, 107.
12. Evans, M.; Iron, L. German Patent 2,043,535. 1971.
13. Evans, M.; Iron, L. German Patent 1,951,247. 1971.
14. Evans, M.; Iron, L; Petty, J. H. P. Biochim. Biophys. Acta., 1971, 243, 259.
15. Feeney, R.F. In "Food Proteins: Improvements Through Chemical and Enzymatic Modifications"; Feeney, R.F.; Whitaker, J.R., Eds.; Advances in Chemistry Series 160, American Chemical Society: Washington, DC, 1977; p 3.
16. Galembeck, F.; Ryan, D.S.; Whitaker, J.R.; Feeney, R.F. J. Agric. Food Chem., 1977, 25, 238.
17. Franzen, K.L.; Kinsella, J.E. J. Agric. Food Chem., 1976, 24, 788.
18. Franzen, K.L.; Kinsella, J.E. J. Agric. Food Chem., 1976, 24, 914.
19. Gandhi, S.K.; Schultz, T.; Boughray, F.; Forsythe, R. J. Food Sci., 1968, 33, 163.
20. Gill, T.A.; Tung, M.A. J. Food Sci., 1981, 43, 1481.
21. Glazer, A.N. "The Proteins"; Neurath, H.; Hill, R., Eds.; Academic Press: New York, NY, 1976; p 1.
22. Goldstein, I.J., Ed.; "Carbohydrate-Protein Interactions"; ACS Symposium Series 88., American Chemical Society; Washington, DC, 1979; 223 pp.
23. Grant, D.R. Cereal Chem., 1973, 50, 417.
24. Groninger, H.S. J. Agric. Food Chem., 1973, 21, 978.
25. Groninger, H.S.; Miller, R. J. Food Sci., 1975, 40, 327.
26. Groninger, H.S.; Miller, R. J. Agric. Food Chem., 1979, 27, 949.
27. Hayes, J.F.; Dunkerley, J.; Muller, L. Aust. J. Dairy Technol., 1969, 24, 69.
28. Hoagland, P.D. Biochem. J., 1969, 7, 2542.
29. Kinsella, J.E. Crit. Rev. Food Sci. and Nutr., 1976, 7, 219.
30. Kinsella, J.E. Chem. and Ind., 1977, No. 5, 177.
31. Kinsella, J.E.; Setty, K.J. In "Functionality and Protein Structure"; Pour-El, Eds., A., ACS Symposium Series 92, American Chemical Society: Washington, DC, 1978; p. 33.
32. Knowles, J.R. In "Chemistry of Macromolecules"; Gutfreund, H., Ed.; Butterworths: London, 1974; p. 149.
33. Lahiry, N.L.; Satterlee, L.D.; Hsu, H.W.; Wallace, G.W. J. Food Sci., 1977, 42, 83.
34. Matoba, T.; Doi, E. J. Food Sci., 1979, 44, 537.
35. McElvin, MD.; Richardson, T.; Amundson, C.H. J. Milk Food Technol., 1975, 38, 521.
36. Means, G.E.; Feeney, R.F, Eds. "Chemical Modification of Proteins". Holden-day: San Francisco, Cal., 1971; 254 pp.
37. Melnychyn, P.; Stapley, R.B. U. S. Patent 3,764,711; 1973.

38. Meyer, E.W.; Williams, L.D. In "Food Proteins-Improvements Through Chemical and Enzymatic Modification"; Feeney, R.F.; Whitaker, J.R., Eds.; Advances In Chemistry Series 160, American Chemical Society: Washington, DC, 1977; p. 52.
39. Muller, L.L. Dairy Sci. Abst., 1971, 33, 659.
40. Oppenheimer, H.; Barany, K.; Harmour, G.; Genton, J. Arch. Biochem. Biophys., 1967, 120, 108.
41. Shetty, K.J.; Kinsella, J.E. J. Food Sci., 1979, 44, 633.
42. Stark, G.R. Adv. Prot. Chem., 1970, 24, 261.
43. Thomas, M.A.; Baumgartner, P.A.; Hyde, K.A. J. Dairy Technol., 1974, 29, 59.
44. Vallee, B.L.; Riordan, J.F. Ann. Rev. Biochem., 1969, 38, 733.
45. Voutsinas, L.P.; Nakal, S. J. Food Sci., 1979, 44, 1205.
46. Wallace, G.W.; Satterlee, L.D. J. Food Sci., 1977, 42, 473.
47. Waugh, D.F. Adv. Prot. Res., 1954, 9, 325.
48. Singer, S.J. Adv. Prot. Chem., 1967, 22, 1.
49. Waslein, C.I. Crit. Rev. Food Sci. Nutr., 1975, 6, 77.
50. Vananuvant, P.; Kinsella, J.E. J. Agric. Food Chem., 1975, 23, 595.
51. Wolf, W.J. "Soybeans: Chemistry and Technology", AVI Publishing Co.: Westport, CT, 1972; p. 112.
52. McCormick, R.D. Food Prod. Dev., 1973, 7, 17.
53. Morr, C.V. In "Functionality and Protein Structure"; Pour-El, A., Ed.; ACS Symposium Series 92, American Chemical Society; Washington, DC, 1979; p. 65.
54. Puigserver, A.J.; Sen, L.C.; Flores, E.G.; Feeney, R.E. J. Agric. Food Chem., 1979, 27, 1098.
55. Waugh, D.F. Discussions Faraday Soc., 1958, 25, 186.
56. Hoagland, P.D. J. Dairy Sci., 1966, 49, 783.
57. Zittle, C.A.; Walter, M. J. Dairy Sci., 1963, 46, 1189.
58. Habeeb, A.F.S.A.; Cassidy, H.G.; Singer, S.J. Biochim. Biophys. Acta., 1958, 29, 587.
59. Beuchat, L.J. J. Agric. Food Chem., 1977, 25, 258.
60. Jencks, W.P. "Catalysis in Chemistry and Enzymology"; McGraw Hill: New York, NY, 1969; 644 pp.
61. Feeney, R.F. "Evaluation of Proteins for Human Food"; Bodell, C.E., Ed.; AVI Publishing Co.: Westport, CT; 1975; p. 233.
62. Frankel-Conrat, H.J. J. Biol. Chem., 1944, 154, 385.
63. Neurath, H.; Pulnam, F.W. J. Biol. Chem., 1945, 160, 397.
64. Deutsch, H.F.; Morton, J.T. Arch. Biochem. Biophys., 1961, 93, 654.
65. Simlot, M.M.; Feeney, R.E. Arch. Biochem. Biophys., 1966, 113, 64.
66. Buttkus, H.; Clar, J.R.; Feeney, R.E. Biochem., 1965, 4, 998.
67. Doscher, M.S. Methods In Enzymol., 1977, 47, 578.
68. Fridkin, M.; Patchormick, A. Ann. Rev. Biochem., 1974, 43, 419.
69. Katsoyannis, P.G.; Ginos, J.Z. Ann. Rev. Biochem., 1969, 38, 881.

70. Merrifield, R.B. Adv. Enzymol., 1969, 32, 221.
71. Carpenter, K.J. Nutr. Abstr. Rev., 1973, 43, 423.
72. Ryan, D.S. In "Food Proteins: Improvements Through
     Chemical and Enzymatic Modification". Advances in Chemistry
     Series 160, American Chemical Society: Washington, DC,
     1977; p 67.
73. Huang, F.K.; Rha, C. Polymer Eng. and Sci., 1974, 14, 81.
74. Tria, E.; Scanu, A.M. "Structural and Functional Aspects of
     Lipoproteins in Living Systems". Academic Press: New York,
     NY; 1969, 661 pp.
75. Karel, M. J. Food Sci., 1973, 38, 756.
76. Pomeranz, Y. Baker's Dig., 1971, 45, 26.
77. Jevons, F.R. "Symposium on Foods: Proteins and Their
     Reactions"; Schultz, H.W., Ed.; AVI Publishing Co.:
     Westport, CT, 1964; p 153.
78. Neuberger, A.; Marshall, R.D. In "Symposium on Foods:
     Carbohydrates and Their Role". Schultz, H.W., Cain, R.F.,
     Wrolstad, R.W., Eds.; AVI Publishing Co.: Westport, CT,
     1969; p 115.
79. Ptitsyn, O.B.; Finelstein, A.V. In "Protein: Structure,
     Function and Industrial Applications", Hofman, E.; Pfeil,
     W.; Aurich, H. Eds.; Permagon Press: New York, NY 1979; p
     105.
80. Wuthrich, K.; Wagner, G.; Richarz, R. In "Protein:
     Structure, Function, and Industrial Application";
     Hofman, E.; Pfeil, W.; Aurich, H. Eds.; Pergamon Press: New
     York, NY, 1978; p 143.
81. Janin, J.; Chothia, C. In "Protein: Structure, Function,
     and Industrial Applications"; Hofman, E.; Pfeil, W.; Aurich,
     H. Eds.; Pergamon Press: New York, NY, 1978; p. 227.
82. Tanford, C.H. "Physical Chemistry of Macromolecules", John
     Wiley and Sons; New York, 1961; 710 pp.
83. Haringome, T.; Kandatsu, M. Agric. Biol. Chem., 1968, 32,
     1093.
84. Srinivasan, D.; Kinsella, J.E. J. Agric. Food Chem., 1980,
     28, 567.
85. Hoon, I.O.; Hoff, J.E.; Armstrong, G.S.; Haff, L.A. J.
     Agric. Food Chem., 1980, 28, 394.
86. Sorenson, J.K.; Richardson, T.; Lund, D.B. In "Functional-
     ity and Protein Structure"; Pour-El, A., Ed.; ACS Symposium
     Series 92, American Chemical Society: Washington, DC, 1979;
     p. 173.
87. Howling, D. Food Chemistry, 1980, 6, 51.
88. Nakai, S.; Powerie, W.D. "Cereals: A Renewable Resource,
     Theory and Practice"; Pomeranz, V.; Munch, L. Eds.; Amer.
     Assoc. Cereal Chem. Symposium, 1981; 736 pp.
89. Whitney, R.M.; Brunner, J.R.; Ebner, K.E.;
     Farrell, H.M.,Jr.; Josephson, R.V.; Morr, C.V.; Swaisgood,
     H.E. J. Dairy Sci., 1976, 59, 785.

RECEIVED June 11, 1982.

Protein Structure and Functional Properties: Emulsification and Flavor Binding Effects

JOHN E. KINSELLA

Cornell University, Institute of Food Science, Ithaca, NY 14853

Proteins are the principal structural and functional components of many food systems; e.g., meat, cheese, gelatin, egg white and many cereal products. In addition, proteins are being used increasingly to fabricate and facilitate the engineering of new foods such as protein beverages and extruded foods. These and other applications depend upon the physicochemical properties of protein ingredients, collectively referred to as the functional properties. Proteins per se as dry powders, have very limited appeal to potential users or consumers. To facilitate their use in foods and their conversion to desirable ingredients they must possess appropriate functional properties following interactions with other food components; e.g., water, carbohydrates or lipids, during processing. Functional properties of proteins are those physicochemical properties of proteins which affect their behavior in food systems during preparation, processing, storage, and consumption, and contribute to the quality and organoleptic attributes of food systems (1). Generally, nutritional properties are not included in this category. Several typical classes of functional properties are shown in Table I.

There are numerous examples of functional properties in traditional foods; e.g., viscoelasticity of wheat gluten, texture, succulence and color of myofibrillar proteins in meats, curd formation of caseins, and whippability of egg white proteins.

Different food applications require different characteristics; e.g., in beverages, proteins should be soluble; in comminuted meats, they should have emulsion stabilizing and gelling properties; and in whipped toppings they must stabilize the foam. The type of functional properties required in a protein or a protein mix varies with the particular food system in question. Thus, in meat systems, water binding, solubility, emulsifying activity, adhesiveness, swelling, viscosity and gelation are typical properties of proteins that determine their usefulness and impact on the final quality. These criteria are critical, and as the number of processed and fabricated foods increase, greater reliance will be placed on the consistent performance of ingredients in specific food formulations.

0097-6156/82/0206-0301$07.50/0

Table I. Typical functional properties performed by proteins in
          food systems.

Functional Property	Mode of Action	Food System Example
Solubility	Protein solvation	Beverages
Water absorption and binding	Hydrogen-bonding of water, Entrapment of water	Meats, Sausages, Breads, Cakes
Viscosity	Thickening, Water binding	Soups, Gravies
Gelation	Protein matrix formation and setting	Meats, Curds, Cheese
Cohesion-adhesion	Protein acts as adhesive material	Meats, Sausages, Baked goods, Pasta products
Elasticity	Hydrophobic bonding in gluten, Disulfide links in gels	Meats, Bakery
Emulsification	Formation and stabilization of fat emulsions	Sausages, Bologna, Soup, Cakes
Fat absorption	Binding of free fat	Meats, Sausages, Donuts
Flavor-binding	Adsorption, entrapment, release	Simulated meats, Bakery, etc.
Foaming	Form stable films to entrap gas	Whipped toppings, Chiffon desserts, Angel cakes

Though the importance of functional properties have been recognized by commodity specialists for many years, the widespread
awareness of their importance to food science and technology is
relatively recent. This has been accelerated by the increased
emphasis on food processing, manufacturing, and formulation. The
universal attempts to utilize less expensive sources of proteins
to fabricate new food analogs, to simulate traditional foods, to
extend traditional foods, and to develop new functional ingredients have accentuated the need for information on functional
properties of proteins.

Factors Affecting Functional Properties

Several factors and processing steps affect the functional
properties of proteins. Thus, the protein source can affect
functionality. In the case of meat the age of the animal, which
is related to the quantity of connective tissue, can markedly
affect the functional properties. The progressive modification
of the collagen molecules via crosslinking increases the toughness

of meat. In the case of egg white there is a progressive thinning of the egg white with storage time. The glutenin content
varies immensely with the different varieties of wheat, and this
affects the functional properties of flours prepared from these
cereals.

Both intrinsic and applied factors influence the observed functional properties of proteins. The inherent molecular
properties of the protein per se (size, shape, conformation,
whether native or denatured), the methods and conditions of isolation (refining, drying, storage), the degree of purification,
(processing alterations) and modification by physical (heat),
chemical (derivatization, hydrolysis) or enzymatic processes, all
influence performance in food systems. The methods employed in
determining a particular functional property, the composition of
the assay system (model or food), the protein concentration, the
pH, temperature, ion concentration and composition, oxidation/re-
duction potential, the type of equipment/machinery used (geometry,
configuration), energy input, temperature controls of the system
and the method of measurement can all affect the observed property (1-6). Thus, multiple interactions are involved, and for
the derivation of useful data all experimental variables should
be defined and controlled. Failure to do so renders the data
redundant for most researchers and vitiates valid comparisons of
data from different laboratories.

The literature is replete with papers on the functional properties of proteins prepared by different methods as measured by a
variety of methods under differing conditions (6). The tendency
has been for each investigator to devise methods and/or conditions to suit a particular situation with limited concern for the
validity, comparative value, or experimental design to facilitate
elucidation of the basis of the physical property being studied.
Thus, much of the data in the literature are of very limited
value and definitely cannot be used in the systematic tabulation
of physical properties in a standardized format. This situation
reflects and is further complicated by the heterogeneous mixture
of proteins being studied; the very complex series of interactions which govern expression of a particular functional property
(and which are not amenable to easy measurement); the multiple
interacting factors which impact on the aggregate effect and the
lack of appropriate measuring devices which can accurately quan-
tify the resultant of these interactions as a function of inherent properties, interactions per se and environmental factors.

Because of the complexity of food systems, the heterogeneity
of the proteins and the variety of functions required in the different food systems for which they are used, it has been difficult to standardize tests for measuring functional properties
(1,6). It is impossible to generalize concerning the available
information and difficult to extrapolate data from one system to
predict the behavior of that particular protein in other systems.

Measurements, in order to have more general applicability,

must be based upon the fundamental properties of the system. In
order to predict the behavior of proteins it is necessary to determine their basic composition and physicochemical properties
under specified conditions. However, to date, there are few
methods that can be validly or universally used to predict the
behavior of proteins in food systems. In measuring the functional
properties of proteins, ideally, equilibrium conditions should
be attained so that thermodynamic criteria are met and so that
the particular measurement reflects the physical properties of
the protein in question. For many functional properties a number
of operational definitions have been used to define those particular properties. Each has its advantages and disadvantages,
and, of course because thermodynamic criteria are rarely attained,
thermodynamic analysis cannot easily be applied (7).

Evaluation of a particular property or application in food
systems usually includes tests in model systems; i.e., systems
that mimic the food system in a limited way and that are selected
to highlight the performance of that particular function.
Ultimately, of course, all functional attributes should be
evaluated in food systems (6).

The method of extraction and refining can have a marked
effect on the functional properties of isolated proteins. This
depends upon the solvent used, the degree of heating, the presence or absence of alkali, pH, ions, mode of precipitation and
drying, and conditions of storage (1-8). Temperature markedly
affects functional properties; e.g., high temperatures which induce denaturation may impair functional properties. Time-temperature relationships used in drying markedly affect the functional
properties; e.g., spray versus drum drying affects the whipping
properties of egg white (9). The exposure of proteins to different ions during preparation can affect functional properties (10).
Moisture levels and storage temperature can affect interactions
during storage and thereby impact on functional properties. The
presence of reactive components such as unsaturated fatty acids
and reducing sugars can result in chemical interactions and
thereby markedly affect the functional properties of proteins.

To facilitate the development of proteins with particular
functional properties it is first necessary to define which physicochemical properties are most important. To achieve this it is
very necessary to develop good techniques for measuring functional properties and devising methods which elucidate relationships between structure and function. Limited advances have been
made in this area though interest has increased. Obviously, in
approaching this challenge one should select a well characterized
protein and use it in a model system wherein functional proteins
are the major active reagents; e.g., protein-based foams, meat
systems, cheese curds, etc. When such relationships are understood it may be feasible to modify inexpensive proteins and impart the necessary functional properties. For this the structural features of the protein, molecular properties, and

interactions required for particular functions need to be understood.

Protein Structure and Functional Relationships

The structure and chemical properties of food components determine their behavior and ultimately the acceptability of many
foods. One challenge to the modern food scientist is to define
the chemical and/or physical attributes that are responsible for
these desirable properties and apply this knowledge toward the
development of functional ingredients and the fabrication of new
foods.
The amino acid composition, the sequence of amino acids, the
degree of modification of the nascent polypeptides during synthesis, the manner in which secondary and tertiary folding occurs
and possible associations between polypeptides all affect the
final structure and physical properties of different food proteins. These structures in turn vary in their response to environmental factors (temperature, pH, ionic strength), which also
affect functional properties (Table II).

Table II. Structure of polypeptides and interactions that
           influence functional properties of food proteins.

Amino acid composition (major groups)
Amino acid sequence (segments/polypeptides)
Secondary/tertiary conformation (compact/coil)
Surface charge, hydrophobicity, polarity
Size, shape (topography)
Quaternary structures
Secondary interactions (intra- and inter-peptide)
    Hydrogen bonding, ionic, Van der Waals, hydrophobic
    and electrostatic interactions.
Disulfide/sulfhydryl content
Environmental conditions (pH, O/R status, salts, temperature)

Amino Acids. The relative amounts of different amino acids
and their disposition in the polypeptide chain affect functional
properties. Because hydrophobic effects are important forces determining the physical behavior of proteins the content and disposition of apolar amino acids (leucine, valine, isoleucine, alanine, proline, phenylalanine, tryptophan, tyrosine and methionine)
significantly affect the structure and function of proteins
(11-13). The shape of protein molecules reflect the internaliza-
tion of the maximum number of apolar amino acids, and the subunits of many oligomeric proteins associate via hydrophobic interactions (14). The content of apolar amino acids (which range
from 25-35% for most proteins) affect conformation, hydration,
solubility, denaturation and gelation properties. Shimada and
Matushita (15) showed that proteins; e.g., soybean, bovine serum

albumin, conalbumin, containing from 26-31% apolar amino acids
formed gels upon heating, whereas proteins with >31%, coagulated.
Kato and Nakai (16) showed a close relationship between the hydrophobicity of proteins and their emulsifying properties.

The presence of charged amino acids enhances electrostatic
interactions which are significant in stabilizing globular proteins and in water binding; e.g., aspartic and glutamic acid bind
6-7 molecules water per charged residue (17). This is important
for many functional properties (hydration, solubility, gelation,
surfactancy). Polar amino groups are chemically reactive and en-
gage in hydrogen-bonding, which influences conformation of α-
helix and β-sheet structures. These may be modified; e.g., acylated, to manipulate the physical properties of food proteins
(18). The hydrophilic amino acids on the surface of globular
proteins impart molecular flexibility (19). Hence, proteins
with a large proportion of hydrophilic residues may be good
functional proteins; e.g., egg white proteins.

Cysteine and cystine significantly affect structure and function of proteins. Sulfhydryl groups may be oxidized to form disulfide bonds (intra- and intermolecular), thiols and disulfides
can undergo interchange reactions which markedly affect structure
and function; e.g., gluten formation (20). Reduction of disulfide bonds can significantly affect molecular properties such as
the unfolding of glutenin and enhance the digestibility of soy
11S. Disulfide crosslinks are important in stabilizing the
tertiary structure of proteins. The formation of the β-lacto-
globulin/κ-casein disulfide linked complexes and improves baking
properties of milk powders (21).

Generally, except in cases where a particular type of amino
acid predominates, knowledge of amino acid composition is of
limited value in predicting tertiary structure or physical properties of proteins; e.g., α-lactalbumin and lysozyme have very similar amino acid composition but differ markedly in properties; the
mere replacement of glutamine with valine changes the conformation and properties of hemoglobin, and casein variants with minor
differences in amino acids show small differences in heat stability.

Protein Structure. The particular sequence of amino acids
in a protein determines its structure, conformation and properties. Knowledge of protein structure and the stability of different conformations is pertinent to understanding functional properties of food proteins (Table II). The structure of protein is
categorized as secondary, tertiary, and quaternary, according to
the spatial arrangement of the polypeptide chains. In an aqueous
environment primary polypeptides tend to coil in a characteristic
fashion to form localized secondary structures; i.e., α-helix or
β-pleated sheet. The driving force for this folding is hydrophobic in origin due to the favorable change in free energy of
solvent upon folding of the polypeptide (11-13). The secondary
structures are stabilized by hydrogen bonding.

The tertiary structure refers to the preferred three-dimensional arrangement of the folded polypeptide chains. In a typical tertiary structure the polypeptides are tightly folded to give a compact molecule in which most of the polar groups of the amino acids are located on the outer surface and are hydrated. Most of the apolar groups are internal in the hydrophobic region from which water is essentially excluded (a few H-bonded water molecules may remain). The proline residues are located at the bends or folds in the polypeptide chain where isoleucine, serine and charged amino acid residues (α-helix disrupting molecules) also tend to be located. These fold regions lack helical structure tending to be random (11, 12). The gross conformation; i.e., tertiary structure of most globular proteins tend to be similar in general organization. The degree of folding and relative proportion of α-helix, β-sheet or random coil varies immensely among proteins. Some proteins are loosely folded and the tertiary structure is stabilized by disulfide bonds, others are compactly folded with many S-S bonds (lysozyme, glycinin). Cytochrome C has no α-helix and many residues are in the extended β-conformation. The polypeptides in fibrous proteins (collagen, myosin, fibrinogen) are mostly α-helix, myoglobin has mostly 70% α-helix; however other globular proteins serum albumin, lysozyme, ribonuclease, are made up of folded polypeptides with extensive regions of unordered structure and proteins like caseins, elastin which possess very little order in manner of polypeptide folding. The compactness and extent of interpeptide interactions or bonding markedly affect the functional properties of proteins (22).

Both entropic and enthalpic factors contribute to the folding of proteins, but it is generally felt that the dominant driving force is the negative entropy effect caused by the exposure of polypeptides to the aqueous solvent phase (10, 11). This results in the unfavorable structuring of the water in the immediate vicinity of the hydrophobic peptide moieties. Consequently, folding of hydrophobic regions out of the water into the protein interior increases the entropy of the previously structured water. Though the enthalpy of this process is usually slightly positive and therefore unfavorable, the large positive entropy greatly favors folding. Hydrophobic interactions are nonspecific because they result from solvent properties rather than attractive bonding between specific groups. Once apolar groups are forced within a critical distance, Van der Waals forces, electrostatic interactions, and other interactions may contribute to stabilization of structure (11, 13, 23). Peturbation of entropic relationships can alter functional properties of proteins (10).

Many polypeptides with molecular weights exceeding 50,000 daltons tend to associate and form quaternary structures. From the amino acid composition of known proteins, it has been observed that proteins containing more than 28 mole percent of

particular amino acids (valine, proline, leucine, isoleucine and
phenylalanine) tend to self-associate (14). Thus, the α and β
chains of hemoglobin associate, whereas myoglobin, which has less
than 28% apolar amino acids, does not self-associate. An explanation for this phenomenon is that below a critical fraction of
hydrophobic residues it is possible to bury all of these completely within the molecule. When this fraction is exceeded the
folding of molecule cannot accomodate these internally and some
are exposed on the surface. When enough such residues appear on
the surface hydrophobic patches will be created, and these can
only be protected or removed from the polar medium by self-asso-
ciation. In many instances forces other than hydrophobic interactions are involved in maintaining multi-subunit structures.
Several major food proteins show quaternary structure. Caseins
associate strongly via hydrophobic forces; the acidic and basic
subunits of soy 11S and the component polypeptides of glutenin
are disulfide linked, and the subunits of myosin associate via
hydrophobic and electrostatic interactions. These structures
markedly affect the functional properties of these proteins.

The food scientist is faced with the situation wherein the
proteins with which he is working possess the tertiary and quaternary structures and conformation that presumably have evolved to
perform specific biological functions and were not necessarily
designed for particular food applications. In most instances the
physical properties of particular food proteins have dictated the
manner in which they are used or consumed. Thus, the peculiar
nature of milk proteins which under appropriate conditions form
curds led to the development of cheese. In the case of soy proteins, their ability to coagulate in the presence of calcium and
heating to form tofu facilitated their preservation, improved
their digestibility and quality attributes. The particular composition, structure and conformation of egg white protein made it
the premier whipping protein and led to the evolution of foam-
based foods and bakery products. The fibrous structure of the
actin and myosin in meats is responsible for the texture of
muscle foods, and the production of leavened breads reflect the
unique viscoelastic properties of wheat gluten. The consumer has
been conditioned to these foods, hence much effort will be required in trying to simulate these properties using other protein ingredients. In order to manipulate these proteins and/or
simulate them with other functional proteins, it is important to
understand their structure and physical properties and to elucidate the relationship between structure and specific functional
properties.

Non-Covalent Forces Stabilizing Protein Structure. Because
functional properties are related to structure and since changes
in structure of food proteins frequently are important in a specific function (surface denaturation of egg white during whipping
or the unraveling of gluten proteins during dough development),

it is important to consider the forces responsible for the native
structure of proteins and factors which effect these in relation
to their function in the native or denatured state.

The forces involved in stabilizing the tertiary structure of
proteins may include hydrogen bonding, Van der Walls forces, dipole and electrostatic interactions, and hydrophobic associations.
Covalent disulfide bonds are important in several proteins. The
nature and importance of these interactions are described in
Chapter 13 of this volume (10). The stability of the quaternary
structure of proteins is governed by the same general classes of
stabilizing interactions, particularly hydrophobic interactions,
ion crossbridging (Ca-casein) and disulfide crosslinking; e.g.,
acidic and basic subunits of 11S proteins and glutenin subunits.
Manipulation of these forces to alter the physical properties of
these proteins provides an approach for improving functional
properties of some food proteins.

Surface Properties/Emulsifying Activity. Surface activity
closely reflects the structural features of proteins and these
affect the usefulness of proteins in emulsions. In protein-based
emulsions the protein functions by forming an interfacial film.
The physical properties of this film matrix and its surface
characteristics determine its capacity to form and stabilize emulsions. Several researchers have studied the film forming properties of proteins, the properties of protein films and tried to
correlate these with emulsifying properties (22, 24-28). During
the formation of an emulsion under ideal conditions, the soluble
protein diffuses to and concentrates at the oil-water interface
once the interfacial electrostatic barrier is overcome. The protein, depending upon its structural stability (compactness, flexibility, charge, disulfide bonds, hydrophobicity) tends to unfold
to establish a new thermodynamic equilibrium. The hydrophobic
segments or loops orient in the apolar oil phase while the polar
and charged segments tend to occupy the aqueous phase. The formation of an interfacial film occurs in sequential stages, all of
which are influenced by the molecular properties of the protein.
The initial diffusion of native molecules to the interface is
concentration dependent and varies with the protein (24). The
newly arrived molecule must penetrate the interface and unfold.
The rate and extent of these events is protein dependent and is
affected by the surface pressure and compressibility of the protein film already formed (25). Finally, rearrangement of adsorbed surface denatured protein molecules occurs to attain the
lowest free energy state.

Solubility of protein is an important prerequisite for film
formation because rapid migration to and adsorption at the interface is critical. This is particularly relevant in fluid emulsions. Different proteins require different times to equilibrate
and form a film. These events are rapid for loose, flexible proteins like β-casein, intermediate for globulins (BSA) and slow

for proteins with a rigid tertiary structure like lysozyme or soy 11S (24). Graham and Phillips (24) related the physical proper-ties of protein films made with different proteins to emulsion formation and stability. They observed that different proteins behave differently at interfaces; e.g., at low concentrations, β-casein spreads out at the interface. As the surface concen-tration progressively increases looping of the apolar regions into the oil phase occurs. Subsequent compression of these loops occurs to yield a condensed film. Further increases in protein concentration does not increase film pressure or reduce surface tension though the thickness increases due to adsorption of multi-layers.

Lysozyme molecules do not unfold as readily as β-casein and take longer to form a film. The film is composed of layers of minimally unfolded molecules. Bovine serum albumin behaves in an intermediate fashion; i.e., it unfolds to a limited extent presumably because the disulfide bonds prevent complete loss of its tertiary structure. Because it possesses a significant num-ber of hydrophobic domains bovine serum albumin is quite surface active. It readily locates at an interface and unfolds to permit polar and apolar segments to protrude into the aqueous and lipid phases respectively while the bulk of the molecule occupies the interface. Adjacent molecules may associate via hydrophobic and electrostatic interactions to form a cohesive film. In model systems the interfacial concentration of protein and the thick-ness of the interfacial film are affected by the type and concen-tration of protein in solution (22, 24). Bovine serum albumin is more effective in stabilizing emulsions than β-casein or lysozyme. It forms a stronger film at lower protein levels, reflecting a greater number of intermolecular interactions.

In the case of food emulsions the rate of interfacial film formation and the type and properties of the film formed are greatly affected by the mechanics of emulsion formation; i.e., shearing, mixing, denaturation, oxidation, and this, perhaps, has more impact on emulsion formation than the properties of the pro-tein per se (4, 29). However, it is important to recognize that the physical and rheological properties of the film, once formed, is critical in governing emulsion stability. Globular proteins with sufficient flexible domains can rapidly form films and de-press interfacial tension. Furthermore, they retain residual tertiary structure which facilitates extensive molecular inter-actions (electrostatic, hydrophobic, disulfide, Van der Waals forces). These contribute to the rheological properties (shear resistance, viscoelasticity, dilatational modulus, incompressi-bility) of the film which enhance film stability. However, their individual contributions to emulsion stability remains unclear.

Generally the stability of protein based emulsions are influenced by the properties of the interfacial film materials and the viscosity of the continuous phase (Table III).

Table III. Factors affecting formation and stability of protein-based emulsions.

PROTEIN
   Intrinsic physical properties of the protein: size, shape,
   compactness, surface charge, hydrophobicity, solubility,
   flexibility, disulfide/thiol groups
PROPERTIES OF ADSORBED PROTEIN FILM
   Thickness; rheological properties (viscosity, cohesiveness,
   elasticity); net charge and distribution; degree of hydra-
   tion (hydrophilic, polar, charged groups)
ENVIRONMENTAL FACTORS
   pH, ions, temperature
PROCESSING PARAMETERS
   Shear forces, temperature, oil composition, droplet size,
   viscosity
CONTINUOUS PHASE
   Viscosity

In a packed emulsion the polyhedral oil droplets are separated by an aqueous lamellar layer. The viscosity of this layer
determines the rate at which adjacent globules can approach each
other. Electrostatic and steric factors retard the approach and
coalescence of fat droplets. The integrity of the protein film
and its stability to shock (thermal, agitation, aging) is governed by its physical and rheological properties (thickness,
viscosity, flexibility and cohesiveness). Thinning of the film
results in weak segments that may upon shock, rupture and coa-
lesce. Hence, the restorative nature of the protein film its
elasticity, dilatational modulus, shear viscosity determines its
intrinsic stability (24-30).
   Coalescence of oil droplets resulting from the breakdown of
the aqueous lamella is a major factor in emulsion instability.
Graham and Philips (24) concluded that the disjoining pressure
stabilizing the lamella is mostly affected by electrostatic and
steric factors contributed by the proteins in the film. Thus,
while capillary suction, Van der Waals forces, and gravity tend
to cause thinning of the aqueous lamella these are counteracted
by electrostatic and steric factors contributed by the surface
layers of the adjacent protein films. Thus, the surface charge,
the size, nature of the degree of hydration of exposed groups all
impair the approach and coalescence of adjacent films. For the
formation of a stable emulsion the interfacial protein film
should be as thick as possible with good cohesive and rheological
properties, be well hydrated, contain exposed polar and charged
groups, and possess a net negative charge (24, 26).

Current Research: Experimental Procedures

   Much current research is concerned with the determination of

the emulsifying properties of food proteins. A wide diversity of
model systems and different conditions have been used to determine emulsification properties of numerous proteins, but useful
comparisons of methods and results are difficult ($\underline{1}$, $\underline{4}$, $\underline{6}$, $\underline{31}$).
Emulsifying activity reflects the ability of the protein to aid
emulsion formation and to stabilize the newly created emulsion.
Emulsifying activity can be easily measured by making an emulsion
and determining the particle size distribution of the dispersed
phase by microscopy or spectroturbidity. In each procedure the
average diameter of the dispersed phase is determined, and from
these data the interfacial area can be calculated. The spectroturbidity method is simple, rapid and theoretically sound and
provides information about the average diameter and particle size
distribution. The method is applicable to emulsions with average
particle size; i.e. diameters between 0.2-8 μ ($\underline{26}$, $\underline{32}$, $\underline{33}$). The
optical density of diluted emulsions is directly related to the
interfacial area; i.e., the surface area of all the droplets, for
coarse emulsions (See Table IV) (34).

Several types of blenders and homogenizers and many sizes
and shapes of containers have been used in emulsion preparation
($\underline{4}$, $\underline{29}$). These instruments vary in their ability to form an
emulsion; i.e., the particle size distribution of the oil drop-
lets vary and frequently ingredient factors affecting emulsifying
activity are overridden by the characteristics of the equipment
used. The Janke-Kunkel homogenizer has consistently proved to be
the best for producing model emulsions in our laboratory ($\underline{29}$).
Using model systems and the procedures described by Pearce and
Kinsella ($\underline{34}$) and Waniska, et al. ($\underline{29}$) some of the structural
factors and interactions affecting emulsion formation and sta-
bility are being studied.

Electrostatic Effects. The importance of electrostatic interactions on the formation and stabilization of protein-based
emulsions is reflected by the effects of charge alteration via
protein modification, pH variation, and ion concentration on
emulsion formation and stability ($\underline{24}$, $\underline{27}$, $\underline{28}$, $\underline{29}$, $\underline{31}$, $\underline{52}$).

Table IV. Emulsifying activity of succinylated yeast proteins.

Modification	Emulsifying Activity ($M^2.g^{-1}$)	
	pH 6.5	pH 8
Yeast protein (unmodified)	8	59
Yeast protein (24% succinylated)	110	204
Yeast protein (62% succinylated)	262	332
Yeast protein (88% succinylated)	322	341

From Pearce and Kinsella ($\underline{34}$).

Succinylation of relatively insoluble proteins markedly improves the emulsifying capacity of plant proteins (35). Yeast protein in which an increasing number of lysine residues were succinylated showed a progressive increase in emulsion activity (Table IV). This is attributed to enhanced solubility and expanded molecular structure which enhanced emulsion formation. However, electrostatic repulsion between the negatively charged, emulsified droplets enhanced stability.

Waniska, et al. (29) studied the effects of pH and structural modification on the emulsifying properties of proteins including bovine serum albumin which possesses excellent emulsifying activity (EA). The EA of bovine serum albumin (BSA) was two to threefold better than other food proteins (Table V) reflecting the favorable molecular characteristics of BSA; i.e., molecular size, ability to unfold to a limited extent and allow polypeptides to reorient at the interface, surface hydrophobicity, disulfide linkages which maintained sufficient tertiary structure and a balance of charged groups (24, 25).

Table V. The relationship between the pH and relative emsusifying capacities of various proteins.

pH of solution	Absorbance (550 nm) of diluted emulsion							
	BSA	R-BSA	S-BSA	Oval-bumin	β-casein	β-Lg	soy isolate	soy 11S
3	0.20	0.15	0.01	0.18	–	0.26	0.02	0.20
4	0.60	0.53	0.03	0.25	0.01	0.25	0.01	0.38
5	0.70	0.56	0.60	0.34	0.10	0.21	–	0.42
6	0.73	0.51	1.20	0.31	0.30	0.21	0.07	0.47
7	0.77	0.48	1.40	0.32	0.30	0.22	0.08	0.45
8	0.80	0.47	1.23	0.33	0.35	0.26	0.08	0.43
9	0.82	0.42	1.24	–	–	–	0.10	0.24
10	0.68	0.30	1.20	–	–	–	–	0.22

The emulsifying capacity was determined by turbidimetry (29, 34); BSA - bovine serum albumin; R-BSA - all disulfides reduced; S-BSA - succinylated BSA; β-Lg - β-lactoglobulin and 11S = soy glycinin.

Between pH 3 and 4 there was a significant increase in emulsifying activity of BSA. At pH 4 BSA undergoes an acid induced molecular expansion, presumably caused by electrostatic repulsion, which results in disruption of some hydrophobic interactions (36). This facilitated emulsion formation. The EA of BSA progressively increased between pH 4 and 9 indicating that as net charge increased the ability to form a film was enhanced. At pH 9.0 BSA undergoes an alkaline expansion (36) which may involve some disulfide cleavage and disruption of hydrophobic attractions (37). The EA of BSA abruptly decreased at pH 9.0 reflecting the altered tertiary structure as disulfide bonds were broken. These data reflect the importance of a certain degree of structural stability (tertiary structure) for optimum emulsion formation. Above pH 9, EA decreased; this may reflect the alkali induced hydrolysis of some disulfide bonds and possibly some crosslinking following β-elimination reactions.

The importance of tertiary and secondary structure is revealed by the almost complete elimination of EA following treatment of BSA with urea (8M). This observation is consistent with the suggestion that while denatured BSA may occupy the interface the lack of ordered structure prevents the formation of a cohesive continuous film. This was corroborated by the EA of BSA in which all the 17 disulfide bonds were reduced by dithiothreitol (0.01M). There was a significant reduction in EA, especially above pH 5.0. Thus, the tertiary conformation stabilized by disulfide bonds are apparently important in the formation of a stable cohesive membrane around the oil droplets. The reduced BSA was more sensitive to pH indicating greater sensitivity to electrostatic repulsions.

Succinylation, which significantly increased the net negative charge and causes molecular expansion markedly improved the EA of BSA between pH 5 and 10. The maximum was observed at pH 6 and subsequently decreased, possibly because the net negativity became excessive and impaired formation of a cohesive film. Presumably succinylation facilitated molecular unfolding at the interface, enhanced electrostatic repulsion between emulsified droplets and the charged groups, being highly hydrated, may have increased the viscosity of the lamellar phase thereby retarding coalescence.

Methylation of the ε-amino groups of lysine which unlike succinylation did not introduce a negative charge, but reduced the net positive charge below pH 8.5, did not significantly alter EA of BSA. These observations indicate that individual effects; i.e., net charge, are modified by molecular structure and flexibility, and optimum functionality is obtained only when these are at an optimum balance.

Ions at relatively low concentrations (0.1 - 0.3 M) influence electrostatic effects in emulsions. Thus, the oil/water interface possesses a concentration of charges which tend to repel approaching charged proteins. This results in a reduced rate of protein adsorption, and unfolding, thus inhibiting the formation of an interfacial film (25). At a particular pH, the net charge on the protein may result in excessive intermolecular repulsion and impair the intermolecular associations necessary for the formation of a cohesive film. Salts at low concentrations function by counteracting such electrostatic effects via counterion binding, thereby permitting adsorption and film formation. Furthermore, this weakens coulombic interactions, especially in aqueous solutions which in turn permits more extensive hydrophobic interactions between proteins at the interface. Waniska et al. (29) observed that both chloride and sulfate anions at concentrations of zero to 0.1 M markedly enhanced the emulsifying

activity of BSA. Chloride enhanced EA slightly up to 0.6 M, but
the 'salting in' effect of sodium chloride reduced the rate of absorption at the interface at salt concentrations above 0.6 M.
Sulfate, a water structure promoting salt ('salting out') favored
the transfer of BSA to the interface at maximum rates at concentrations from 0.1 to 1M. Thus, ions in addition to influencing
coulombic interactions, may also affect conformational stability
of proteins, especially at higher concentrations via the chaotropic effects (10).
     Proteins with different molecular sizes and structures
showed markedly different EA under identical conditions. Soy 11S
protein and arachin are oligomeric proteins, around 330,000 daltons, with over 40 disulfide crosslinks and very little (~12%)
helical structure. These are structurally very stable proteins
yet 11S had a much greater EA compared to arachin. This observation is being further explored. β-Casein is a highly amphipathic molecule that is surface active and rapidly forms films
:nd facilitates emulsification, but because films of β-casein are
weak, emulsion stability is poor. Ovalbumin is reasonably hydrophobic, but tends to denature easily and coagulate resulting in
weak interfacial films.

     Hydrophilic Effects. The hydrophilic character of glycopro-
teins may improve their surface activity. Conceivably the carbohydrate moieties because of their hydration provide steric stabilization in emulsions. To study the effect of hydrophilic
groups we covalently acylated different carbohydrate groups to
β-lactoglobulin which contains 16 free amino groups and 25 carboxyl groups available for modification. β-Lactoglobulin has a
low intrinsic viscosity (0.034 dl/g) reflecting its globular nature (38). It has an average hydrophobicity of 1060 compared to
1000 and 980 for BSA and ovalbumin, respectively (39). β-Lactoglobulin is reasonably surface active; i.e. surface tension = 60,
interfacial tension 11.0 and emulsifying activity index 151;
bovine serum albumin has corresponding values of 58, 10.3 and
166, respectively (16).
     Activated cyclic carbonate derivatives of maltose and β-
cyclodextrin were linked to β-lactoglobulin via the free amino
groups on the protein (R-NH-CO-maltose). Glucosamine was attached to β-lactoglobulin following carbodiimide activation of
the free carboxyl groups (R-CO-NH-glucose). The number of groups
attached to the β-lactoglobulin was varied by manipulating reaction conditions (27).
     From studies of the molecular properties (viscosity, circu-
lar dichroism, fluorescence and UV difference spectroscopy)
glycosylation of β-lactoglobulin with maltose caused greater hydration, molecular expansion and elongation as the number of carbohydrate residues added was increased. Fluorescence and UV
spectra together with rates of proteolysis indicated a more open
tertiary structure and some disordering of the molecule; i.e.,

reduced amount of α helix. As more maltose groups were added the
surface hydrophobicity was decreased and net negative charge was
increased. Modification with glucosamine caused little conformational changes except at high levels of modification; i.e.,
when 22 residues were glycosylated some loss of secondary structure became evident (27).

Some surface active properties of the glycosylated β-lactoglobulin were studied to determine if glycosylation affected
these. Because glycosylation, especially with maltose, caused a
more expanded structure the gain in free energy during protein
adsorption should be lower and the rate of adsorption at an oil
water interface should be less. However, glycosylation may also
facilitate unfolding and reorientation at the interface and alter
surface area occupied by each molecule. Initially the properties
of films of β-lactoglobulin and modified β-lactoglobulin were
studied.

The soluble globular β-lactoglobulin (Lg) adsorbed rapidly
and partly unfolded to form a fairly tightly packed condensed
film. Maximum surface pressure of β-Lg was observed at pH 5.0
slightly below the isoelectric point (5.25) of β-Lg. Thus, when
net charge was close to zero more protein was adsorbed because
β-lactoglobulin had a more compact structure. As pH was increased the electronegative charge caused repulsion and probably
fewer protein segments occupied the interface. Systematic studies of films revealed that glycosylation of β-lactolobulin with
maltose, glucosamine or cyclodextrin reduced the rates of adsorption and rearrangement at the interface. This was attributed
to lower free energy gain, the increased viscosity, the steric
hindrance by carbohydrates, the stabilizing effect of the carbohydrate moiety against unfolding, alteration of net charge,
and reduced surface hydrophobicity of the modified β-lactoglobulin (27). Though glycosylation resulted in a reduction of
electrostatic repulsion the carbohydrates apparently restricted
close molecular packing in the interfacial film, inhibited unfolding of the hydrophobic regions to the apolar phase and
significantly increased the amount of work required for each
molecule to penetrate the interface.

The general conclusion of these studies were that while
glycosylation enhanced the hydrophilic character of β-Lg it concurrently decreased the number of ionizable groups and perhaps
more importantly, it restricted the expression of the hydrophobic properties at the interface thereby reducing the surface
activity of β-lactoglobulin as indicated by properties of interfacial films.

Because the hydrophilic glycosyl-substituents would be expected to enhance emulsion formation and stability via hydration
and steric effects, the emulsifying properties of glycosylated
β-lactoglobulin were examined (Table VI).

Table VI. Effects of glycosylation of β-lactoglobulin on emulsifying properties at different pH values.

Protein Preparation	Emulsion Activity (OD 1:10³ Dilutions)			Stability Oil Separation (%)		
	pH			pH		
	5	6.5	7.5	5	6.5	7.5
βL	0.20	0.23	0.26	9.0	13.0	11.6
B	0.24	0.30	0.40	7.5	8.2	6.1
C	0.23	0.29	0.31	6.8	11.5	13.0
D	0.21	0.26	0.26	8.0	11.4	10.8
E	0.25	0.21	0.34	9.5	12.8	8.5
F	0.26	0.32	0.29	12.3	7.9	9.8
G	0.12	0.10	0.23	9.9	10.0	9.8

Emulsions were prepared with 0.5% protein and measured as reported by Waniska et al. (29).
Legend: β-Lactoglobulin (βLg) with amino or carboxyl groups modified with B 7 maltosyl; C 11 maltosyl; D 3 β-cyclodextrin; E 6 glucosamine; F 16 glucosamine and G 12 glucosamineoctaosyl residues, respectively.
Isoelectric points βLg 5.2; B = 4.8; C = 4.3; D = 4.3; G = 5.5.

Glycosylation of β-lactoglobulin improved emulsion formation and stability. This was in contrast to its general effects on the properties of interfacial films. However, the energy input in emulsion preparation may have overcome some of the energy barriers encountered in the film studies where equilibrium conditions prevailed. Overall the data are consistent with the suggestion that the increased hydrophilic nature of the protein enhances emulsion properties by stabilizing the aqueous interface through enhanced hydration and perhaps steric hindrance preventing coalescence of the oil droplets. However, derivatization altered electrostatic and hydrophobic interactions within and between the polypeptides and these may have impeded attainment of maximum emulsifying properties.

Emulsifying activity increased with increased pH indicating that as the net electronegativity increased electrostatic repulsion facilitated formation and stabilization of a greater number of oil droplets. The number and size of hydrophilic groups introduced affected both EA and emulsion stability. The attachment of seven maltose groups into β-lactoglobulin was superior to 11 groups in improving emulsifying properties. Conceivably the lower number of maltose groups provided less steric hindrance to ionic and hydrophobic interactions between the polypeptides in the interface. The β-cyclodextrin had little impact on emulsifying properties. The emulsifying properties of β-lactoglobulin, modified with glucosamine residues, were generally improved.

Near the isoelectric range of the derivatized protein, emulsion
formation was good. Proteins have a low net charge in this pH
range and maximum hydrophobic interactions are possible (40).
Therefore, it is possible that small glucose derivatives (even
when 16 carboxyl groups were derivatized) did not completely
impede hydrophobic interactions at this pH. Derivatization with
a large amino sugar, i.e., glycosamineoctaose, impaired emulsion
formation, but stability of the emulsion was good.

These data indicate the importance of net charge, hydrophilic interactions and steric effects in emulsion formation and
emulsion stabilization and confirm the positive effects of hydrophilic groups. However, large, bulky hydrophilic residues may
interfere with hydrophobic interactions causing excessive steric
hindrance and actually reduce emulsifying properties.

Hydrophobic Effects. The ionic, hydrophilic and hydrophobic
characteristics of protein affect their surface activity and
functional properties. In an emulsion one 'side' of the protein
film, i.e., that exposed to the oil phase, should be preponder-
antly hydrophobic. The hydrophobicity influences adsorption and
orientation of protein at the interface (22). The side of the
film that is exposed to the continuous aqueous phase has most of
the polar and charged groups exposed. Hydrophobic interactions
stabilize the conformation of proteins in the native state in
solution. At an interface the normal thermodynamic equilibrium
is disturbed. The apolar segments of the polypeptides tend to
unfold, reorient and occupy the less polar oil phase. This phenomenon depends upon the amphipathic attributes of the protein,
a property required for surface activity.

The hydrophobicity of proteins is related to their contents
of apolar amino acids. The relative hydrophobicities of these;
i.e., trp, ileu, tyr, phe, pro, leu, val, lys, meth, cys/2, ala,
arg are 3.00, 2.95, 2.85, 2.65, 2.60, 2.40, 1.7, 1.5, 1.3, 1.0,
0.75, 0.75 Kcal/residue respectively (13). The average hydrophobicity (HQ) representing the total hydrophobicity divided by the
total number of residues was calculated for several proteins by
Bigelow (39). Proteins have HQ values ranging from 1000-12000
cal/residue (Table VII). While general relationships are apparent, it should be remembered that the disposition of the apolar
residues (sequence and whether located internally, on the surface,
or on flexible segments) affect physical properties more than the
average hydrophobic amino acid content.

Leslie (41) related the strong NMR signals of the apolar
residues in soy proteins to the emulsifying properties and Kato
and Nakai (16) reported good correlations between molecular
hydrophobicity and surfactant properties. They successfully
quantified effective hydrophobicity; i.e., true surface hydrophobicity, using cis-parinaric acid, an excellent fluorescent
probe for measuring hydrophobicity of food proteins. Proteins
were reacted with cis-parinaric acid. The conjugates were

excited at 325 nm and fluorescence intensity measured at 420 nm.
The initial slope (S_O) obtained by plotting fluorescence against
protein concentration indicates surface hydrophobicity. The
energy transfer efficiency of conjugates at zero protein concentration (T_O) also reflect hydrophobicity. These two values were
highly correlated and corresponded to data obtained using hydrophobic chromatography (42).

Table VII. The average hydrophobicity of some food proteins.

	MW 10^3	HQ	NPS	P	Charge
Tropomyosin	–	870	0.20	1.94	0.47
Ovomucoid	27	920	0.25	1.50	0.30
Collagen (soluble calf)	80	960	0.19	0.89	0.16
Lysozyme	15	970	0.26	1.18	0.16
Myosin	500	1020	0.28	1.43	0.35
Actin	57	1050	0.31	1.15	0.27
Gliadin	120	1080	–	–	–
Conalbumin	87	1080	0.31	1.30	0.32
Myoglobin	17	1090	0.32	1.12	0.34
Ovalbumin	46	1110	0.34	0.92	0.24
BSA	65	1120	0.32	1.22	0.33
α–Lactalbumin	16	1150	0.34	1.11	0.28
α–Casein	–	1200	0.38	1.27	0.24
β–Lactoglobulin	17	1230	0.37	0.96	0.28
Zein	50	1310	–	–	–

MW = mol. weight; HQ = total hydrophobicity divided by total
number of residues; NPS = sum of Trp, Ile, Tyr, Phe, Pro, Leu,
and Val residues as a fraction of total residues; P = polarity
ratio roughly all polar side chains/all nonpolar side chains and
charge is sum of Asp, Glu, His, Lys, Arg as fraction of total
amino acids. From Bigelow (39).

     Bovine serum albumin, κ–casein and β–lactoglobulin showed
high S_O values (Table VIII) reflecting the binding of cis-parinaric acid to their hydrophobic regions. The surface hydrophobicity showed strong correlations with surface tension, interfacial tension and emulsifying activity of these proteins.
     This was further demonstrated by studies indicating that the
progressive thermal denaturation of lysozyme increased surface
hydrophobicity and the EA was concurrently enhanced (3). These
data indicate that hydrophobic interactions are important, particularly for the intermolecular interactions and associations
which enhance the cohesiveness within the protein films surrounding fat droplets in emulsions.
     The above sections demonstrate some relationships between
the molecular properties of food proteins and their emulsion
properties. However, much research is needed to determine the
optimum quantitative relationships between individual molecular

properties (charge, hydrophobicity and hydrophilicity), environmental parameters, and particular types of food emulsions.

Table VIII. Relative Hydrophobicity (fluorescence intensity, S_0)
             Surface Tension, Interfacial Tension and Emulsifying
             Activity Index Values of Various Proteins.

	S_0	Surface Tension (Dynes/cm)	Interfacial Tension (Dynes/cm)	Emulsifying Activity Index $(m^2 g^{-1})$
Bovine serum albumin	1400	57.9	10.3	166
k-Casein	1300	54.1	9.5	185
β-lactoglobulin	750	59.8	11.0	151
Trypsin	90	64.1	12.0	93
Ovalbumin	60	61.1	11.6	57
Conalbumin	70	63.7	12.1	105
Lysozyme	100	64.0	11.2	55

From Kato and Nakai (16)

Protein Ligand Interactions - Flavor Binding

      Proteins being surface active can bind lipids, fatty acids,
aldehydes, ketones, tannins, chlorogenic acid, etc., and adversely affect functional and nutritional properties (43,44). The
binding of low amounts of surface active compounds and lipids to
food proteins may alter their stability and thermal properties.
Thus, the binding of lauric acid to egg white proteins enhances
their thermal stability (45) and low molecular weight alcohols or
carbonyls at low concentrations stabilize bovine serum albumin
(46).
      The binding of flavors to food components, especially pro-
teins is of particular importance, and it is a problem that has
been dramatized as food technologists have tried to fabricate
analogs and new foods from novel proteins (47). More fundamental
information concerning the nature of flavor-protein interaction is
needed to help solve problems of off-flavor binding, to aid the
development of more successful flavoring procedures and understand the mechanism and thermodynamics of flavor release when
food is chewed.
      Because the perceived flavor is important in determining
food acceptability, the phenomenon of flavor binding and release
is extremely significant. The off-flavors that become associated
with proteins limits their application. This is exemplified by
many soy protein preparations which, though possessing good

functional properties, have limited use because of the impact of bound off-flavors (48). To develop effective methods for the removal of off-flavors we have been studying flavor-protein interactions in simple model systems composed of soy proteins and carbonyl compounds and determining binding using equilibrium dialysis (49).

Binding isotherms indicated a progressive increase in binding of carbonyls with their concentration. There were four binding sites per 100,000 molecular weight soy protein for 2-heptanone, 2-octanone and 2-nonanone, respectively. The binding affinities increased with chain length of the carbonyls indicating that hydrophobic interactions are involved (Table IX). There was a threefold increase in Keq for each increment of methylene group ($-CH_2$) in the chain length of the carbonyl which corresponded to a change in free energy of -600 calories/CH_2 group. Similar results were obtained with bovine serum albumin except that in the case of soy protein and BSA the Keq for nonanone was 930 and 1800 M^{-1} respectively. This reflects the difference in structure of these proteins and the greater surface hydrophobicity in BSA which possesses six binding sites for these carbonyls (50).

Table IX. Thermodynamic constants for interactions of carbonyls with soy proteins in model systems.

Ligand	Protein	Temp °C	n	K_{eq} (M^{-1})	ΔG (Kcal/mole)
2-Heptanone	native	25	4	110	-2.781
2-Octanone	native	25	4	310	-3.395
2-Nonanone	native	25	4	930	-4.045
5-Nonanone	native	25	4	541	-3.725
1-Nonanal	native	25	4	1094	-4.141
2-Nonanone	(succinylated)	25	2	850	-3.992
2-Nonanone	native	25	4	930	-4.045
2-Nonanone	native	5	2	2000	-4.221
2-Nonanone	(heated 90°C)	25	4	1240	-4.215
2-Nonanone	7S	25	4	930	4.040
2-Nonanone	11S (0.03M Salt)	25	-	-	-
2-Nonanone	11S (0.5M Salt)	25	8	290	-

These data strongly support hydrophobic binding and the ΔG value for the binding of each CH_2 group, i.e., -550 and -600 cal/-CH_2 group for BSA and soy protein closely approximate the -540 cal/mole-CH_2 observed for the transfer of a CH_2 group from water to an apolar solvent (51). Further evidence of hydrophobic interaction is obtained by comparing the binding of nonanol, 2-nonanone and 5-nonanone to soy protein (Table IX). The binding constant K decreased as the keto group was moved toward the middle of the chain; i.e., Keq was 1094, 930 and 541 M^{-1} for these three carbonyls reflecting the steric hindrance caused by the keto group, especially when located near the center of the hydrocarbon chain. The binding affinity of succinylated soy protein for 2-nonanone was slightly reduced compared to the native protein, and the number of binding sites was reduced to two reflecting major conformational changes in the soy protein in the neighborhood of the binding sites. Though the two binding sites have almost the same intrinsic binding constant as the native soy protein, the binding capacity (moles ligand bound per mole protein) of the succinylated soy was decreased. Therefore, succinylation decreased flavor binding and improved the flavor of succinylated soy protein as noted previously (52).

At low temperature (5°C) the number of binding sites was reduced to two, but these had significantly greater binding constants reflecting changes in the conformation of soy protein at low temperatures. The binding constant was significantly increased (30%) following heating of soy (90°C, 1 hr), but there were still four binding sites. The enhanced binding upon heating corroborates   the data of Arai et al. (53) and indicates that the thermal processing of soy protein may exacerbate the problem of off-flavor binding by increasing the binding affinity of the soy protein for carbonyls.

The effects of heating of soy protein on ligand binding seemed inconsistent with the knowledge that soy 11S dissociates, and therefore heating should increase the number of binding sites. Therefore, we studied the binding of carbonyls to 11S and 7S, the major protein components of soybean. Significantly, there was negligible binding of nonanone to the 11S fraction, whereas the binding to 7S was comparable to that for whole soy protein. This selective interaction obviously reflects the differences in the molecular structure, conformation and surface properties of these two proteins. The 7S protein (17,500 daltons) is comprised of three subunits of different sizes and amino acid composition (54). The 11S protein (320,000 daltons) contains six acidic and six basic subunits (55). The spatial arrangement of the subunits in these two proteins may be such that the 7S protein has hydrophobic regions which are accessible for ligand binding, whereas in the case of 11S protein such hydrophobic regions may be buried inside the protein or be at the points of contact or association between subunits, and hence may not be accessible to the ligand.

Because the weak interaction between 11S and 2-nonanone may reflect its unique quaternary structure, then changes in this structure may affect the binding affinity for 2-nonanone. The oligomeric structure of soy 11S is influenced by ionic strength; i.e., at 0.5 M ionic strength the protein has a sedimentation coefficient of 11S, below 0.1 M ionic strength it dissociates into 'half 11S' units (56). At the ionic strength used in our initial study the 11S was in half 11S form and perhaps had a

relatively high surface charge which impaired binding. At higher
ionic (0.5 M) strength when the electrostatic repulsion is weakened, the intact 11S is formed, and this is accompanied by a
marked increase in nonanone binding. This observation is significant because if the 11S exists in an environment of low
ionic strength in the soybean seed or if low ionic strength pre-
vails during milling of the seed, the 11S fraction which is ∿35%
of soy protein, should possess less off-flavors.

   Because hydrophobic interactions are the predominant forces
involved in binding of apolar organic ligands to protein it may
be feasible to remove off-flavors by chaotropic agents which
weaken hydrophobic interactions (57). Thus, we determined the
effects of increasing urea concentrations on flavor binding
characteristics of soy protein. There was a progressive decrease
in binding affinity for nonanone as urea was increased from 0 to
4.5 M. It is known that the denaturing action of urea involves a
hydrophobic mechanism that favors exposure of nonpolar residues
in the protein to solvent environment (58). As these conformational changes occurred in the soy proteins the binding affinity
concomitantly decreased. At low concentrations of urea the binding affinity of soy for nonanone was decreased significantly;
i.e., by ∿50 and 70% at 1.5 and 3.0 M urea, respectively. This
observation may have practical implications. Thus, reversible
dissociation of soy protein subunits; e.g., using urea or other
dissociating agent may facilitate the removal of off-flavors
such as carbonyls, because binding is an equilibrium process.
Thus, ultrafiltration of dilute urea solutions of undenatured
soy proteins may be useful for the preparation of bland soy protein.

Conclusion

   The functional properties of proteins reflect the inherent
molecular properties of the proteins (size, shape, conformation,
molecular flexibility, susceptability to denaturation), the manner in which they interact with other food constituents and how
they interact with the prevailing environmental conditions. Thus
numerous intrinsic and extrinsic factors acting in concert determine the final effect. This situation makes it extremely chal-
lenging to the food scientist to develop standard methods based
on the physicochemical behavior of protein in a simple system.
Because of the complexity of the systems involved, it is obvious
that the conventional empirical methods for comparing the functional properties of proteins will continue to be used. Recognizing this, food scientists should strive to standardize the
methods as much as possible so that published data can be validly
used to compare the relative functionality of different protein
preparations.

   There is a continuing need for dedicated basic research
(Table X) directed toward the elucidation of the physicochemical

bases of specific functional properties; e.g., gelation, surface
activity, film formation. While it may not be possible to define each functional property in mathematical terms, the knowledge gained from fundamental research will provide a clearer
understanding of the physical and chemical relationships involved.
This information will enable the food scientist to select the
most appropriate protein for a specific functional property or
application and also enable the scientist to modify the proteins
to optimize desirable physical properties.

Table X.   Some research needs in the area of functional
           properties.

Standardizing the definition of functional properties
Development of standardized methods based on physical properties
Relating functional properties to structural features and
    secondary interactions
Manipulation of protein structure by chemical or physical
    alterations
Testing in food systems

Acknowledgments

    The author gratefully acknowledges the support and dis-
cussions of Drs. J. Shetty, S. Damodaran, and R. Waniska, some
of whose data are presented in this paper. Financial support
from the National Science Foundation Grant #CPE 80-18394 is
gratefully acknowledged.

Literature Cited

1. Kinsella, J. E.   CRC Crit. Rev. Food Sci. Nutr., 1976, 7, 219.
2. Cherry, J. P., Ed. "Protein Functionality in Foods", American
    Chemical Society Symp. Series #147: Washington, DC, 1981.
3. Hermansson, A. M.   In "Problems in Human Nutrition"; Porter,
    J. & Rolls, B., Eds., Acad. Press, New York, 1973; p. 407.
4. Tornberg, E.; Hermansson, A. M. J. Food Sci., 1977, 42,
    468; 1978, 43, 1553.
5. Pour-El, A., Ed. "Functionality and Protein Structure";
    American Chemical Society Symp. Series #92: Washington, DC,
    1980.
6. Pour-El, A. In "Protein Functionality in Foods"; Cherry,
    J. P., Ed.; American Chemical Society Symp. Series #147:
    Washington, DC, 1981; p. 1.
7. Shen, J. In "Protein Functionality in Foods"; Cherry, J. P.,
    Ed.; American Chemical Society Symp. Series #147: Washington,
    DC, 1981.
8. Smith, A.; Circle, A. J. "Soybeans: Chemistry and
    Technology"; Avi Publ. Co.: Westport, CT, 1978.

9. Galyean, R.; Cotterill, O. J. Food Sci., 1979, 44, 1365.
10. Damodaran, S.; Kinsella, J. E. This volume Chap. 13.
11. Kauzman, P. Advances Protein Chem., 1959, 14, 1.
12. Anfinsen, C.; Scheraga, H. Adv. Protein Chem., 1975, 29, 205.
13. Tanford, C. "The Hydrophobic Effect"; Wiley, J., Ed., New York, 1973.
14. Van Holde, N. In "Food Proteins"; Whitaker, J. & Tannenbaum, S., Eds.; Avi Publ. Co.: Westport, CT, 1971; p. 1.
15. Shimada, K.; Matushita, S. J. Agr. Food Chem., 1980, 28, 413.
16. Kato, A.; Nakai, S. Can. Inst. Food Sci. & Technol. J., press, 1981.
17. Bull, H.; Breese, K. Arch. Biochem. Biophys., 1968, 128, 488.
18. Feeney, R.; Whitaker, J., Eds.; "Food Proteins: Improvement Through Chemical and Enzymatic Modification"; Advances in Chem. Series 160, American Chemical Society: Washington, DC, 1977.
19. Williams, R. J. Biol. Rev., 1979, 54, 389.
20. Ewart, J. J. Sci. Food Agr., 1972, 23, 687.
21. Morr, C. J. Dairy Sci., 1975, 58, 977.
22. Phillips, M. C. Food Technol., 1981, 35, 50.
23. Brandts, J. In "Structure and Stability of Macromolecules"; Timasheff, S. & Fasman, G., Eds., Marcel-Dekker: New York, 1969, Chap. 3.
24. Graham, D. E.; Phillips, M. C. In "Theory and Practice of Emulsion Technology"; Smith, A. L., Ed., Acad. Press: New York, 1976; p. 75.
25. MacRitchie, F. Adv. In Protein Chem., 1978, 32, 283.
26. Halling, P. J. Crit. Rev. Food Sci. and Nutr., 1981, 12, 155.
27. Waniska, R.; Kinsella, J. E. Unpubl. research, 1982.
28. Tornberg, E. In "Functionality and Protein Structure"; Pour-El, A., Ed.; American Chemical Society Symp. Series #92: Washington, DC, 1976, p. 105.
29. Waniska, R.; Shetty, J.; Kinsella, J. E. J. Agr. Food Chem., 1981, 29, 826.
30. Kinsella, J. E. Food Chem. (Lond.) 1982 (press).
31. McWatters, K; Cherry, J. P. In "Protein Functionality in Foods"; Cherry, J. P., Ed.; American Chemical Society Symp. Series #147: Washington, DC, 1981, p. 217.
32. Mulder, H.; Walstra, P. "The Milk Fat Globule: Emulsion Science as Applied to Milk Products"; Pub. Commonwealth Agr. Bureau: Bucks, England, 1974.
33. Walstra, P.; Ourtwijn, H.; deGroaf, J. Neth. Milk Dairy J., 1969, 23, 12.
34. Pearce, K.; Kinsella, J. E. J. Agr. Food Chem., 1976, 26, 716.

35. Kinsella, J. E.; Shetty, J. In "Functionality and Protein
     Structure"; Pour-El., Ed.; American Chemical Society Symp.
     Series #92: Washington, DC, 1979, p. 37.
36. Leonard, W.; Foster, J. J. Biol. Chem., 1961, 236, 2262.
37. Aoki, K.; Murata, M.; Hiramatus, K. Anal. Biochem., 1974,
     59, 16.
38. McKenzie, H. "Milk Proteins: Chemistry and Technology";
     Acad. Press: New York, 1971; p. 257.
39. Bigelow, C. J. Theor. Biol., 1967, 16, 187.
40. Kuntz, L.; Kauzmann, W. Adv. Protein Chem., 1974, 28, 239.
41. Leslie, R. B. J. Am. Oil Chem. Soc., 1976, 56, 290.
42. Kato, A.; Makai, S. Biochem. Biophys. Acta., 1980, 624, 13.
43. Kinsella, J. E. In "Criteria of Food Acceptance"; Solms, J.
     & Hall, R., Eds.; Forster Verlag A. G.: Zurich, Switzerland,
     1981, p. 140.
44. Blouin, F.; Zarins, Z.; Cherry, J. In "Protein Functional-
     ity in Foods"; Cherry, J. P., Ed.; American Chemical Society
     Symp. Series #147: Washington, DC, 1981, p. 21.
45. Hegg, P. O.; Lofquist, B. J. Food Sci., 1974, 39, 1231.
46. Damodaran, S.; Kinsella, J. E. J. Biol. Chem., 1980, 255,
     8503.
47. Kinsella, J. E. Crit. Rev. Food Sci. Nutr., 1978, 10, 147.
48. Kinsella, J. E. J. Am. Oil Chem. Soc., 1979, 56, 242.
49. Damodaran, S.; Kinsella, J. E. J. Agr. Food Chem., 1981,
     29, 1253.
50. Damodaran, S.; Kinsella, J. E. J. Agr. Food Chem., 1980,
     28, 567.
51. Abraham, M. J. Am. Chem. Soc., 1980, 102, 5910.
52. Franzen, K.; Kinsella, J. E. J. Agr. Food Chem., 1976, 24
     914.
53. Arai, S.; Noguchi, M.; Kato, H.; Fujimaki, M. Agr. Biol.
     Chem., 1970, 34, 1420
54. Thanh, V.; Shibasaki, K. J. Agr. Food Chem., 1978, 26, 692.
55. Badley, R.; Atkinson, D.; Oldani, D.; Green, J.; Stubbs, J.
     Biochem. Biophys. Acta., 1975, 412, 216.
56. Koshiyama, L. Int. J. Peptide & Protein Res., 1972, 4, 167.
57. Damodaran, S.; Kinsella, J. E. J. Biol. Chem., 1981, 256,
     3394.
58. Wetlaufer, D.; Malik, S.; Stoller, L.; Coffin, R. J. Am.
     Chem. Soc., 1964, 86, 508.

RECEIVED June 1, 1982.

Effects of Ions on Protein Conformation and Functionality

SRINIVASAN DAMODARAN and JOHN E. KINSELLA

Cornell University, Institute of Food Science, Ithaca, NY 14853

The unique three-dimensional structure of a protein molecule is the resultant of various attractive and repulsive interactions of the protein chain within itself and with the surrounding solvent. The various types of interactions and forces which contribute to the overall stability of the native structure of a protein may be broadly categorized as hydrogen bonding, electrostatic and hydrophobic interactions. For the protein to assume a particular ordered structure, the favorable free energy change from the above interactions in the ordered form should be more than the unfavorable free energy increase resulting from the configurational entropy of the chain. In other words, the favorable change in the free energy for the stability of the folded native protein may be expressed as Equation 1:

$$\Delta G_n = \Delta G_h + \Delta G_e + \Delta G_\phi - T\Delta S_{configuration}$$

where ΔG_h, ΔG_e and ΔG_ϕ are the free energy contribution from hydrogen bonding, electrostatic and hydrophobic interactions, respectively, and ΔS is the configurational entropy of the protein chain and T is the temperature.

For many proteins the net free energy favoring the formation and stability of the native structure is only about 10–20 Kcal/mole (1). Hence any perturbation leading to decrease in the free energy of stabilization even by few kilocalories would profoundly affect the structural stability of the native protein and hence its function. Such changes may be brought about by agents like heat, organic solvents and chaotropic salts. The major emphasis of this chapter will be to discuss the effect of ionic solutes on protein conformation and function, and to elucidate the mechanisms of these interactions. Before venturing into this it may be appropriate to understand the nature and magnitude of some of the major forces responsible for the stability of protein structure, viz. hydrogen bonds, electrostatic and hydrophobic interactions.

0097-6156/82/0206-0327$09.00/0

Major Forces in Protein Structure Stability

    Hydrogen bonds. The existence and importance of hydrogen
bonds in proteins is well documented. In fact, the α-helical and
the β-sheet structures in proteins are based on the formation of
hydrogen bonds between the peptide groups. However, there has
been considerable disagreement about the degree of stabilization
they provide to the protein structure. The reason for this is
the fact that in aqueous solutions, the solvent itself can compete with both the acceptor and donor groups in proteins for
hydrogen bond formation. In other words, if the formation of
hydrogen bonds in aqueous solution is expressed as Equation 2:

$$\text{protein-H}\cdots\text{H}_2\text{O} + \text{H}_2\text{O}\cdots\text{H-protein} \rightleftharpoons$$
$$\text{protein-protein} + \text{H}_2\text{O}\cdots\text{H}_2\text{O}$$

then, in order to form intrapeptide hydrogen bonds the change in
the free energy of the above reaction should favor the forward
reaction. With N-methylacetamide as a model compound, which is
comparable to groups in proteins, it has been shown that the
change in the free energy for the formation of hydrogen bonds
between N-methylacetamide molecules is +0.75 Kcal/mole (2). But
the data for helix-coil transition in synthetic polypeptides
shows that the free energy contributed by the hydrogen bonds for
the stability of proteins is only about -0.15 Kcal/mole residue
(3). One can, therefore, assume that the energy contributed by
hydrogen bonds to the overall stability of protein structure is
very small. It should be borne in mind that the hydrogen bond is
primarily ionic in character due to the dipole moment of the peptide group. Although the hydrogen bonds give stability to the α-
helical structure, the driving force for the formation of hydrogen bonds is the hydrophobic interaction between the sidechain
groups (4). Conversely, any perturbation which causes destabilization of these hydrophobic interactions in the protein would
also destabilize the hydrogen bonds in this vicinity.

    Electrostatic interactions. The presence of charged groups
like glutamic, aspartic, lysine and arginine residues in proteins
form the basis of electrostatic interactions. But little is known
about the thermodynamic contribution they provide for the overall
stability of the protein structure. Near the isoelectric pH the
electrostatic interaction between the oppositely charged groups
may be significant and may provide some stabilization energy to
the protein structure (5). Above or below the isoelectric pH
range the protein will have a net negative or positive charge and
hence repulsive interactions will exist within the protein. But
binding of counter ions may decrease these repulsive forces and
provide stability to the protein structure. Since most of the
charged groups are present on the surface of the protein, the
charge repulsion or attraction between these groups may be minimal

because of the high dielectric constant of the surrounding solvent. But the magnitude of such interactions may become significant if these charges are partially or completely buried inside
the protein where the dielectric constant is lower. In such
cases, each of the buried charges will destabilize the native
protein structure to the order of 20 Kcal/mole (6). In the presence of an oppositely charged residue in the interior of the protein, a salt bridge may be formed which will provide considerable
energy for stabilizing protein structure.

    Hydrophobic interactions. It is recognized that the major
force responsible for the stability of protein structure is the
hydrophobic interaction between the nonpolar sidechains in proteins. The driving force for such interactions arise not from
the inherent attraction of nonpolar sidechains for each other,
but in the energetically unfavorable effect they have on the
structure of the water molecules around them. The basis for this
fundamental concept is the thermodynamic data concerning the
transfer of various nonpolar compounds from apolar solvents to
water (7-10). When a hydrocarbon is transferred from a nonpolar
solvent to water, the changes in volume and enthalpy are negative
and there is a very large negative excess entropy change over the
entropy of ideal mixing. This leads to a large positive change
in free energy and hence, low solubility (11). These abnormal
changes are considered to be the result of an ordering of water
molecules around the hydrocarbon chain forming clathrate or cage-
like structures. The above mentioned non-ideal thermodynamic
behavior of hydrocarbon solutions is due to the formation of
these clathrate structures and to the orientational specificity
of the water molecules in these clathrate structures compared to
that in the free bulk water in the absence of nonpolar solutes
(12). The requirements on the orientational specificity of water
molecules for the formation of clathrate-like structures around
hydrocarbons is more specific (Figure 1; 12). Figure 1A represents the water pair geometry in ice or idealized bulk water.
This is the geometrically possible orientation of two hydrogen
bonded water molecules which permit the maximum entropy possible.
The structure shown in Figure 1B represents the water-water
orientation in clathrate hydrates. The difference between these
two orientations is that in the second configuration one of the
water molecules is rotated through a tetrahedral angle on the
hydrogen bond. This structure will have a higher energy state
because of the closeness of the hydrogen atoms and the lone pairs
of electrons which will enhance the repulsive interaction (12,
13). This would explain the positive standard free energy change
observed for the transfer of hydrocarbons into aqueous solution.
Furthermore, such ordering of the water molecules around the hydrocarbon would restrict their freedom of rotational and translational motions and thus decrease the entropy of the system (14).
Such a state would be highly unstable and would tend to go back

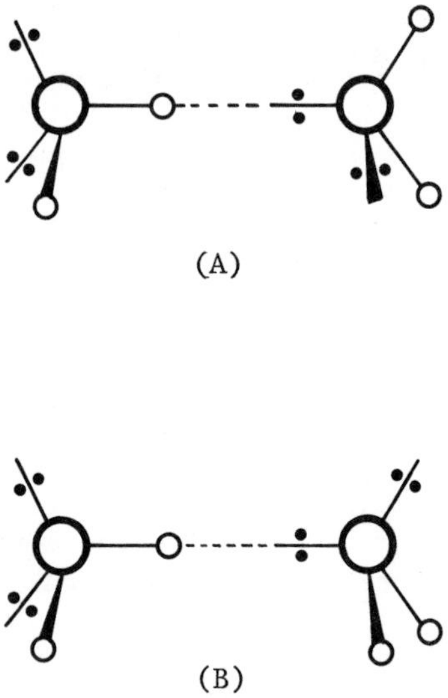

Figure 1.   Mutual orientation of two water molecules in ice (A) (and liquid water), and in clathrate hydrates (B). Dots indicate the direction of lone electron pair orbitals, small circles refer to hydrogen atoms and big circles refer to oxygen atoms. Configuration B is obtained from A by rotating one water molecule through a tetrahedral angle about the axis of the hydrogen bond. (Reproduced, with permission, from Ref. 12. Copyright 1975, Academic Press Inc. (London) Ltd.)

to the original state of higher entropy. To do so it is obliga-
tory for the system to expel the hydrocarbon from the solution.
This is partially accomplished by forcing two hydrophobic molecules together, which allows formation of a single sphere of hydration around them with release of some of the water molecules
to their entropically favorable state (Figure 2). Such grouping
of hydrophobic molecules in aqueous solution is known as hydrophobic interaction. In other words, it is the interaction between the hydrophobic molecules forced by the thermodynamically
unfavorable structural state of water in the presence of these
molecules. Since proteins contain such nonpolar residues, the
hydrophobic interaction between these residues would result in
the formation of hydrophobic regions which would impose a particular three dimensional conformation on the protein.

From the solubility measurements of amino acids in water and
in ethanol, Nozaki and Tanford (15) calculated the free energy of
transfer of amino acid residues from water to ethanol. Assuming
that the transfer from water to ethanol is comparable to transfer
of hydrophobic residues in protein from water environment to a
nonpolar region in the protein the data obtained by Nozaki and
Tanford (15) suggest that on an average the hydrophobic free
energy contribution for the protein folding and stability by each
hydrophobic residue is about -2 Kcal/mole residue. Since many
proteins contain about 40 to 60% content of hydrophobic amino
acid residues, it may be deduced that the major force responsible
for the stability of the protein conformation is the hydrophobic
interaction.

As a corollary, if the entropy driven hydrophobic free energy is responsible for the native structure of the protein, then
it should be possible to manipulate the structure of the protein
by decreasing this hydrophobic free energy by changing the solvent conditions. Such changes can be induced by chaotropic ions.

Before attempting to study the ionic effects on hydrophobic
interaction, and hence on the protein conformation and function,
it is pertinent to understand the effects of ions on the structure of water. The type of interaction between an ion and water
in solution is the strong ion-dipole compared to the weak dipole-
induced dipole interaction in hydrocarbon solutions. Such ion-
dipole interaction would certainly disrupt the regular tetrahe-
drally hydrogen bonded structure of water while concomitantly
imposing a new kind of order around each ion. A detailed discussion on the structure of water will not be presented since
this is done elsewhere (13, 16).

The orientation of water molecules around an ion depends on
the type of ion; i.e. cation or anion. In the hydration shell of
a cation the hydrogen atoms of water are directed out, whereas
with the anion they are directed in (Figure 3). This qualitative
difference in the orientation of the water molecules around ions
itself profoundly affects the polarity of the bulk water (17) and
in this respect the anions decrease the polarity of water to a

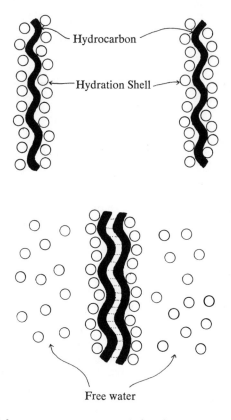

Figure 2.   *Scheme of water structure induced association of hydrocarbons.*

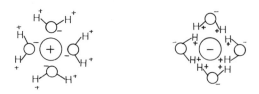

Figure 3.   *Scheme of orientation of water molecules in the hydration spheres of cations and anions.*

greater extent than cations. The degree to which the water structure is affected depends on the size and charge density of the anion. In general for anions, the extent of structure breaking effect on water progressively follows the order $F^- < CH_3COO^- < SO_4^= < Cl^- < Br^- < I^- < NO_3^- < ClO_4^- -SCN^- < Cl_3CCOO^-$; this series is known as the lyotropic or chaotropic series [18].

One of the thermodynamic manifestations of the effect of ions on water structure is the partial molar entropy (Table I). For monovalent ions, a progressive increase in entropy accompanies with increasing ionic radius, suggesting that the entropy gain may arise primarily from the steric hindrance to the formation of the water framework.

Table I. Effects of anions on water structure.*

Anion	$S_iO(M)$ [a]	S_i^b hydration
F^-	-2.3	-31.2
Cl^-	13.5	-18.2
Br^-	19.7	-14.4
I^-	25.3	-10.1
SCN^-	36.0	- 8.1

[a]Partial molar entropy of aqueous ions; [b]partial molar entropy of hydration.
*Taken from Hatefi and Hanstein [18].

It may also be noted that with decreasing entropy of hydration, the partial molar entropy of the solution also decreases. This shows that the entropy of ionic solutions is closely related to the orientation and to magnitude of water molecules in the hydrogen sphere of the ions. Small ions with greater charge density on the surface tend to have the least effect on the entropy of bulk water, whereas large ions with lower charge density on the surface tend to break the hydrogen bonded structure of water. Reviews on the effect of ions on the physical properties of water were presented by von Hippel and Schleich [16] and Dandliker and De Saussure [19].

It is reasonable at this point to question what effect the chaotropic ions would have on the hydrophobic interactions in proteins. Since the hydrophobic interaction is intimately related to the structure of water, chaotropic ions should have profound effects on the strength of the hydrophobic interaction. As explained in the previous sections, the conformation of a protein in aqueous solution is mainly induced by the characteristics of the solvent such as water. Although the grouping of the nonpolar sidechains into a protein's hydrophobic region involves a decrease in the conformational entropy, the high gain in the entropy of the solvent in the process overrides the decrease in the conformational entropy and thus thermodynamically stabilizes the folded native structure (5). If the hydrogen bonded structure of water is altered by chaotropic ions, the driving force for the hydrophobic interactions within and between protein molecules (association and polymerization reactions) will decrease and hence would destabilize their structures.

Effect of salts on the solubility of hydrocarbons and model peptides. One possible approach to understanding the effect of salts on protein conformation and function is to study the solubility of appropriate model compounds in salt solutions. The classical study was on the solubility of a model peptide, acetyltetraglycylethyl ester (ATGEE), in salt solutions (20). This study showed that the logarithm of the activity coefficient (log S_0/S) of ATGEE in salt solutions exhibited a linear relationship with the salt concentration. That is Equation 3:

$$\text{Log } S_0/S = K_s C_s$$

where S_0 is the solubility of ATGEE in water, S is the solubility in a salt solution, C_s is the molar concentration of the added salt and K_s is the salting-out constant. Here it should be mentioned that an increase in the solubility of ATGEE in a salt solution corresponds to a decrease in the activity coefficient (S_0/S). Ions such as I^-, ClO_4^-, SCN^- and Cl_3CCOO^- exhibited negative K_s values, while $SO_4^=$, $H_2PO_4^-$ and citrate had positive K_s values. Chloride and bromide exhibited neither salting-in nor salting-out properties; i.e., their K_s values were almost zero.

In a similar approach Schrier and Schrier (21) studied the solubilities of N-methylacetamide and N-methylpropionamide in various salt solutions. While the order of effectiveness of various neutral salts as activity coefficient affectors were the same as that of ATGEE, the level at which K_s changes from a positive to a negative value for the salts was different. Furthermore, increase in the chain length of the amide increased the salting-out coefficient indicating that the effect of ions on the solubility behavior of these model compounds was mainly on the nonpolar moieties in these model compounds. The salting-out constant for N-methylacetamide was 0.099 in NaCl and −0.023 in NaSCN. The change in the free energy for transfer of one mole of N-methylacetamide from water to 1 M salt solution can be calculated from the K_s values by Equation 4 (22):

$$\Delta G_{tr} = 2.3 \, RTK_s C_s$$

where C_s is the molar concentration of the salt, R is the gas constant and T is the temperature. At 25°C, the free energy of transfer of N-methylacetamide from water to 1 M NaCl is unfavorable to the magnitude of about +133 cal/mol whereas the transfer to 1 M NaSCN is favorable to the order of −30.9 cal/mole.

The effects of neutral salts on solubility of many apolar

solutes such as amides, benzene, purines and pyrimidines, etc., in aqueous solutions were reported (22, 23). These studies with simple compounds reveal that the neutral salts affect both the polar and apolar moieties in these compounds and certain ions exhibit positive effects while other ions have negative effects on solubility. In general the salting-out effect of the anions and cations follow the order

$$SO_4^= > Cl^- > Br^- > NO_3^- > ClO_4^- > SCN^- > Cl_3CCOO^-$$

$$NH_4^+ > K^+, Na^+ > Li^+ > Mg^{2+} > Ca^{2+} > Ba^{2+}$$

Since proteins have both polar and apolar groups, more comparable to the above model compound, the salting-in or salting-out effect of ions on these groups would have profound affect on the conformational stability of proteins. Since the hydrophobic interaction between the apolar residues is the major stabilizing force of the native conformation of the protein, the ions which salt-out the hydrocarbons should enhance these hydrophobic interactions in the protein and provide more stabilizing energy. In the presence of ions which exhibit salting-in effect on hydrocarbons, the hydrophobic interactions in the protein would be decreased and hence destabilize the protein structure.

The simplest way to study the effects of ions on the conformational stability of proteins is to follow the change in α-helical content of a protein as a function of temperature at different salt concentrations. von Hippel and Wong (24) studied the effects of CaCl$_2$ on the melting temperature of collagen at pH 7.0. The melting temperature, T_m, which is defined as the temperature at which the transition from helical conformation to random coil is 50% complete, decreased linearly with increase in CaCl$_2$ concentration. In a similar approach von Hippel and Wong (25) studied the temperature transition of ribonuclease as a function of salt concentration. The plots of T_m versus salt concentration for many salts exhibited linear relationships. In general the relationship between T_m and the salt concentration may be functionally expressed as Equation 5 (16):

$$T_m = T_m^o + K_s C_s$$

where T_m^o is the transition temperature in the absence of salt, C_s is the molar concentration of the salt and K_s is the molar effectiveness of the salt in perturbing T_m. In the study with ribonuclease, PO_4^{3-}, $SO_4^=$ and Cl^- exhibited positive K_s values indicating that these ions stabilized the protein. Br^- and SCN^- exhibited negative K_s values indicating that they destabilized the protein structure. The relative effectiveness of both anions and cations in perturbing the stability of ribonuclease followed the same chaotropic or lyotropic order observed with model compounds.

One can assume that the observed effects of ions on the

stability of proteins cannot be due to simple nonspecific interactions with their charged groups. As mentioned in the previous
section, the free energy contributed by the electrostatic interactions to the overall stability of protein is very small and
hence cannot account for the magnitude of the salt effect
on protein conformation. In most proteins the electrostatic interactions are essentially suppressed at about 0.1-0.15 M salt
concentration (26). Hence, the lyotropic effect of ions on the
protein conformation, which is observed at relatively higher concentrations, may be due to their effect on the hydrophobic interactions mediated via changes in the water structure.

If the conformational changes in protein in the presence of
various ions are due to stabilization or destabilization of the
hydrophobic regions that are mediated via changes in the water
structure, then such changes will be reflected in the binding
ability of hydrophobic ligands to the protein. Damodaran and
Kinsella (27) studied the effect of anions on the binding of 2-
nonanone to bovine serum albumin. Binding of 2-nonanone to bovine serum albumin exhibited positive cooperativity which was
interpreted as being due to initial binding induced conforma-
tinal changes in the protein which increased the binding affinities at the subsequent binding sites (27). In other words, the
positive cooperativity may be due to stabilization of the hydrophobic binding sites in bovine serum albumin upon initial binding. As a corollary, if the exhibition of positive cooperativity
is due to binding induced stabilization of the binding sites,
then in the presence of structure stabilizing or destabilizing
agents the degree of such stabilization should be increased or
decreased, respectively.

The binding isotherms for the interaction of 2-nonanone with
bovine serum albumin in the presence of various anions is shown
in Figure 4. In the presence of Cl^-, Br^-, SCN^- and Cl_3CCOO^-,
the curves are sigmoidal, whereas in the presence of F^- and $SO_4^=$,
they are hyperbolic. When the same data are presented in the form
of Scatchard plots, the binding exhibited positive cooperativity
in the presence of all anions except F^- (Figure 5). The initial
positive slopes at low molal ratios of binding varied distinctly
with the type of anion. The effect of anions in increasing the
initial positive slope followed the classical Hofmeister series
or chaotropic series (18); i.e., $F^- > SO_4^= > Cl^- > Br^- > SCN^- >$
Cl_3CCOO^-. These ions, in the same order, have been known to
affect the water structure (18) as well as stability of globular
proteins (16). Apart from changing the slopes of the Scatchard
plots, these ions also induced a shift in the position of the
maximum in the Scatchard plot away from the ordinate. A shift
away from the ordinate may reflect increased cooperativity in
binding. In this respect, the results show that the binding of
2-nonanone to bovine serum albumin is highly cooperative in the
presence of Cl_3CCOO^- and SCN^- in contrast to its behavior in the
presence of $SO_4^=$ and F^-.

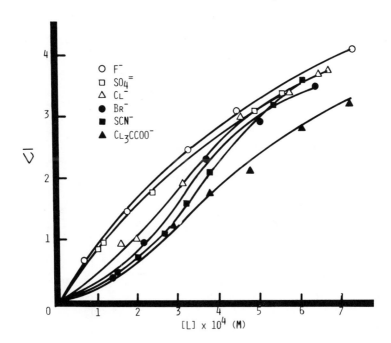

Figure 4.   Binding isotherms for the interaction of 2-nonanone to bovine serum albumin (BSA) in the presence of various anions at 0.15 M ionic strength. $\overline{\nu}$ is the number of moles of 2-nonanone bound per mole of BSA and [L] is the free 2-non-anone concentration. (Reproduced from Ref. 27.)

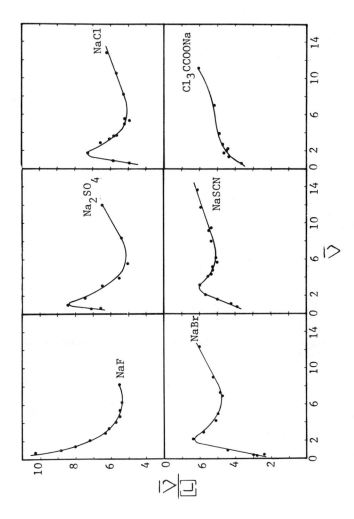

Figure 5. Scatchard plots for the binding of 2-nonanone to BSA in the presence of various anions at 0.15 M ionic strength. The symbols are the same as in Figure 4. (Reproduced from Ref. 27.)

The above changes in the binding behavior of 2-nonanone to bovine serum albumin in the presence of various anions may be intimately related to changes in the protein structure. In other words, the initial positive slopes (Figure 5), which may be considered to be proportional to the net stabilization of the hydrophobic sites upon ligand binding, are the result of two opposing forces; i.e., the destabilizing effect of anions and the stabilizing effect of 2-nonanone upon binding. Assuming that the stabilization energy provided by 2-nonanone binding is constant, the increase in the destabilizing energy provided by different anions would obviously decrease the initial positive slope in the order of the lyotropic series.

If the destabilization/stabilization of the protein structure by anions is via changes in the water structure, then one would expect a correlation between the changes in the binding parameters and the partial molar entropy of the solutions of various anions. Figure 6 shows the correlation between the free energy of interaction, ΔG_{int}, (27), which is calculated from the initial positive slopes, the Hill coefficient (28) which indicates the degree of cooperative changes, and the partial molar entropy of solution. It is obvious that as the partial molar entropy of the solution is increased (which indicates increased disorderliness of the water structure), the hydrophobic side-chains in the binding sites would be dissociated mainly by a decrease in the hydrophobic interaction between them. In such an environment, the binding induced stabilization of the binding sites will be relatively unfavorable. This is reflected by an increase in the free energy change (more positive) with increase in the partial molar entropy of the solution. However, the co-operativity of 2-nonanone binding to bovine serum albumin increases with increase in the partial molar entropy of the solution as revealed from the Hill coefficient. In the presence of F^- ion which has salting-out effect on hydrocarbons, it is possible that the structural state of the binding sites would be compact so that the initial binding of 2-nonanone would not result in further stabilization of the binding sites. In the presence of Cl^-, Br^- and SCN^- which have positive partial molar entropy values (Table I; 18), the binding sites may have some degree of freedom for further strengthening of these regions upon 2-nonanone binding. This is reflected in the cooperativity of binding.

It may be pointed out that the above binding studies with the various ions were done at 0.15 M ionic strength. The results show that even at this ionic strength the anions exhibit lyotropic effects on the hydrophobic forces in the protein, and such effects follow the same Hofmeister series as one would expect at much higher salt concentrations. This is further confirmed from the solubility studies of 2-nonanone at 0 to 0.15 M concentration of NaCl and NaSCN (Figure 7). Even at this low concentration, NaCl tends to salt-out 2-nonanone whereas NaSCN tends to salt-in the hydrocarbon. This indicates that even at this low

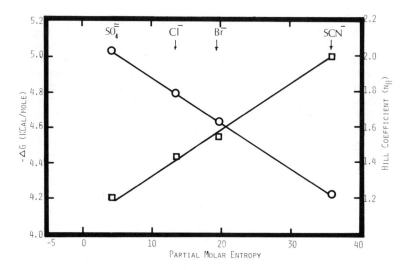

Figure 6.   Relationship between partial molar entropy of solutions of various anions and the free energy of interaction, ΔG_{int} (O – O), and the Hill coefficient (□ – □). (Reproduced from Ref. 27.)

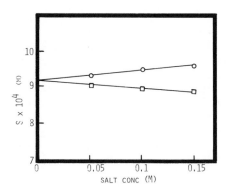

Figure 7.   Solubility of 2-nonanone in NaSCN (O – O) and NaCl (□ – □) solutions.

concentration the anions induce changes in the water structure
and affect the solubility of hydrocarbons in the order of
Hofmeister series. Hence, the effect observed on the 2-nonanone
binding to bovine serum albumin at 0.15 M concentration of various anions may be attributed to changes in the hydrophobic regions in the protein mediated via changes in water structure.

While the above studies; i.e., on the solubility of model
compounds, helix $\rightleftharpoons$ coil transitions in protein and hydrophobic
ligand binding experiments, provide a general view of the effects
of salts on protein conformation, they fail to predict the actual
thermodynamics of salt-protein interactions. In a protein structure which is stabilized by hydrophobic, hydrogen bonding and
electrostatic forces, the thermodynamics of the solubilization of
nonpolar residues from the protein's interior into the aqueous
environment may not be the same as those derived from the representative model compounds.

To understand the magnitude of the effects of neutral salts
on the hydrophobic forces in proteins, Damodaran and Kinsella
(29) used a novel method involving a protein ligand-binding
system. Let us consider this system as Equation 6:

$$P + L \underset{}{\overset{K}{\rightleftharpoons}} PL$$

where a certain concentration of the free ligand, [L], is in
equilibrium with the ligand hydrophobically bound to the protein,
P, with an intrinsic binding constant, K. The intrinsic binding
constant is a function of the conformational state of the binding
sites in the protein. Any change in the hydrophobic interactions
within the protein would alter these binding sites and affect the
binding constant. In the presence of ions which stabilize these
hydrophobic regions, the binding equilibrium would favor the
right-hand side of the above equation with considerable increase
in the binding constant. In other words, a decrease in the free
ligand concentration would occur compared to the system lacking
the stabilizing salt, other conditions being constant. Conversely, in the presence of salts which destabilize the hydrophobic
regions in the protein, the equilibrium would be pushed to the
left-hand side of the above equation, with a decrease in the
binding constant and a concomitant increase in the free ligand
concentration. The important criteria in this approach is
that the free ligand concentration in the absence of salts
should be selected in such a way that it is well below the maximum solubility at the concentrations of all salts studied. By
monitoring the changes in the free ligand concentration (keeping
the total ligand concentration constant) as a function of salt
concentration, it is possible to use a hydrophobic ligand bound
to the protein to elucidate the effect of neutral salts on the
hydrophobic interactions within the protein.

On this basis, Damodaran and Kinsella (29) studied the thermodynamics of various ion effects on hydrophobic regions in

bovine serum albumin using 2-nonanone as the hydrophobic probe.
They treated the data according to Equation 7:

$$\log [L]_w/[L]_s = K_s C_s$$

where $[L]_w$ is the free ligand concentration in water (buffer),
$[L]_s$ is the free ligand concentration in salt solutions, C_s is
the salt concentration in molarity and K_s is the salting-out constant. The ratio $[L]_w/[L]_s$ is defined as the activity coefficient of the free ligand in salt solutions.

The effects of salts on the activity coefficient of the free
2-nonanone in a 2-nonanone-bovine serum albumin binding system is
shown in Figure 8. In the presence of both Na_2SO_4 and NaCl, the
activity coefficient of the free ligand increases with salt concentration indicating that the ligand was salted-out into the
protein phase from salt solution. Since the ligand concentration
in these experiments were well below the maximum solubility at
all concentrations of the salts studied, the decrease in the free
ligand concentration cannot be attributed to a solubility phenomenon. On the other hand, the salting-out nature of Na_2SO_4 and
NaCl on the hydrophobic sidechain may stabilize the hydrophobic
binding sites and hence may increase the binding affinity for the
ligand, with concurrent decrease in the free ligand concentration.
In the presence of NaSCN and $Cl_3CCOONa$, the activity coefficient
of the free ligand decreases with salt concentration indicating
that some of the ligand initially bound to albumin was salted-in
to the salt solution from the protein phase. It is known that
both NaSCN and $Cl_3CCOONa$ destabilize protein structure. Such destabilization of the hydrophobic regions in the protein would
decrease the binding affinity for the ligand, resulting in an
increase in the free ligand concentration.

From the salting-out constant, K_s, which is obtained from
the slopes in Figure 8, one can calculate the various thermodynamic parameters for the transfer of one mole of 2-nonanone from
protein phase to 1 M salt solution. The free energy change,
ΔG_{tr}, can be calculated from the Equation 4 ($\Delta G = 2.3\ RTK_s C_s$).
By studying the changes in K_s value as a function of temperature,
one can obtain the enthalpy change, ΔH_{tr}, from the slopes of
plots of K_s vs $1/T$ for each salt. Assuming that ΔH_{tr} is independent of temperature, the entropy change, ΔS_{tr} can be obtained
from Equation 8:

$$\Delta G_{tr} = \Delta H_{tr} - T\Delta S_{tr}$$

The thermodynamic parameters for the transfer of 1 mole of
2-nonanone from protein phase to 1M salt solution are given in
Table II. It may be noted that the ΔH_{tr} values for all salts
studied are positive and increase in the order of the Hofmeister
series. This suggests that the transfer of hydrophobic groups
from the protein phase to salt solution is an energy-consuming
process and may involve breaking of certain interactions either

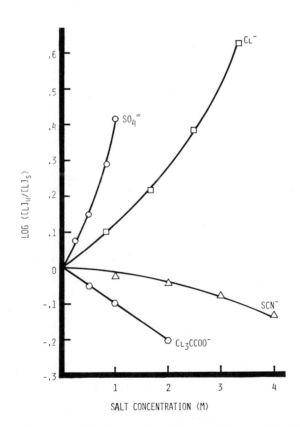

Figure 8.   Effect of neutral salts on the activity coefficient of free 2-nonanone in the BSA-2-nonanone binding system. (Reproduced from Ref. 29.)

Table II: Thermodynamic parameters for the transfer of 1 mole
of 2-nonanone from protein (BSA) phase to 1M salt
solutions.*

Salt	T(C)	K_s	ΔG_{tr}	ΔH_{tr}	ΔS_{tr}
Na_2SO_4	5	+0.27	+0.343		
	25	+0.27	+0.367	0	−1.23
	40	+0.27	+0.386		
NaCl	5	+0.165	+0.209		
	25	+0.120	+0.163	+0.025	−0.46
	40	+0.120	+0.171		
NaSCN	5	0	0		
	25	−0.028	−0.038	+0.053	+0.31
	40	−0.083	−0.012		
$Cl_3CCOONa$	5	−0.070	−0.089		
	25	−0.100	−0.136	+0.199	+0.855
	40	−0.166	−0.237		

*Thermodynamic parameters are calculated as described in the text.
All ΔG_{tr} and ΔH_{tr} values are in Kcal/mole, and ΔS_{tr} values in
entropy units (cal/mole/deg).

in the protein or in the solvent. It may also be noted that,
despite the positive ΔH_{tr} values in the case of NaSCN and
$Cl_3CCOONa$, the ΔG_{tr} values are negative, indicating that the unfavorable enthalpy change is more than overcome by greater
changes in the entropy of the system. This is presumably because
of the breakdown in the water structure which acts as the driving
force for the dissociation of the ligand from the protein phase.
Such an argument is further supported by the relationship between
the above thermodynamic changes and the partial molar entropy of
the solutions (Figure 9). The ΔG_{tr} values become more negative
as the partial molar entropy of the solution is increased, where-
as the ΔH_{tr} and ΔS_{tr} values increase with partial molar entropy of
the solution. These observations strongly suggest that the
mechanism by which ions induce lyotropic effects on protein conformation and function may be via changes in the water structure
caused by these ions.

Ionic effects on protein solubility. In addition to their
effects on the conformational transitions in globular proteins,
chatropic ions exert striking effects on the solubility as well
as association-dissociation equilibria of proteins. These macroscopic effects which are manifestations of specific protein-protein interactions, may or may not be the result of conformational
changes discussed in the previous section. For example, while
the conformational transitions in collagen and ribonuclease lead

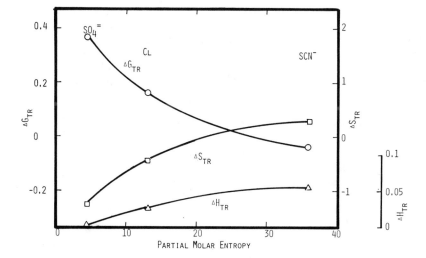

Figure 9.   Relationship between partial molar entropies of salt solutions and the thermodynamic parameters for the transfer of one mole of 2-nonanone from protein phase to 1 M salt solutions. (Reproduced from Ref. 29.)

to precipitation (25, 30), crystallization of myoglobin in ammonium sulfate did not produce the same results (31).

When the solubility of proteins is studied as a function of
salt concentration, two macroscopic effects are observed. First,
the protein solubility increases at low salt concentrations
(salting-in)* and second, after reaching a maximum the solubility
begins to decrease again (salting-out) (32). These two effects
on protein solubility in salt solutions is considered to be the
net result of electrostatic and hydrophobic interactions (33).
The salting-in process is considered to be due to the nonspecific
interaction between the ions and the charged groups on the protein. These interactions result in a net decrease in the activity coefficient of the protein and thus increases the solubility
(34, 35). The salting-in process depends mainly on the ionic
strength and not the type of ion. Above a particular ionic
strength the magnitude of this electrostatic effect becomes constant and the hydrophobic effect predominates, which results in
salting-out of the protein. Based on the above assumptions,
Melander and Horvath (33) proposed that the solubility of a
protein in salt solutions may be expressed as Equation 9:

$$\ln(w_o/w) = \beta + M(\Omega\sigma - \Lambda)$$

where w_o and w are the solubilities of the protein in water and
salt solutions respectively, m is the molar concentration of the
salt, Ω is related to surface hydrophobicity of the protein, σ is
the molal surface tension increment of the salt solution, Λ is
related to the electrostatic interactions in the protein and β is
a constant. The salting-out constant K_s is defined as in
Equation 10 (33):

$$K_s = \Omega\sigma - \Lambda.$$

In other words, the salting-out constant for a salt is the summa-
tion of the salting-out coefficient, $\Omega\sigma$ which is related to hydrophobic interactions, and the salting-in coefficient, Λ, which
is related to the electrostatic interactions in proteins (33).

The classical example of the above phenomena is the solubility of carboxyhemoglobin in ammonium sulfate solutions (Figure 10; 35). At low salt concentrations, the solubility of the
protein increases due to the electrostatic term, Λ. At higher
concentrations the electrostatic term becomes constant, whereas
the hydrophobic term $\Omega\sigma$ becomes predominant and hence the protein
is salted-out.

If one takes a cursory glance at the above equation, it may
be noted that the salting-out coefficient, $\Omega\sigma$, is proportional to
the molal surface tension of the salt solution. For a given protein, one may assume that the hydrophobic term, Ω, and the elec-
static term, Λ, are constant. Hence, according to the above
equation the variation in the solubility at higher salt concentrations is a function of the molal surface tension increment of

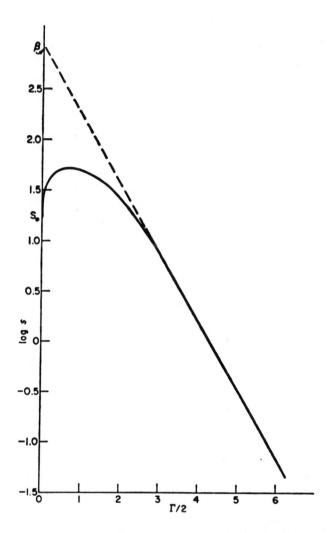

Figure 10.   Relationships between logs (solubility) and ionic strength for carboxy-hemoglobin and ammonium sulfate. (Reproduced, with permission, from Ref. 32.)

the salt solution. Melander and Horvath (33) reported that the molal surface tension increment exhibit positive values for all the salts studied. In practical terms this means that the hydrophobic salting-out coefficient will always be positive for all the salts and hence the solubility of a protein in any salt solution should always decrease compared to water. This is in disagreement with many experimental results on the solubility of hydrocarbons and model peptides in salt solutions (20, 29). As shown earlier, while the salting-out constant of 2-nonanone in NaCl is positive, it was negative in NaSCN (Figure 7). This indicates that the salting-out effect cannot be solely explained in terms of the surface tension of the salt solution, as claimed by Melander and Horvath (33).

Recently, Poillon and Bertles (36) studied the effect of various lyotropic salts on the solubility of deoxygenated sickle hemoglobin. The difference between normal hemoglobin and sickle hemoglobin is that sickle hemoglobin has valine residue at the β_6 position instead of glutamate residue in normal hemoglobin. The substituted valine at β_6 position interacts hydrophobically with leucine and phenylalanine at β_{88} and β_{85} positions of another HbS molecule. A chain of such interactions between HbS molecules lead to formation of polymers. In other words, the polymerization of sickle hemoglobin is solely due to hydrophobic interactions as shown below (37).

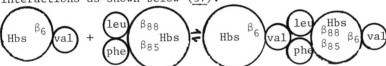

Hbs = Sickle hemoglobin
leu = Leucine
phe = Phenylalanine
β_6, β_{85}, β_{88} = Amino acid positions in the chain.

In such cases, any salt which promotes solubility of sickle hemoglobin must do so by decreasing the hydrophobic interaction between the sickle hemoglobin molecules.

The effect of various ions on the solubility of sickle hemoglobin is shown in Figure 11. In the presence of various sodium salts, the solubility of sickle hemoglobin increased with salt concentration. Among the anions studied (Figure 11), the salting-in effectiveness followed the Hofmeister series, $SCN^- > ClO_4^-$, $I^- > Br^- > Cl^-$. Since apparently no electrostatic interactions are involved in the polymerization of sickle hemoglobin, the salting-in effect by salts cannot be due to electrostatic shielding effects. Even if electrostatic interactions are involved, a solution about 0.15 M ionic strength should be sufficient to suppress these interactions (26). The observed salting-in effect even at concentrations as high as 1 M indicates that the

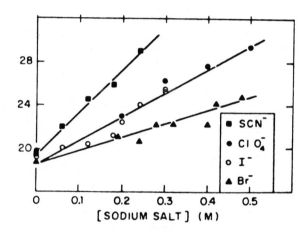

Figure 11.   Salting-in effects of anions on the solubility of deoxysickle hemoglobin. (Reproduced from Ref. 36.)

solubilization of sickle hemoglobin was due to decreased
hydrophobic interactions between sickle hemoblobin molecules.
This is incompatible with the notion that the increase in surface
tension of the solution should always result in salting-out of
the protein (33). Also, it is generally believed that the effect
of ions on the hydrophobic interactions occur only at higher salt
concentrations (16). The data on sickle hemoblobin and our results on the solubility (Figure 7) and binding of 2-nonanone to
bovine serum albumin (27) indicate that the lyotropic effect of
ions on hydrophobic interactions can occur at relatively low
concentrations.
      The argument that the decrease in electrostatic interactions
at low salt concentrations is responsible for the salting-in
effect (34, 35) may be valid for simple globular proteins. But,
complex protein systems, like proteins having multisubunits or
systems having more than one fraction, may behave differently at
low salt concentrations. For example, as the salt concentration
is increased to about 0.2 M ionic strength, the solubility of soy
protein decreases, and then increases at higher concentrations
(Figure 12; 38, 39). In this respect both NaCl and $CaCl_2$ exhibited the same effects (39). Similar results were obtained with
solubility of wheat gluten in various salt solutions (40). Thus,
if the electrostatic effect at low salt concentration has salting-
in effects, then how does the solubility of soy protein decrease
at low salt concentration. This may be due to complex interactions involving both electrostatic and hydrophobic forces between the subunits and different protein fractions (glycinin and
conglycinin) in soy. Although the electrostatic shielding of the
charged groups by salts may increase protein solubility, the decrease in the electrostatic repulsion between proteins may enhance the hydrophobic interaction between their nonpolar surfaces
and lead to formation of aggregates. This is further supported
by the fact that the observed solubility is minimum at about 0.1-
0.15 M ionic strength (Figure 12), at which the electrostatic
interaction in proteins is suppressed (26). The data also sup-
ports our contention that even at low concentrations the ions
have lyotropic effects on the hydrophobic interactions. If the
ionic strength is solely responsible for the decrease in the solubility at low salt concentrations, then the solubility level at
0.15 M ionic strength should be the same irrespective of the nature of the ion. But the data (Figure 12) show that the minimum
solubility level at 0.15 M ionic strength varies with the nature
of the anion and the increase in solubility at this ionic
strength apparently follows the lyotropic series; i.e., $Cl^- = Br^- <$
$NO_3^- < I^-$. This clearly indicates that even at this low concentration the anions exert lyotropic effects on the hydrophobic interactions. But, the magnitude of this lyotropic (solubilizing)
effect at this low salt concentration may be smaller than the
magnitude of the favorable hydrophobic interaction between molecules and hence result in insolubility. As the salt concentration

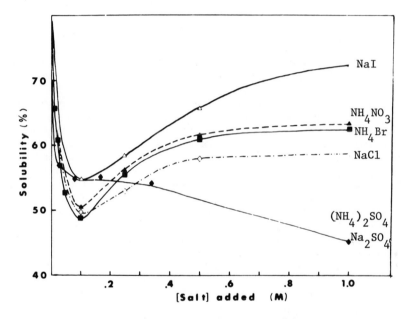

Figure 12.   Effect of various sodium salts on the solubility of soybean protein. (Reproduced from Ref. 39. Copyright 1974, American Chemical Society.)

is increased, the magnitude of these hydrophobic interactions may
be either increased or decreased depending on the lyotropic nature of the ion and thus result in decreased or increased solubility, respectively.

To support the above argument, we studied the binding of
8-anilinonaphthalene-1-sulfonate (ANS) to soy protein in the
presence of various salts. The rationale in this method is that
when ANS is hydrophobically bound to soy protein, its quantum
yield of fluorescence is increased. Thus by monitoring the
fluorescence of ANS in a soy protein-ANS system, it is possible
to measure the changes in the hydrophobic interaction as a
function of salt concentration.

The effects of salts on the fluorescence intensity of ANS
bound to soy protein is shown in Figure 13. The ANS concentration was 1.54×10^{-6} M and the protein concentration was 0.0027%.
At this protein concentration no precipitation was observed in
salt solutions at the concentration range reported here. In the
presence of all the salts studied, the fluorescence intensity of
ANS increased up to about 0.15 M salt concentration. The increase in the fluorescence intensity at this low ionic strength
is due to an increase in the hydrophobic interaction between ANS
and soy protein. The fluorescence intensity undergoes a transition at about 0.15 M salt concentration and above 0.15 M the
fluorescence intensity either increases or decreases depending
on the lyotropic effect of the ion on hydrophobic interactions.
In this respect $SO_4^=$ and F^- further increased the fluorescence,
indicating increased hydrophobic interaction between ANS and soy
protein, and Br^-, I^- and SCN^- ions decrease the fluorescence
intensity. These results clearly demonstrate that the observed
decrease in the solubility of soy protein (Figure 11) at low salt
concentrations is due to enhancement of hydrophobic interactions
between protein molecules. In other words, the salting-in and
salting-out profile of proteins by salts at various concentrations is the net result of the maximization or minimization of
electrostatic and hydrophobic forces at different salt concentrations. It would also depend on the ratio of nonpolar surface
to charge density on the surface of the protein. If the protein
has larger nonpolar surface areas the decrease in the electrostatic interaction at low salt concentrations would promote hydrophobic aggregation of such proteins resulting in decreased
solubility. At higher salt concentrations, further decrease or
increase in solubility depends on the lyotropic nature of the
ion. In soy protein, $SO_4^=$ further decreases the solubility
whereas other anions increase the solubility in the order of
$Cl^- < Br^- < NO_3^- < I^-$ (Figure 12). On the other hand, if the
hydrophobic surface area on the protein is less and the charge
density is high, the protein will be salted-in at lower salt
concentrations and at higher salt concentrations the solubility
would follow the lyotropic effect of different ions. Such complex interplay of different forces may be the reason for the

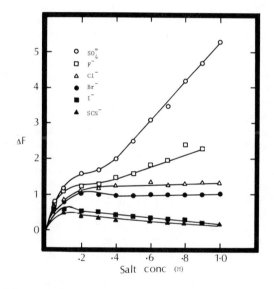

Figure 13. Effect of sodium salts on the binding of 8-anilinonaphthalene-1-sulfonate (ANS) to soybean protein in 30 mM tris-HCl buffer, pH 8.0. Protein and ANS concentrations were 0.04% and 1.5 × 10⁻⁶ M, respectively. ANS binding was monitored as the increase in fluorescence intensity at 485 nm with excitation at 360 nm.

observed differences in the solubility behavior of various proteins in salt solutions (36, 38, 40, 41).

Mechanisms of Salt Effect on Protein Structure

The effects of ions on protein conformation and function may
be divided into two categories: First, their electrostatic interaction with the charged groups and polar groups (e.g., amide
groups) in proteins, and second, their effect on hydrophobic
forces via influence on the structure of water. From the
foregoing arguments the electrostatic shielding effect which
reaches a maximum at 0.15 M ionic strength may not have great
influence on the conformational transitions in proteins that are
more pronounced at high salt concentrations. The salt may interact with the peptide groups in proteins (42, 43). Mandelkern
et al. (44) have suggested that salt interaction with the peptide
groups may alter the partial double bond character of the peptide
bond. This permits free rotation of the peptide backbone which
might lead to conformational changes in proteins. Although the
possibility of such interactions between salts and peptide groups
exist, it has been shown that the hydration of the ions greatly
prevents formation of such complexes (45). Hence in aqueous
solutions, concentrations at which the ions induce conformational
changes in proteins, it is unlikely that the formation of the
complex between ions and peptide groups is the major cause for
the conformational changes in proteins. Although it is thought
that such a mechanism exists (43), 46, 47) there are no direct
evidence to support this hypothesis.

The conformational changes in proteins in the presence of
ions may be mainly via ionic effects on water structure (16, 19).
As mentioned earlier, the native structure of a protein is im-
posed by the state of the water structure because of its thermodynamically unfavorable interaction with nonpolar sidechains.
The changes in the hydrogen bonded structure of bulk water in the
presence of salts, due to ion-dipole interaction, would alter the
degree of hydration as well as the orientation of the water
molecules around the nonpolar sidechains. In this respect, the
ions which enhance the hydrogen bonded structure of the bulk
water (e.g., F^-, $SO_4^=$) presumably have the tendency to increase
the degree of hydration of the nonpolar groups in proteins. The
ions which increase the entropy of the bulk water (e.g., Br^-, I^-,
ClO_4^-, SCN^-) tend to decrease the hydration of the hydrophobic
sidechains. As mentioned earlier, the driving force for the
hydrophobic interactions between the nonpolar molecules is the
entropically unfavorable water-water orientation in the hydration sphere of the hydrocarbon (Figure 1B). Such relative
orientation of the water molecules around the hydrocarbon exerts
an interfacial tension of about 51 ergs/cm^2 of water-hydrocarbon
contact area (48). In other words, separation of 1 cm^2 area of
water-hydrocarbon interface to form 1 cm^2 of water-water and

1 cm^2 of hydrocarbon–hydrocarbon interfaces will be favored by about –102 ergs. It is this free energy which is responsible for the formation of hydrophobic regions in proteins. The ions which stabilize or destabilize these hydrophobic regions may do so by increasing or decreasing the interfacial tension, respectively. If we assign an interfacial tension of 51 ergs/cm^2 for the water–water orientation in the hydration shell as shown in Figure 1B, (13), the ions which increase the interfacial tension above 51 ergs/cm^2 may do so by altering the water–water geometry in the hydration shell in such a way that the repulsion between the lone pair of electrons and the protons is increased. Conversely, the ions which decrease the interfacial tension below 51 ergs/cm^2 may do so by changing the water–water geometry such that the repulsive interactions between the lone pair of electrons and the hydrogen atoms are decreased. However, such changes in the water–water orientation in the hydration shell of hydrocarbon by ions is only through the changes in the bulk water structure, which can be defined in terms of the partial molar entropy. One can experimentally determine the decrease in the interfacial tension between water and hydrocarbon in the presence of various ions. From these estimations and assuming that 51 ergs/cm^2 corresponds to water–water orientation as shown in Figure 1B (12, 13), it may be possible to calculate the relative water–water orientation around the hydrocarbon with a statistical-mechanical approach for various interfacial tensions in the presence of different ions. This would provide an understanding of the relationships between the partial molar entropy, interfacial tension and the hydrophobic effect and the influence of these effects on protein conformation and function.

Conclusion

The primary effect of ions on protein structure at low concentration is their electrostatic interaction with the counter charges on the protein. Such charge neutralization may induce dissociation of the subunits in oligomeric proteins which are held together by electrostatic interactions. At sufficiently higher concentrations the ions either weaken or strengthen the hydrophobic interactions in proteins, depending on the nature of the ion. Such effects on hydrophobic interactions are mediated via changes in the water structure in the presence of ions. However, some of the evidences suggest that even at low concentrations (0.15 M) anions affect hydrophobic interactions. The effectiveness of various anions in destabilizing hydrophobic interactions follow the Hofmeister series. The changes in electrostatic and hydrophobic forces in proteins by ions induce conformational changes in proteins and hence affect their functional behavior. Systematic investigation of such effects may provide better understanding of the factors affecting the functional properties of proteins in food systems.

Acknowledgment

Support from the National Science Foundation Grant #CPE 80-18394 is gratefully acknowledged.

Literature Cited

1. Pain, R. H. In "Characterization of Protein Conformation and Function"; Franks, F., Ed.; Symposium Press: London, 1978, p. 19.
2. Klotz, I. M.; Franzen, J. S. J. Am. Chem. Soc., 1962, 84, 3461.
3. Schachman, H. K. Cold Spring Harbor Symp. Quant. Biol., 1963, 28, 409.
4. Ross, R. D.; Subramanian, S. Biochemistry, 1981, 20, 3096.
5. Tanford, C. "Physical Chemistry of Macromolecules"; Wiley: New York, 1961; p. 480.
6. Lumry, R.; Biltonen, R. In "Structure and Stability of Biological Macromolecules"; Timasheff, S. N. & Fasman, G. D., Eds.; Marcel-Dekker, Inc.: New York, 1969; p. 106.
7. Kauzmann, W. Adv. Protein Chem., 1959, 14, 1.
8. Nemethy, G.; Scheraga, H. A. J. Chem. Phys., 1962, 36, 3382.
9. Nemethy, G.; Scheraga, H. A. J. Chem. Phys., 1962, 36, 3401.
10. Miller, K. W.; Hildebrand, J. H. J. Am. Chem. Soc., 1968, 90, 3001.
11. Frank, H. S.; Evans, M. W. J. Chem. Phys., 1945, 13, 507.
12. Franks, F. In "Water Relations of Foods"; Duckworth, R. B., Eds.; Academic Press: New York, 1975; p. 3.
13. Stillinger, F. H. Science, 1980, 209, 451.
14. Lewin, S. "Displacement of Water and Its Control of Biochemical Reactions"; Academic Press: New York, 1974.
15. Nozaki, Y.; Tanford, C. J. Biol. Chem., 1971, 276, 2211.
16. von Hippel, P. H.; Schleich, T. In "Structure and Stability of Biological Macromolecules"; Vol. 2; Timasheff, S. N. & Farman, G. D., Eds.; Marcel-Dekker: New York, 1969; p. 417.
17. Greyson, J. J. Phys. Chem., 1967, 71, 2210.
18. Hatefi, Y.; Hanstein, W. G. Proc. Natl. Acad. Sci., 1969, 62, 1129.
19. Dandliker, W. B.; De Saussure, V. A. In "The Chemistry of Biosurfaces"; Hair, M. L., Ed.; Marcel-Dekkar, Inc.: New York, 1971; p. 1.
20. Robinson, D. R.; Jencks, W. P. J. Am. Chem. Soc., 1965, 87, 2470.
21. Schrier, E. E.; Schrier, E. G. J. Phys. Chem., 1967, 71, 1951.
22. Hamabata, A.; Chang, S.; von Hippel, P. H. Biochemistry, 1973, 12, 1271.
23. Robinson, D. R.; Grant, M. E. J. Biol. Chem., 1966, 241, 4030.
24. von Hippel, P. H.; Wong, K. Y. Biochemistry, 1963, 2, 1387.

25. von Hippel, P. H.; Wong, K. Y. J. Biol. Chem., 1965, 240, 3909.
26. Eagland, D. In "Water Relations of Foods"; Duckworth, R. B., Ed.; Academic Press: New York, 1975; p. 73.
27. Damodaran, S.; Kinsella, J. E. J. Biol. Chem., 1980, 255, 8503.
28. Dahlquist, F. W. Methods in Enzymology, 1978, 48, 270.
29. Damodaran, S.; Kinsella, J. E. J. Biol. Chem., 1981, 256, 3394.
30. Puett, D.; Garmon, R.; Ciferri, A. Nature, 1966, 211, 1294.
31. Kendrew, J. C. Science, 1963, 139, 1259.
32. Dixon, M.; Webb, E. C. Adv. Protein Chem., 1961, 16, 197.
33. Melander, W.; Horvath, C. Arch . Biochem. Biophys., 1977, 183, 200.
34. Czok, R.; Bucher, T. Adv. Protein Chem., 1960, 15, 315.
35. Dixon, V. P.; Webb, E. C. Adv. Protein Chem., 1961, 16, 197.
36. Poillon, W. N.; Bertles, J. F. J. Biol. Chem., 1979, 254, 3462.
37. Benesch, R.; Benesch, R. Nature, 1981, 289, 637.
38. Shen, J. L. In "Protein Functionality in Foods"; Cherry, J. J. P., Ed.; American Chemical Society Symp. Series #147: Washington, DC, 1981; p. 89.
39. van Megan, W. H. J. Agric. Food Chem., 1974, 22, 126.
40. Preston, K. R. Cereal Chem., 1981, 58, 317.
41. Sawyer, W. H.; Puckridge, J. J. Biol. Chem., 1973, 248, 2429.
42. Bello, J.; Bello, H. R. Nature, 1961, 190, 440.
43. Jencks, W. P. Federation Proceedings, 1965, 24, S-50.
44. Mandelkern, L.; Halpin, J. C.; Posner, A. S. J. Am. Chem. Soc., 1962, 84, 1383.
45. Diorio, A. F.; Lippincott, E.; Mandelkern, L. Nature, 1962, 195, 1296
46. Gordon, J. A.; Jencks, W. P. Biochemistry, 1963, 2, 47.
47. Herskovits, T. T. Biochemistry, 1963, 2, 335.
48. Tanford, C. Proc. Natl. Acad. Sci., 1979, 76, 4175.

RECEIVED June 1, 1982.

Effect of Disulfide Bond Modification on the Structure and Activities of Enzyme Inhibitors

MENDEL FRIEDMAN, OK-KOO K. GROSJEAN, and JAMES C. ZAHNLEY

United States Department of Agriculture, Western Regional Research Center, Agricultural Research Service, Berkeley, CA 94710

Enzyme (trypsin, chymotrypsin, α-amylase, carboxypeptidase, etc.) inhibitors appear in many agricultural products (1-4) such as legumes (4), corn (5), wheat (6), and potatoes (7,8).

Soybeans and other legumes containing active trypsin inhibitors depress growth in rats compared to analogous feeding of inhibitor-free soybeans. Growth inhibition and its accompanying pancreatic hypertrophy are presumably partly due to the antitryptic activity of the inhibitor(s) (4, 9-13).

Since naturally occurring protease inhibitors such as trypsin and chymotrypsin inhibitors are proteins that require their own structural integrity in order to inactivate proteolytic enzymes by complex formation (14-19), any alteration of the inhibitor's structure will inactivate it if interaction with the enzyme is thereby prevented. A reasonable strategy is to cleave the inhibitor's disulfide bonds, which often control protein structure (20). Reduced trypsin inhibitors are expected to be inactive.

In addition to the cited primary benefit, i.e., inactivation of protease inhibitors, reductive cleavage of disulfide bonds of both inhibitors and structural proteins in legumes by, for example, cysteine or other thiols, should also offer an important secondary benefit, namely, improvement in the digestibility of the treated proteins, because disulfide cleavage is expected to decrease crosslinking thus permitting more rapid digestion.

In this chapter we examine effects of 1) reductive S-pyridylethylation and S-quinolylethylation of disulfide bonds on analytical, structural, and inhibitory aspects of enzyme inhibitors; and, 2) cooperative and synergistic effects of heat and thiols in inactivating soybean and lima bean inhibitors.

Experimental Procedures

        Enzymes, inhibitors, and substrates were obtained from
Worthington Biochemical Corporation or Sigma Chemical Co. Soy
flour was a gift of Archer Daniels Midland Corporation (Decatur,
IL). Lima bean flour was prepared from raw beans purchased in
a local store. 2-Vinylpyridine was obtained from Aldrich Chemical
Co. and was redistilled before use. 2-Vinylquinoline was from
Eastman Kodak and was vacuum-distilled before use. Tributyl-
phosphine was obtained from Aldrich Chemical Co. and was stored
under nitrogen. S-β-(2-Pyridylethyl)-L-cysteine (2-PEC) was
synthesized as described by Friedman and Noma (21) and S-β-
(2-quinolylethyl)-L-cysteine (2-QEC) according to the procedure
of Krull et al. (22). Other chemicals were reagent grade.

        Instrumental. Absorption spectra were recorded on a Cary
15 spectrophotometer; and trypsin was assayed by absorbance
measured with a Beckman DB spectrophotometer. A Radiometer pHM
26 meter and Beckman 39030 thin-probe combination electrode were
used for most pH measurements. Titrimetric (pH-stat) experiments were carried out using a Radiometer TTT1b Titrator, SBR
2c Titrigraph, SBU1a syringe burette, and a TTA31 microtitration
assembly, equipped with a G2222c glass electrode and a K4112
calomel electrode. Constant temperature was maintained with a
Haake Model FE circulator and a thermostatting jacket (Radiometer
V526). Thermal denaturation measurements were made with a DuPont
Model 990 differential scanning calorimeter (DSC), at a heating
rate of 10°/min and a nominal sensitivity of 100 μcal s^{-1}.
Calibration and operation of the DSC and the interpretation of
results have been described (23,24,25). Fluorescence spectra
were recorded on a Turner Model 210 spectrofluorometer.

        Solutions. Buffers were prepared at room temperature by
adding HCl or NaOH to adjust the pH to the desired value.
Distilled, deionized water was used. Unless otherwise indicated,
native and S-pyridylethylated trypsin inhibitors were dissolved
in 0.05 $\underline{M}$ glycine adjusted with 1 $\underline{N}$ HCl to pH 3.0. Trypsin was
dissolved in 2 m$\underline{M}$ HCl/20 m$\underline{M}$ CaCl$_2$, and kept at 0-4°C. Benzoyl-
$\underline{DL}$-arginine $\underline{p}$-nitroanilide (BAPNA) stock solutions (0.1 $\underline{M}$) were
made with dimethyl sulfoxide (DMSO), and N-acetyl-$\underline{L}$-tyrosine
ethyl ester (ATEE) stock solutions (0.2 $\underline{M}$) were made with 100%
ethanol.

        Modification of inhibitors. 200 mg of Kunitz soybean trypsin
inhibitor (STI) was dissolved in 10 ml of pH 7.6 Tris-8$\underline{M}$ urea
buffer. To this solution were then added 10 ml ethanol, 0.2 ml
tri-n-butylphosphine and 0.2 ml of 2-vinylpyridine. Nitrogen was
bubbled in for one minute. The reaction mixture was shaken and

5 ml samples were removed at various times. The samples were
dialyzed about two days in water- 2% methanol, and lyophilized.
The procedure with ovomucoid (OMI) was identical to that described for STI. Reductive alkylation with lima bean protease
inhibitor (LBI) was carried out with 50 mg of inhibitor, 5 ml
pH 7.6-8M urea buffer, 5 ml of ethanol, 0.05 ml of n-Bu₃P, and
0.05 ml 2-vinylpyridine. Control experiments were done in pH
7.6 buffer without urea and in pH 7.6 buffer without n-Bu₃P
and 2-vinylpyridine.
   A similar procedure was used to transform half-cystine
residues in STI to 2-QEC side chains (26).

   Reaction with thiols. The following procedure with soy
flour illustrates the general method used to inactivate STI
(18). Soy flour (300 mg) and N-acetyl-L-cysteine (NAC)
(67.4 mg, 0.4 millimoles) in 30 ml of 0.5 M Tris buffer were
placed in a water bath at a specified temperature for 1 hr.
The reaction mixture was then cooled in ice-water. Portions
(0.05 ml) of the suspension were diluted to 0.5 ml with pH 8.2,
0.05 M Tris buffer. Control experiments without NAC were carried
out concurrently. Pure STI was studied similarly except that
dilutions were 1:300 with the pH 8.2 Tris buffer. The remaining
trypsin inhibitory activity was then assayed as described below.
All experiments were carried out in duplicate. The reproducibility is estimated to be ± 3% or better.

   Amino acid analysis. Amino acid analyses of aliquots of
hydrolyzed weighed samples dissolved in pH 2.2 buffer were
carried out with a single-column Durrum Amino Acid Analyzer,
Model D500 under the following conditions: single-column ion-
exchange chromatographic method (17); Resin, Durrum DC-4A,
citrate buffer pH, 3.25, 4.25, 7.90; Photometer, 590 and 440
nm; Column, 1.75 mm x 48 cm; analysis time, 105 minutes.
Norleucine was used as an internal standard (Figures 1-6).
Extent of reaction was deduced from the amount of 2-PEC found.

   Ultraviolet spectroscopy of 2-PEC and 2-pyridylethylated
(2-PE) inhibitors. Extent of reaction of 2-vinylpyridine with
the inhibitors was also determined from the difference between
A_{263} for equimolar concentrations of the modified and native
inhibitors, using ε for 2-PEC of 7070 at pH 3.0. The value
of ε at pH 1, 7200, was slightly higher than previously found
in 6 N HCl (21). To determine ε for 2-PEC, a 2.7 mM stock
solution in 2 mM HCl was made and diluted 1:20 in appropriate
buffers. Duplicate dilutions in 0.05 M glycine/HCl (pH 3.0)
gave ε values consistent within 1%.
   Spectra for 2-pyridylethylated inhibitors were also
analyzed simultaneously for concentrations of 2-PE side
chains and protein by using a two-wavelength two-component
analysis (27). One wavelength was 263 nm, the absorption

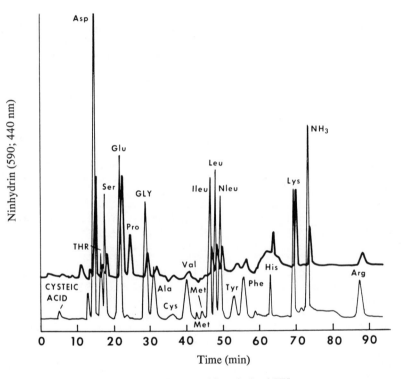

Figure 1.   Amino acid analysis of STI.

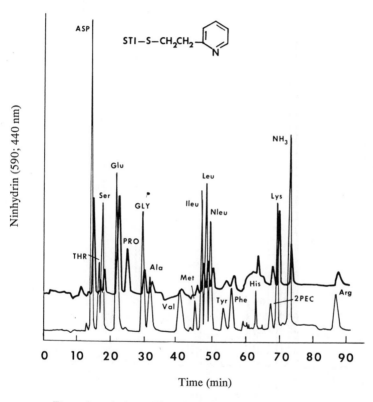

Figure 2.   Amino acid analysis of 2-PE-STI. Note 2-PEC peak.

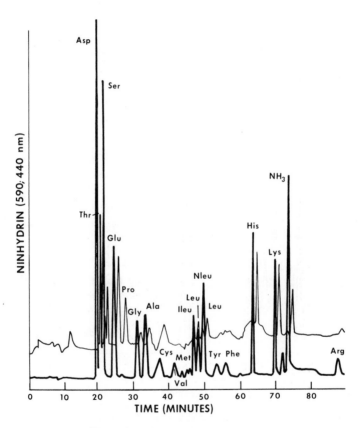

Figure 3. Amino acid analysis of LBI.

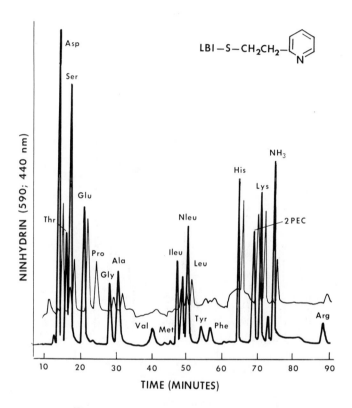

Figure 4. *Amino acid analysis of 2-PE-LBI.*

366 FOOD PROTEIN DETERIORATION

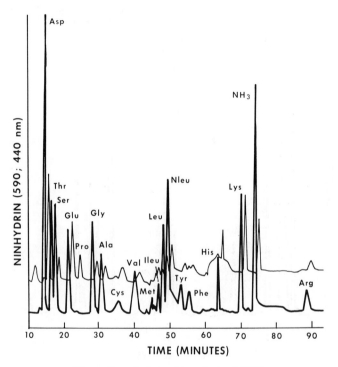

Figure 5. *Amino acid analysis of OMI.*

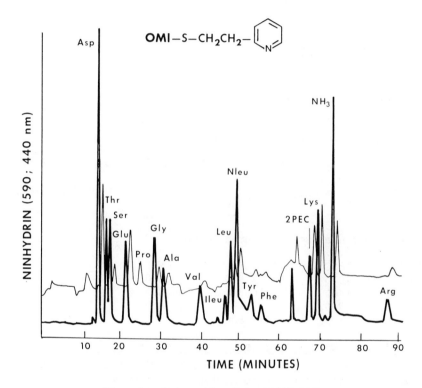

Figure 6. Amino acid analysis of 2-PE-OMI.

maximum for 2-PEC at pH 3, since $\varepsilon_{PEC}/\varepsilon_{protein}$ is maximal at this wavelength. The second wavelength, at which $\varepsilon_{protein}/\varepsilon_{PEC}$ is maximal or nearly so, was 285 nm for PE-OMI, 280 nm for PE-LBI, or 289 nm for PE-STI. Values of $\varepsilon_{protein}$ and ε_{PEC} for wavelengths for which ε is not listed were calculated from spectra using the formula

$$\varepsilon_\lambda = \frac{A_\lambda}{A_{max} \ (or \ A_{280})} \times \varepsilon_{max} \ (or \ \varepsilon_{280})$$

Two simultaneous equations of the form:

$$A = \varepsilon_{protein} \times C_{protein} + \varepsilon_{PEC} \times C_{PEC}$$

were solved in each case.

For example, equations for PE-STI were:

$$A_{263} = \varepsilon_{STI_{263}} \times C_{STI} + \varepsilon_{PEC_{263}} \times C_{PEC} = 14{,}900 \ C_{STI} + 7070 \ C_{PEC}$$

$$A_{289} = \varepsilon_{STI_{289}} \times C_{STI} + \varepsilon_{PEC_{289}} \times C_{PEC} = 12{,}600 \ C_{STI} + 395 \ C_{PEC}$$

The ε values used are tabulated below:

λ(nm)	ε_{PEC}	ε_{STI}	ε_{OMI}	ε_{LBI}
263	7070	14,900	10,850	2480
280	1460	20,300	14,890	2390
285	700	---	12,860	
289	395	12,600		

Ultraviolet spectroscopy of 2-quinolylethyl-STI (2-QE-STI).

Since 2-QE-STI was insoluble in nondenaturing aqueous solvents, it was first dissolved in formamide at 5-15 mg/ml and then diluted in aqueous solutions. Aliquots of stock solutions of 2-QEC or 2-QE-STI were diluted to $A_{max} \leq 1.0$. Diluents were: 0.1 and 0.2 M HCl; 0.1 N H_2SO_4: 0.1 and 0.2 M acetic acid; 0.2 M glycine--HCl, pH 3.0; 0.2 M acetate buffers, pH 3.6 and pH 5.0; 0.2 M 3-(N-morpholino)propanesulfonic acid (MOPS)--NaOH buffer, pH 7.0; 0.2 M glycine--NaOH buffer, pH 9.0. Spectra were recorded against solvent blanks.

Fluorescence spectroscopy.

2-QE-STI stock solutions were normally diluted in water or 0.1 N H_2SO_4 to A_{max} (1-cm light path) of 0.04 or less, and the pH was adjusted as needed with 1 N H_2SO_4 and 1 N or 4 N NaOH. Fluorescence was excited at 316 nm and 25°C. Yields were determined from areas of the emission

spectra, and intensities were measured as net peak heights at
maximum emission.

Trypsin inhibitor assay. Many units are used to define the
inhibitory activity of protease inhibitors (28). The following
method, based on the defined stoichiometry of enzyme, inhibitor,
and substrate, was adapted from Mikola and Suolinna (29).
Inhibition of trypsin activity was measured at pH 8.2 and 37°C,
with BAPNA as substrate. Inhibitor was incubated with 25 μg
of twice-crystallized bovine trypsin in a total volume of 0.75
ml for 5 min at 37°C. BAPNA (30 mg in 1.00 ml of DMSO) was
diluted to 100 ml with 0.05 $\underline{M}$ Tris-HCl, pH 8.2, containing 0.02
$\underline{M}$ CaCl$_2$. The diluted BAPNA was kept at 37°C and used within
1 hr. To start the reaction, 3.0 ml of substrate-buffer mixture
was added rapidly to the trypsin-inhibitor mixture. The reaction was stopped after 5 or 10 min by adding 0.50 ml of 30%
acetic acid.

Determination of active trypsin. The concentration of
active trypsin was determined by active site titration with
$\underline{p}$-nitrophenyl $\underline{p}$'-guanidinobenzoate (NPGB), essentially according to Chase and Shaw (30). Absorbance at 410 nm was followed in
a Cary 15 spectrophotometer, with the 0.1 absorbance slidewire.
NPGB (0.01 M) was dissolved in dimethylformamide-acetonitrile
(1:4). Trypsin was dissolved in 2 m$\underline{M}$ HCl/0.02 $\underline{M}$ CaCl$_2$ at 5-6
mg/ml, then diluted 1:10 in the same solvent. Each cuvette
contained 0.05-0.09 $\underline{M}$ Na barbital buffer (pH 8.35 or 8.39 at
23°C), 0.02 $\underline{M}$ CaCl$_2$, 0.1 $\underline{M}$ KCl. Trypsin (3-5 nmol of active
enzyme) was added to the sample cuvette as 0.20 ml of the 1:10
dilution or 25 μg of the stock solution. After the baseline
was determined, 5 μL of NPGB was added to each cuvette. Total
volume was 1.00 ml. In our instrument, 1 A_{410} = 61.1 mol of
$\underline{p}$-nitrophenolate (31). The trypsin used was 61% ac$\neg$ enzyme,
based on its protein content as determined by abs$\neg$:e at
280 nm and an active site titration.

Calculations. The following sample calculation was used
to estimate the trypsin inhibitor activity per mg of active
enzyme, where

A_{410} = absorbance measured at 410 nm
ε_M = molar extinction coefficient
V_o = rate of BAPNA hydrolysis by trypsin in the absence of
inhibitor
V_i = rate of BAPNA hydrolysis by trypsin in the presence of
inhibitor
TU = trypsin unit, defined as the amount of trypsin that
catalyzes the hydrolysis of 1 μmol of BAPNA per min at
37°C and pH 8.2
TIU = trypsin inhibitor unit, defined as reduction in activity
of trypsin by one trypsin unit (TU).

Using ε_M for p-nitroaniline = 8800 1/mol-cm (32), and a total volume of 4.25 ml, 1 TU corresponds to an increase in A_{410} of 2.07 min^{-1}.

We found trypsin activity of 2.51 TU per mg of active enzyme (see sample calculation). Using molecular weights of 23,400 for trypsin and 21,500 for STI, we calculated 2.92 TIU per mg of STI.

For trypsin only, 0.358 A_{410}/5 min; Rate, V_o = 0.0716/min; for trypsin and inhibitor, 0.389 A_{410}/10 min.

$$V_o - V_i = \frac{A_{410} \text{ (no inh.)}}{t_o} - \frac{A_{410} \text{ (inh.)}}{t_i} = \frac{0.358}{5} - \frac{0.389}{10}$$

$$= 0.0716 - 0.0389 = 0.0327 \ A_{410} \ min^{-1}$$

Since 2.07 A_{410} min^{-1} is equivalent to 1 trypsin unit (TU), the amount of inhibitor present is:

$$\frac{0.0327}{2.07} = 0.0158 \text{ TIU}$$

For pure STI the inhibitor content of the sample used is:

$$\frac{0.0158 \text{ TIU}}{2.92 \text{ TIU/mg STI}} = 5.41 \times 10^{-3} \text{ mg STI} = 5.41 \text{ μg STI}$$

Since 1 mg of active trypsin = 2.51 TU, 0.0158 TIU represents

$$\frac{0.0158 \text{ TU}}{2.51 \text{ TU/mg trypsin}} = 6.29 \times 10^{-3} \text{ mg active trypsin inhibited in}$$

the pure STI sample.

If inhibitors other than STI are also present (as is apparently the case in most plant foods), a more precise estimate is:

$$\frac{\text{TIU in sample}}{\text{wt. of sample}} = \text{TIU/mg (or g).}$$

The following average values demonstrate the reproducibility of the assay. Pure commercial STI inhibited 1.23 ± 0.03 mg of active trypsin per mg STI (eight replicates). (This value is higher than unity because the ratio of molecular weights of trypsin to STI is greater than one and because of possible minor contributions of lower molecular weight inhibitors to the total enzyme inhibitory process.) For untreated soy flour, the average for four determinations was found to be 109.51 ± 1.52 TIU or 43.63 ± 0.605 mg trypsin inhibited per gram of flour. Similarly, the trypsin inhibitor assay for LBI gave the following average values for four determinations: 2.76 ± 0.041 TIU/mg commercial LBI and 51.33 ± 0.994 TIU/g lima bean flour.

Chymotrypsin inhibition assay. Inhibition of chymotrypsin
was determined titrimetrically, with 0.01 M ATEE as substrate;
the procedure was adapted from that of Wilcox (33). Inhibitor
was incubated with chymotrypsin (25 μl of 0.44 mg/ml solution)
at room temperature, then equilibrated at 25°C, in 1.9 ml of
5 mM Tris/HCl, 0.1 M in CaCl₂, pH 8.0. At least 10 min was
allowed for enzyme and inhibitor to react. After the baseline
uptake of titrant (0.1 N NaOH) was recorded, 100 μl of 0.2 M
ATEE was injected into the solution. Initial rates were determined from the linear portion of the plot of base uptake against
time. Chart speed was 2 cm/min. Measurements of the nonenzymatic breakdown of ATEE and the effect of order of adding ATEE
and chymotrypsin (no inhibitor) were also made and used to
correct assays of inhibition.

Proteolytic digestion assay. Treated soy flour was dialyzed for three days against distilled water and lyophilized.
Samples were dissolved or dispersed (5 mg/ml) in 0.05 M KCl with
stirring and adjusted to pH 8.5. To determine digestibility by
trypsin, aliquots (5 mg of protein) were pipetted into plastic
titration vessels (Radiometer V 524) and diluted to 4.0 ml with
0.05 M KCl. Samples were incubated under N_2 to minimize CO_2
absorption. In a typical run, the substrate was brought to pH
8.5 and 37°C, and the baseline rate of uptake of titrant (0.02
N NaOH) was determined. Then, 20 μL of bovine trypsin (5 mg/ml
in 2 mM HCl-0.02 M CaCl₂) was added. Further additions of enzyme
were made when the uptake of alkali became very slow. When the
reaction was essentially complete, the net uptake was determined.
Since liberation of hydrogen ions by proteolysis is incomplete at pH values where some new α-amino groups are ionized
(34), calculation of the number of peptide bonds cleaved requires
a value for the average pK of the newly formed peptide α-amino
groups. If we assume a pK_a of 7.5 (35), free hydrogen ions will
be produced in 90% of the peptide bond cleavages at an assay pH
8.5. To calculate the actual average number of peptide bonds
cleaved per chain corrected for incomplete liberation of hydrogen
ions, we also assumed an average molecular weight of 23,000 for
the soy protein (36).

Results

The 2-PE-derivatives of all three inhibitors showed absorption maxima at 263-264 nm in acidic (pH 3.0) solution (Figures
7-10), coinciding with that for free 2-PEC. Reaction was rapid
in buffers containing 8 M urea, especially in the cases of STI
and chicken OMI. All 4 half-cystine groups of STI and 12 of
18 half-cystine groups of OMI reacted within 30 min under these
conditions, based on an increase in A_{263} (Figure 8 and Table I).
As expected, LBI reacted more slowly (Figure 10 and Table I).
Only 5 to 6 of the 14 half-cystines, on the average, reacted

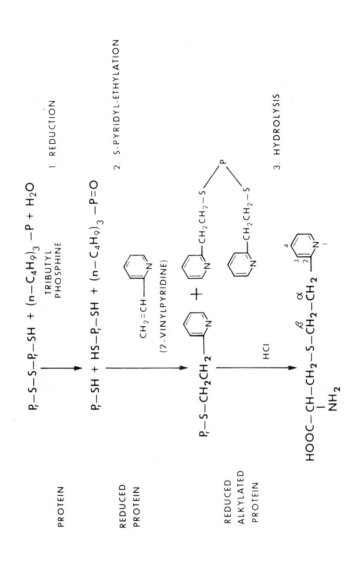

Figure 7.   Reductive S-pyridylethylation of a protein.

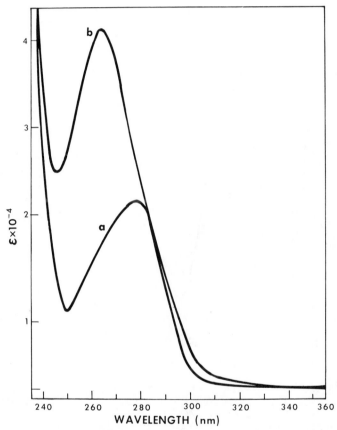

Figure 8. UV spectra of trypsin inhibitors and their 2-PEC derivatives at pH 3.0. Lyophilized proteins were dissolved in 0.1 M KCl by adjusting the pH to 3.2 at 7.8 mg/mL (native) or 3.1–3.7 mg/mL (modified). Aliquots were diluted to 0.5 mg/mL (2.3×10^{-5} M) in 0.05 M glycine/HCl (pH 3.0). Key: a, native STI; and b, modified STI.

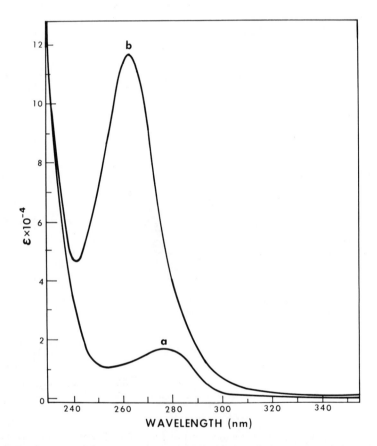

Figure 9.   UV absorption spectra of chicken ovomucoid inhibitors (OMI). Protein samples were dissolved and diluted to 0.25 mg/mL. Key: a, untreated OMI: 8.2 × 10^{-6} M; and b, treated OMI: 7.6 × 10^{-6} M.

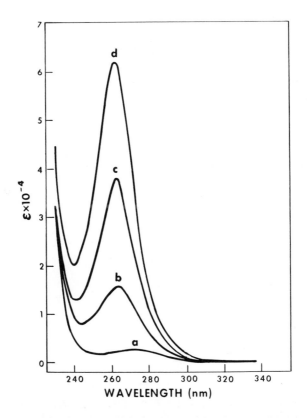

Figure 10. The UV absorption spectra of lima bean trypsin inhibitors (LBI) is shown. Control samples in the presence or absence of urea gave almost identical spectra; therefore, only one is shown. Key to protein concentrations (w/w): a, untreated, 2.28×10^{-5} M; controls (urea), 2.36×10^{-5} M; b, treated 30 min without urea, 2.17×10^{-5} M; c, treated 30 min with urea, 2.10×10^{-5} M; and d, treated 240 min, 2.48×10^{-5} M.

Table I. 2-PEC (half-cystine) contents of reduced alkylated STI, LBI, and OMI determined by amino acid analysis and ultraviolet spectroscopy.

Protein	Time of Reduction (min)	Moles 2-PEC/mole protein[1]	
		Amino acid analysis	UV spectra
2-PE-STI	30	4.0	4.37^2
	60	4.6	4.38^2
	120	4.4	4.47^2
	240		4.24^2
	Mean ± S.D.	4.3±0.3	4.36±.10
2-PE-OMI	30	11.4	13.27^2
	60	12.3	13.63^2
	240	15.3	14.89^2
2-PE-LBI	30 no urea		2.0^3
	30 8M urea		5.6^3
	240	7.4	7.1^3

[1] Molecular weights = STI = 21,500; OMI = 29,600; LBI = 9000.
[2] Determined by simultaneous equations (see text).
[3] Determined by direct subtraction of 2-PE LBI and LBI absorbances at 263 nm (peak maximum for 2-PEC).

within 30 min in the presence of urea. Even after 240 min, only about 7 of 14 had reacted. Omitting urea from the reaction medium slowed the alkylation of LBI, suggesting that the rate of reaction and possibly the selectivity for different disulfide bridges could be modified by appropriate choice of reaction conditions.

Table II shows the effect of extent of reduction and alkylation on trypsin inhibition. Essentially complete loss of activity of LBI was observed when 7 half-cystines per molecule, on the average, were modified. This result is consistent with the findings of Ferdinand et al. (37). Loss of activity was disproportionate to the extent of modification at lower levels. When an average of 1 disulfide bridge per LBI molecule was lost, approximately 2/3 of the inhibitory activity was lost. When modification was almost 40% complete (average of 3 disulfide bridge per LBI molecule) almost 90% of the inhibitory activity was lost. The controls show that little or no inactivation of inhibitor was due to the procedure itself and that, under these conditions, the effect of urea is reversible.

Essentially complete reduction and alkylation of STI and OMI abolished their trypsin inhibitory activities, as shown in Table II. These results are consistent with those obtained with other reducing agents.

Since some variants of LBI also inhibit chymotrypsin (38), the effect of modification on inhibitory activity against chymotrypsin was also determined to see whether the sensitivity of antichymotrypsin activity to reduction and alkylation differed from that of trypsin inhibition. As shown in Table II, the losses of inhibitory activity against both enzymes were similar. Less than 30% of the LBI consisted of variants that inhibit chymotrypsin.

Free 2-PEC (0.01 $\underline{M}$) inhibited chymotrypsin weakly (13%) at a mol ratio of ~ 5 x 10^4. PE-OMI (0.8 nmol) did not inhibit 0.4 nmol of chymotrypsin (Mol ratio of 2-PEC moiety to enzyme 30-40). Thus, the 2-PEC does not appear to bind effectively to chymotrypsin.

Where little or no inhibitory activity is found, free trypsin or chymotrypsin can attack the inhibitor, if sites of potential cleavage are accessible. Thus, the modified inhibitors may well show increased susceptibility to proteolysis. Table III shows that the 2-PE-inhibitors are attacked more readily by trypsin or chymotrypsin than are the native inhibitors, even those that do not inhibit chymotrypsin. Digestion was incomplete, however, under the conditions used. (Both primary and secondary specificities of chymotrypsin were counted in deducing the potential sites.) In the case of LBI, susceptibility to digestion by trypsin or chymotrypsin was not increased until three or more disulfide bridges were broken. These results and those in Table II show that inhibitory activities of LBI are more easily destroyed by this modification procedure than are the structural elements responsible for resistance to proteolysis.

Table II. Effect of modification on inhibitory activities against trypsin and chymotrypsin.

Inhibitor	Time (min)	S-S bonds cleaved (%)	Reductive treatment: % Initial specific inhibitory activity against:	
			Trypsin[6]	Chymotrypsin[7]
STI			100	
2-PE-STI	All	100	0	
LBI			100	100
LBI-controls[1]			92	
2-PE-LBI	30[2]	14[4]	32	19
	30[3]	40	10	15
	240	51[5]	$\leqslant 1$	0
OMI			100	
2-PE-OMI	30	68[5]	0	
	60	72[5]	0	
	240	83[5]	0	

[1] Carried through procedure with or without urea or without reduction-alkylation.
[2] Reduced-alkylated without urea.
[3] Reduced-alkylated with urea.
[4] Using ultraviolet spectroscopy.
[5] Average from ion-exchange chromatography and ultraviolet spectroscopy.
[6] BAPNA assay.
[7] ATEE assay.

Table III. Susceptibility of native and 2-PE inhibitors to proteolysis[1]

Inhibitor	Reductive treatment Time (min)	S-S bonds cleaved[5] (%)	Trypsin Digestion Time (min)	Final mg inhibitor per mg trypsin (approx.)	Bonds cleaved per molecule (avg.)	Chymotrypsin Digestion Time (min)	Final mg inhibitor per mg chymotrypsin (approx.)	Bonds cleaved per molecule (avg.)
STI			90	57	0.7	105	25	3.1
2-PE-STI	60	100	60	56	6.0	105	26	16.9
LBI			60	45	0.7	120	21	0.8
2-PE-LBI	30[2]	14	60	40	1.1	110	16	1.7
	30[3]	40	90	43	0.8	105	19	4.5
	240	51	90	58	3.8	120	17	5.9
	240	51	150	35	5.1			
OMI			110	59	0.3	100	19	2.5
2-PE-OMI	30	68	120	68	12.1	110	13	17.2
Casein[4]			120	46	9.2	30	103	9.7

1 Digestion conditions: pH 8.0, 25°C, under N_2, 4 ml of 0.1M KCl/0.02M $CaCl_2$. The average number of bonds cleaved per molecule is equal to the net titrant consumed corrected for incomplete dissociation (51).

2 Reduced-alkylated without urea.

3 Reduced-alkylated in the presence of urea.

4 Included for comparison.

5 STI, 2 S-S bonds/21,500 molecular weight; LBI, 7 S-S bonds/9,000 molecular weight; OMI, 9 S-S bonds per 29,600 molecular weight.

Another way of analyzing structural alterations produced by reduction and alkylation is comparison of thermal stabilities of native and modified inhibitors by DSC. In this technique, sample and reference solutions are heated in sealed pans at a constant rate and the difference in applied heat required to maintain the same rate of temperature increase is plotted as a function of temperature (thermogram). This endothermic heat flow or excess heat uptake by the sample as protein denaturation ("melting") occurs is seen as an endothermic peak in the thermogram. The extreme is referred to as the denaturation temperature (T_d); its position depends on heating rate. The area of the endotherm is proportional to the enthalpy of denaturation of the protein.

Thermal stability of the inhibitors was drastically altered by modification. A typical example is shown in Figure 11. The DSC thermogram of STI at pH 7.0 shows a large endothermic peak with T_d at 73°C, or 3° below that found by Donovan and Beardslee (23) at pH 6.7. That of modified STI, however, shows only a small peak (near 100°C), corresponding to a decrease of about 85% in the enthalpy of denaturation. Thus, the modified inhibitor appears to lack most of the organized structure normally disrupted near 73°C. The reason for the high T_d of the modified inhibitor is not clear, but it may represent conversion of part of the structure to a more stable conformation.

Similar reductions were obtained with OMI and LBI at pH 7. Partially reduced and alkylated LBI samples at pH 7 showed lower T_d and enthalpies of denaturation than native LBI. The enthalpy of denaturation of native LBI is very small (39). Consequently the even smaller areas of endotherms for modified LBI were too small to measure accurately.

Other samples were heated at pH 3, where the modified inhibitors are more soluble. Results were generally similar to those obtained at pH 7. The thermogram of OMI, however, became indistinct at pH 3 (cf. 40) because of effects of low pH on conformation.

Amino acid chromatograms of native and modified inhibitors are shown in Figures 1-6.

The structure of 2-QEC and ultraviolet and fluorescence spectra of 2-QE-STI are shown in Figures 12-15 and the influence of structurally different thiols on the enzyme-inhibitory activities of STI and LBI, both in pure form and in flours, are summarized in Figures 16-19.

Discussion

Reductive S-pyridylethylation (17). Most protein inhibitors of trypsin or other proteases contain disulfide bonds, some or all of which are needed to maintain active conformations (12-18). Partial loss of disulfide bridges may or may not entail selective reduction of specific bonds. In some cases, remaining inhibitory

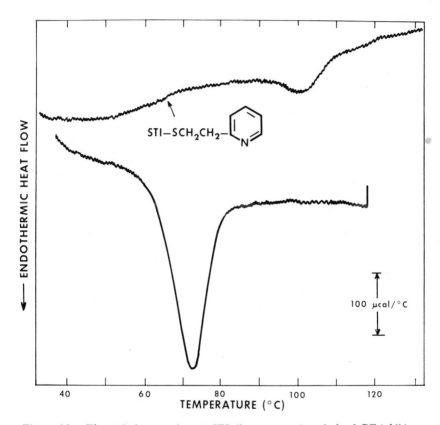

STI—SCH₂CH₂—

Figure 11. Thermal denaturation of STI (bottom scan) and the 2-PE-inhibitor (top scan) in 0.1 M Mops/NaOH, pH 7.02. Samples were heated at 10°/min in the DSC as described in the text. Native inhibitor was 16.07 mg of 5.5% (w/w) solution and the modified (30 min reaction) inhibitor was 16.18 mg of 5.3% (w/w) solution.

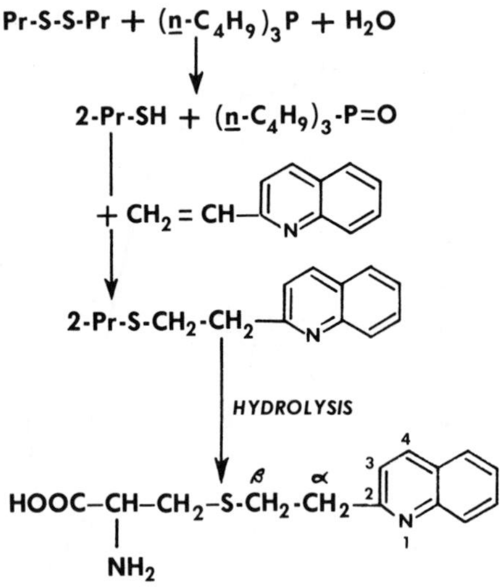

$$Pr\text{-}S\text{-}S\text{-}Pr + (\underline{n}\text{-}C_4H_9)_3P + H_2O$$

$$\downarrow$$

$$2\text{-}Pr\text{-}SH + (\underline{n}\text{-}C_4H_9)_3\text{-}P{=}O$$

$$+ \; CH_2 = CH-\text{(quinoline)}$$

$$\downarrow$$

$$2\text{-}Pr\text{-}S\text{-}CH_2\text{-}CH_2-\text{(quinoline)}$$

HYDROLYSIS

$$\downarrow$$

$$HOOC\text{-}CH\text{-}CH_2\text{-}S\text{-}CH_2\text{-}CH_2-\text{(quinoline)}$$
$$\phantom{HOOC\text{-}CH\text{-}CH_2\text{-}S\text{-}CH_2\text{-}CH_2}$$
$$NH_2$$

Figure 12.   Reductive S-quinolylethylation of a protein disulfide bond.

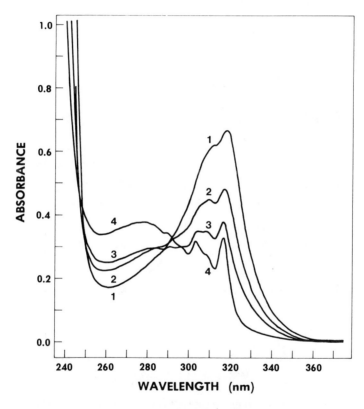

Figure 13. Effect of pH on the UV absorption spectra of 2-QE-STI. Key to pH: 1, 1.5; 2, 3.0; 3, 3.5; and 4, 7.0.

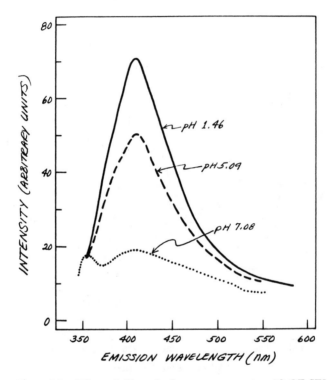

Figure 14.   Effect of pH on the fluorescence spectra of 2-QE-STI.

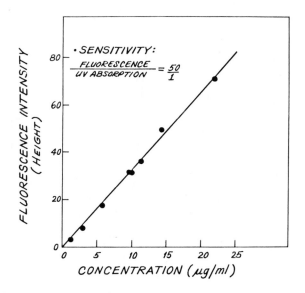

Figure 15.   Beer's law plot of fluorescence intensity vs. concentration of 2-QE-STI.

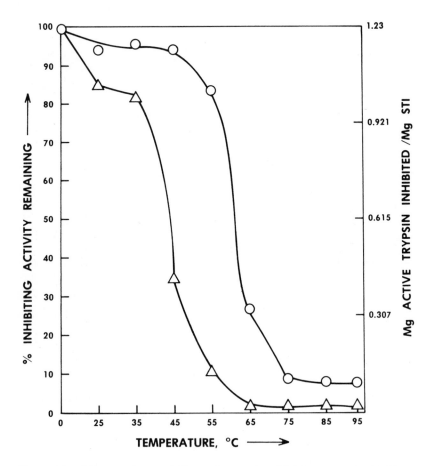

Figure 16.   Effects of heat and N-acetyl-L-cysteine *(NAC) on trypsin inhibitory activity of pure STI. The 100% value is for untreated inhibitor. Key:* △*, with NAC; and* ○*, without NAC.*

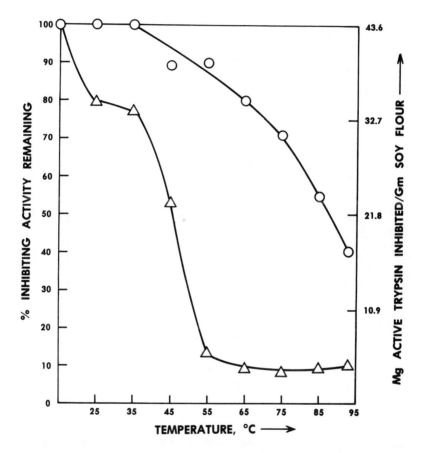

Figure 17.   Effect of heat and N-acetyl-L-cysteine (NAC) on trypsin inhibitor activity of STI in soy flour. The 100% value is for untreated soy flour (18). Key is the same as in Figure 16.

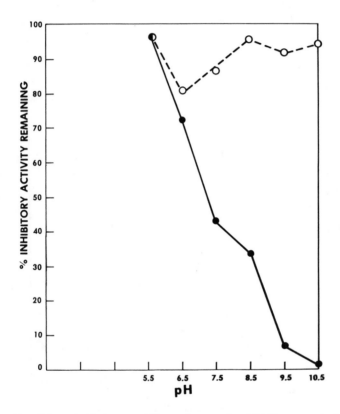

Figure 18.   Effect of pH, heat, and N-acetyl-L-cysteine *(NAC) on trypsin inhibitor activity of commercial LBI (18). Key:* ○, *heat only; and* ●, *heat and NAC.*

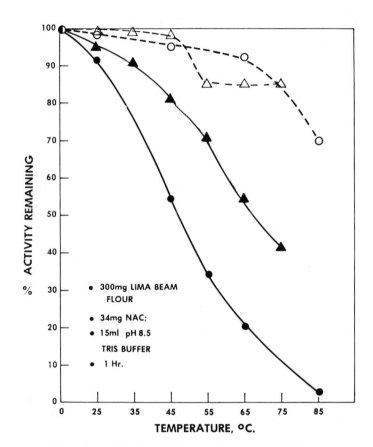

Figure 19. Effect of temperature, heat, and N-acetyl-L-cysteine (NAC) on trypsin and chymotrypsin inhibitor activity in lima bean flour (18). Key: ○, trypsin inhibitors, heat only; △, chymotrypsin inhibitor, heat only; ●, trypsin inhibitor, heat plus NAC; and ▲, chymotrypsin inhibitor, heat plus NAC.

activity could be attributed to native inhibitor mixed with modified inhibitor (15,41). In other cases, selective reduction of more susceptible disulfide bonds was shown to occur without significant loss of inhibitory activity (42-46). Incomplete reduction may essentially abolish inhibitory activity of Bowman-Birk inhibitor from soybean (15) or lima bean protease inhibitor (37,47) when over half of the disulfide bonds have been lost. Recovery of inhibitory activity on reoxidation may vary for double-acting (double-headed) inhibitors; loss or regain of chymotrypsin-inhibiting and trypsin-inhibiting activities may differ (41,48).

Since protease inhibitors vary widely in disulfide content, selected inhibitors provide useful means for applying reduction with tri-n-butylphosphine plus alkylation with 2-vinylpyridine to investigate the relationship of inhibitory activity and chemical structure. The present study was carried out with STI (low S-S), OMI (intermediate S-S), and (high S-S), as representative examples.

Unlike alkylating agents that complicate measurement of trypsin activity by introducing negatively charged groups into the region of the inhibitor reactive site (14,43), the 2-pyridylethyl moiety bears no charge at the pH of the trypsin assay, and appears not to interfere with determination of inhibitor activity.

Reductive alkylation of the trypsin inhibitors was carried out in a mixture of equal volumes of ethanol and $8\underline{M}$ urea in pH 7.6 Tris - HCl buffer. The stoichiometry of the reduction is given in Figure 7.

An ideal reagent for the selective modification of SH groups should meet the following requirements: 1) it should selectively modify the SH groups under mild conditions, 2) modified cysteine residues should survive acid hydrolysis, 3) cysteine derivatives should be eluted in a convenient position as a well-resolved peak in standard amino acid analysis, and 4) the derivative should contain a chromophore that can be determined independently, namely by ultraviolet or fluorescence spectroscopy. Our studies suggest that 2-vinylpyridine and 2-vinylquinoline appear to meet these criteria.

A two-wavelength method already outlined was used, in addition to direct subtraction of peak maxima of native and modified protein, to calculate the 2-PEC content of modified trypsin inhibitors. The results obtained with the two-wavelength procedure were more reproducible and agreed better with those obtained from amino acid analysis, particularly, where, as in STI and OMI, protein absorption contributes appreciably to the observed A_{263}.

Direct subtraction, however, gave results in better agreement with those from ion-exchange chromatography for LBI (Table I). In this case, the two-wavelength method appears to overestimate the unmodified protein concentration. Although the

reason for this difference is not apparent, this effect may
arise from aromatic stacking of the large number of pyridine
groups. Such stacking would result in lower absorptivity of
the pyridine chromophores (cf. 49).

   Results in Table I and Figures 1-11 show that advantages of
the cited procedure for reductive alkylation of disulfide bonds
in trypsin inhibitors include: 1) the cystine content can be
estimated by two independent techniques; i.e. ion exchange
chromatography and ultraviolet spectroscopy; 2) the use of ion-
exchange chromatography permits estimation of the other amino
acids as well as cystine in a single determination; and 3) the
method is especially useful for determining the extent of
partial reduction of disulfide bonds.

   The relative reactivities of disulfide bonds of the three
protein inhibitors deserve further comment. STI has two
cystine residues per mole. Both disulfide bonds are completely
reduced in about 30 minutes with the cited procedure. The
disulfide content of OMI is about three times as high as that
of STI. It has 9 cystine (18 half-cystine) residues per mole
(50). The susceptibility of OMI disulfide to reduction appears
to be less than that of STI since the data in Tables 1 and 2
show that only about 68% of OMI disulfide bonds are reduced in
thirty minutes. Finally, LBI is a highly disulfide cross-linked
protein with 7 disulfide bonds per mole (Molecular weight,
9,000) or a disulfide content about eight times as high as that
of STI. (The disulfide content of LBI approaches that of keratin
proteins). The extent of reduction and alkylation of LBI S-S
bonds is only about 51% after a 4-hr reaction time. No further
attempt was made in those studies to effect complete reductive
alkylation.

   Measurements of in vitro digestibilities of native and
modified trypsin inhibitors determined by potentiometric
titration (Table III) show the following: 1) reductive S-
pyridylethylation of disulfide bonds in STI results in an
increase in bonds cleaved by trypsin per mole of inhibitor
from 0.7 to 6.0 and, by chymotrypsin, from 3.1 to 16.9;
2) partial reductive alkylation of LBI has only a minor effect
on its digestibility by either trypsin or chymotrypsin; and
3) in all cases, digestion approached a limiting value below
the theoretical limit. Thus, although reductive alkylation
under the conditions used appears not to expose all of the
potential sites of attack to trypsin and chymotrypsin, the
number of sites cleaved is similar to that observed with
unmodified casein (Table III) cf., 50).

   These results clearly confirm that loss of inhibitory activity by the three protease inhibitors follows partial or
complete modification of their disulfide bonds.

   The loss of inhibitory activity following reduction of
disulfide bonds presumably arises from the change in the native
conformation of the protein inhibitors. Because of their

altered geometry, the modified inhibitors very likely do not
readily combine with the active site of trypsin or chymotrypsin;
or if combination does occur, dissociation is rapid because of
the change in binding forces needed to maintain the trypsin
complexes. These conclusions are supported by results from
differential scanning calorimetry (Figure 11), which show that
2 PE-STI has lost its native conformation. The small peak
exhibited by 2-PE-STI suggests that the modified protein is not
totally denatured but shows a small but measurable amount of
structural order, possibly related to the introduced hydro-
philic-hydrophobic character of the S-pyridylethyl side chains.

Since the structure of 2-PEC is somewhat similar to that
of S-aminoethylcysteine and lysine, we wished to establish
whether trypsin also hydrolyzed peptide bonds of S-pyridylethyl
residues in the modified trypsin inhibitors. Similarly, we
tested proteolysis by chymotrypsin, inasmuch as the 2-PEC side
chain includes an aromatic (pyridine) group, that might interact with the active site of chymotrypsin. As reported above,
interaction appears weak. Moreover, no evidence was found that
S-pyridylethylation produces new sites for chymotryptic or
tryptic attack, even at a pH (5.5) near the pK_a of the pyridine
nitrogen.

Reductive S-quinolylethylation (26). The 2-QE-STI differed
from its 2-PE-STI analog in solubility and spectral properties.
When the ring nitrogen of quinoline was protonated, maximum
absorption was at 318 nm. The molar extinction coefficient for
2-QEC at pH 1.5 was 11,550 M^{-1} cm^{-1}. The quinoline absorption
at 315-318 nm decreases with increasing pH, with apparent pK_a
values of about 5 for 2-QEC and about 3 for 2-QE-STI. This
large difference in pK_a suggests that the microenvironment of
the quinoline chromophores differs significantly in the model
compound and in the protein. The UV spectra at pH 1.5 yielded
a value of 4 mol of 2-QEC per mol of STI, in good agreement with
the amino acid composition.

Since 2-QEC-STI has an absorption peak where STI chromophores do not absorb, the quinoline fluorescence can be excited
independently of intrinsic protein fluorescence. Fluorescence
emission of 2-QE-STI (excitation at 316 nm) decreased with
increasing pH, as did the UV absorption. The fluorescence
efficiency was influenced by the prior treatment of the sample,
and possibly by the sequence of protonation and deprotonation.
Maximum fluorescence was obtained by acidification of a neutral
solution in Mops buffer to pH 1.5, using dilute H_2SO_4. (Since
Cl^- quenched the fluorescence, HCl was avoided.) Fluorescence
was about 50 times as sensitive as absorption, to $< 4 \times 10^{-7}$
M 2-QEC residues, and was linear with concentration up to at
least 8×10^{-6} M 2-QEC.

These results suggest that reductive 2-quinolylethylation
is a specific and sensitive method for measuring the extent of

modification of protein disulfide bonds. In the case of STI, the method yields both analytical and structural information.

Effect of thiols (18). Figure 16 shows the effect of heating STI in the absence and in the presence of NAC from 25°C to 93°C. The data are drawn in Figure 16 both in absolute terms (number of actual trypsin units titrated) and in percent decrease in STI activity as a function of temperature. Figure 17 illustrates an analogous study for STI in soy flour.

Generally, the effect of NAC starts at 25°C and is maintained throughout the temperature range studied. The thiol is especially effective in enhancing inactivation in the temperature range 35°C to about 65°C. Thus, for pure STI at 45°C, heat alone destroys only about 6% of STI compared to about 65% for heat plus NAC. Corresponding values without and with NAC at 55°C are 16% and 89%, and at 65°C, 27% and 98%, respectively. The corresponding values for STI in soy flour without and with NAC are 0% and 23% at 35°C; 11% and 47% at 45°C; and 20% and 91% at 65°C.

Several additional parameters expected to influence inactivation were also investigated. Varying the concentration of soy flour from 25 to 300 mg and keeping the concentration of NAC constant did not influence the extent of inactivation at 45°C above the 50 mg soy flour level.

Comparison of four structurally different thiols in Tables IV and V shows that all four catalyzed inactivation to about the same extent, with the exception of L-cysteine, which was less effective in inactivating the inhibitor in soy flour but not in pure STI.

The effect of pH on inactivation in the presence and absence of NAC at 45°C is shown in Table VI. Adding thiol greatly increases inactivation at every pH. Essentially all STI activity was destroyed above pH 10. A complicating factor in this pH region is that disulfide bonds are destroyed at high pH (51,52) in the absence of thiol so that the observed inactivation at high pH could result from several effects: 1) reductive cleavage of protein disulfide bonds; 2) thiol-exchange cleavage; and 3) alkaline degradation.

Additional studies revealed that the thiol effect is more efficient with larger samples of soy flour and also appears to operate in the solid state in an autoclave (Tables VII and VIII). The generality of the method is confirmed by our observation that thiols also facilitate inactivation of the highly cross-linked LBI, as illustrated in Figures 18 and 19 and Tables IX to XII. These results suggest that inactivation of enzyme inhibitor by thiols may be useful for large-scale, economical application.

Table IV.   Effectiveness of four different thiols in
            inactivating commercial STI.[1]

Thiol	mg active trypsin inhibited per mg STI	Percent STI remaining
L-Cysteine	0.11	8.9
N-Acetyl-L-cysteine	0.16	13.0
Reduced glutathione	0.11	8.9
Mercaptopropionyl-glycine	0.12	9.8
Untreated control	1.23 ± 0.03	100.0

[1]Conditions: 36 mg STI; 0.588 millimoles thiol; 10 ml
pH 8.5 Tris buffer, 55°C; 1 hour.

Table V.   Relative effectiveness of four thiols on STI
           in soy flour:[1]

Thiol	mg active trypsin inhibited per gram soy flour	Percent inhibitory activity
Cysteine	7.68	17.6
Reduced glutathione	4.04	9.2
Mercaptopropionyl-glycine	3.77	8.6
N-Acetyl-L-cysteine	3.64	8.3
Untreated control	43.63 ± 0.605	100.0

[1]Conditions: 100 mg soy flour; 0.4 millimoles thiol; 30 ml
pH 8.5 Tris buffer; 75°C; 1 hour.

Table VI.   Effect of pH and N-acetyl-L-cysteine on STI
activity of <u>soy flour.</u>[1]

pH	mg active trypsin inhibited per gram soy flour		Percent inhibitory activity remaining	
	Heat	Heat plus NAC	Heat	Heat plus NAC
2.5	33.7	32.83	72.4	70.4
4.5	31.02	31.40	66.5	67.3
5.4	31.26	30.94	67.0	66.3
6.5	26.88	20.89	57.6	44.8
7.5	29.38	15.10	63.0	32.4
8.5	38.48	22.03	82.5	47.2
9.5	36.74	10.28	78.9	22.0
10.5	36.04	2.72	77.3	5.8

Untreated control: 43.63 ± 0.605

[1]Conditions: 300 mg soy flour; 67.4 mg (0.4 millimole) NAC;
30 ml appropriate buffer; 1 hr.; 45°C. pH 2.5: 0.2M glycine•HCl;
pH 4.5-5.4: 0.5M citrate-phosphate; pH 6.5-8.5: 0.5M phosphate;
pH 8.5: 0.5M Tris; pH 9.5-10.5: 0.05M borate.   The final pH was
was adjusted with 0.5N HCl or 1.0N NaOH.

Table VII. Effect of N-acetylcysteine on STI activity in <u>soy</u>
            flour.[1]

NAC (g)	Time (hr)	% Inhibitory activity remaining
0	1	44.2
1	1	11.4
2	1	4.6
3	1	0.0
4	1	0.0
5	1	0.0
1	2	0.0

[1] Conditions: 100 grams soy flour in 500 cc (slurry) pH 8.5 Tris
    buffer; 65°C. Numbers are averages of three assays
    for each sample.

Table VIII. Effect of N-acetylcysteine on STI activity
             in <u>soy flour</u>.[1]

Time (min)	Soy flour (mg)	NAC (mmoles)	% Inhibitory activity remaining[2]
15	300	0.0	58.7
15	300	0.1	15.4
15	300	0.2	20.1
15	300	0.4	9.6
15	300	0.8	1.5

[1] Conditions: Solid mixtures in an autoclave (121°C).
[2] Results from single experiments.

Table IX.   Effectiveness of four thiols at two concentration
            levels on activity of commercial LBI.[1]

Thiol	Concentration (micromoles)	% LBI remaining
L-Cysteine	72	82.9
	290	33.7
N-Acetyl-L-cysteine	72	92.3
	290	39.9
Reduced gluta-thione	72	72.0
	290	30.5
Mercaptopropio-nylglycine	72	79.8
	290	30.5
Untreated control:		100.0

[1] Conditions: 9 mg LBI; 5 ml pH 8.5 Tris buffer; 45°C; 1 hour.

Table X. Effectiveness of four thiols at three temperatures[1]
          on trypsin inhibitor activity in lima bean flour.

Thiol	Temperature (°C)	% Inhibitory activity remaining
L-Cysteine	45	100.0
	55	68.7
	65	41.7
N-Acetyl-L-cysteine	45	78.2
	55	35.8
	65	11.8
Reduced glutathione	45	71.9
	55	35.5
	65	7.5
Mercaptopropionylglycine	45	77.2
	55	31.9
	65	17.4
None	45	109.0
	55	108.0
	65	84.1

[1] Conditions: 100 mg flour; 0.2 mmoles thiol; 15 ml pH 8.5
                 Tris buffer; 1 hour.

Table XI. Effect of NAC concentration on trypsin inhibitor activity in lima bean flour.

Time (min)	Lima bean flour (mg)	NAC (mmoles)	% Inhibitory activity remaining[2]
15	300	0.0	50.10
15	300	0.1	20.43
15	300	0.2	15.15
15	300	0.4	0.00

[1] Conditions: Solid mixtures in an autoclave.

[2] Results from single experiments.

Table XII. Effect of time and NAC on trypsin inhibitor activity of lima bean inhibitor in lima bean flour.[1]

Time (min)	Lima bean flour (mg)	NAC (mmoles)	% Inhibitory activity remaining[2]
5	300	0.0	73.2 ± 8.08
5	300	0.4	28.6 ± 4.32
15	300	0.0	50.1 ± 1.20
15	300	0.4	7.8 ± 2.63
30	300	0.0	23.7 ± 4.72
30	300	0.4	26.1 ± 4.28

[1] Conditions: Solid mixture in an autoclave (121°C).

[2] Average from four experiments.

If one represents the inactivation as a thiol-disulfide interchange, then excess thiolate anion is expected to force the equilibrium to the right (Table XIII). Since STI has two disulfide bonds per molecule, sulfhydryl-disulfide interchange can generate a large number of rearranged disulfide or mixed forms to produce proteins that are structurally different from the native STI. Although the inhibitor may tend to regenerate its original conformation, this process may be hindered or suppressed by excess thiol, or the presence of mixed disulfides. Because of altered geometry, the modified inhibitors do not readily combine with the active site of trypsin or chymotrypsin or if combination does occur, dissociation is rapid because of the change in binding forces needed to stabilize the trypsin complexes.

It should be emphasized that from both chemical and nutritional viewpoints, inactivation of a protease inhibitor by an added thiol <u>via</u> a sulfhydryl-disulfide interchange should be distinguished from the related reductive cleavage of disulfide bonds followed by alkylation of the generated SH groups. The former treatment produces new disulfide bonds whereas the latter eliminates the original ones from the protein. In the case of soy flour, possibilities for sulfhydryl-disulfide interchange include interaction between inhibitor SH and S-S bonds with soybean protein SH and S-S groups, respectively, to form a complex network that may make it difficult for the original disulfide bonds to be reformed (Table XIII). The relative concentration of SH and S-S bonds in the thiol and protein may dictate whether the thiol is considered a catalyst or stoichiometric reagent in the inactivation. For these reasons, optimum conditions for the inactivation of disulfide-containing protease inhibitors by thiols have to be established in each case.

Table XIII. Sulfhydryl-disulfide interchange pathways:

R-SH   Added thiol (Cysteine, NAC, GSH, etc.)
In-S-S-In   Inhibitor (In) disulfide bonds
Pr-S-S-Pr   Protein (Pr) disulfide bonds

1. R-SH + In-S-S-In $\rightleftharpoons$ R-S-S-In + HS-In
2. R-SH + Pr-S-S-Pr $\rightleftharpoons$ R-S-S-Pr + HS-Pr
3. In-SH + Pr-S-S-Pr $\rightleftharpoons$ In-S-S-Pr + HS-Pr
4. In-SH + HS-Pr $\rightleftharpoons$ In-S-S-Pr
5. In-SH + HS-In $\rightleftharpoons$ In-S-S-In

Net Effect: network of new disulfide bonds.

It should also be emphasized that although the observed inactivation of the inhibitor can be rationalized by the postulated sulfhydryl-disulfide rearrangement mechanism, other less obvious possibilities cannot be ruled out. Additional physicochemical studies are needed to unequivocally establish to what extent the thiol treatment rearranges native disulfide bonds or facilitates unfolding inhibitor protein chains.

Modifying disulfide bonds of both inhibitors and structural proteins with which they are associated may, besides inactivating trypsin inhibitors, also improve digestibility and nutritional quality since disulfide cleavage is expected to decrease protein crosslinking. The mixed-disulfide proteins should be more accessible to proteolytic enzymes, thus permitting more rapid digestion.

This expectation was realized since measurement of in vitro digestibilities of NAC-treated soy flour by trypsin (Table XIV) show that although NAC had no effect at 25°C, its beneficial influence at higher temperatures was striking. Treatment with NAC at 45°C results in an increase in bonds cleaved by trypsin per mole of soy protein from about 4 to 10; treatment at 75°C, from about 6 to 18. This last value is nearly equal to all the available trypsin-susceptible arginine and lysine residues in soy proteins (36).

The theoretical rationale for measuring the extent of peptide bonds cleavages is based on potentiometric titration of hydrogen ions liberated when peptide bonds are hydrolyzed (51) (Equation 1).

Equation 1.

$$P-CO-NH-P' \xrightarrow[H_2O]{Trypsin} P-COO^- + (H_3N^+-P' \underset{}{\overset{K}{\rightleftharpoons}} H^+ + H_2N-P')$$

P, P' = Protein side chains.

Above pH 6.5, the number of hydrogen ions liberated is equal to the number of dissociated α-amino groups produced (34). Since the average pK of these new amino groups is near 7.5 (34,35), about 90% of the splits will produce titratable hydrogen ions at the pH (8.5) used in the present study. Since pH is kept constant during the titration, the degree of ionization of lysine ϵ-amino groups (pK near 10) does not change. Similarly, the slight contribution of hydrogen ions from the single terminal α-amino group can also be neglected.

In a previous study (54), digestibility measurements on the pH-stat were carried out at pH 7.5 and the correction for incomplete ionization of the liberated amino groups was

Table XIV. Effect of temperature and N-acetyl-L-cysteine on trypsin inhibitor activity and trypsin digestion of dialyzed soy flour.[1]

Temperature (°C)	N (%)	Percent inhibitory activity remaining		Number of peptide bonds cleaved by trypsin/mole of protein[2]	
		Heat	Heat plus NAC	Heat	Heat plus NAC
25	10.8	98.7		5.6	
25	11.7		84.4		4.9
45	12.1	82.5		3.9	
45	10.4		47.2		10.6
75	10.9	53.2		6.4	
75	10.5		0.0		18.2

[1] Conditions: 300 mg soy flour; 67.4 mg (0.4 millimole) NAC; 30 ml pH 8.5 Tris buffer; 1 hour at the indicated temperatures.

[2] Based on calculated protein content (N X 6.25) and an average molecular weight per protein chain of 23,000 (36).

inadvertently overlooked; therefore, column headings in 54 reading "number of peptide bonds cleaved per mole casein" should read "moles hydrogen ions liberated per mole casein".
        The relation between the number of peptide bonds cleaved and hydrogen ions produced is given by Equation 2:

Equation 2.

$$pH - pK = \log \frac{NH_2}{NH_3^+} = \log \frac{H^+}{NH_3^+}$$

The number of enzyme-catalyzed peptide bonds cleaved per unit (grams or moles) of protein is given by Equation 3:

Equation 3.

$N = H^+ [1 + 10^{(pK - pH)}]$, where

N = number of peptide bonds cleaved per unit of protein.

H^+ = number of hydrogen ions liberated and titrated during the enzymatic digestion per unit weight of protein.

pK = average pK values of the new α-amino groups produced during the enzymatic hydrolysis (equation 1). This value generally ranges between 7.5 and 7.6 (34,35,56).

pH = pH value at which the enzymatic digestion was carried out.

Thus, at pH 7.5 cleavage of two peptide bonds liberates one hydrogen ion since N = H$^+$ (1 + 1). At pH 8.0 and 8.5, the corresponding values are 1.32 and 1.1, respectively.

Finally, forming mixed disulfides with cysteine or NAC would also increase the cystine content of the proteins. This is a desirable objective because methionine is a limiting amino acid in legumes (56, 57,59) and cystine has a sparing effect on methionine. This expectation was also realized since preliminary studies show that the half-cystine content of treated commercial LBI and STI and of lima bean and soy flours are significantly higher than the corresponding values for the starting materials (Table XV).

Table XV. Half-cystine content of treated and untreated materials.

Material	Treatment	Half-cystine (nmoles/mg)[1]
LBI	dialyzed in H$_2$O	855.0
LBI	36 mg LBI + 96 mg NAC; pH 8.5; 45°C; 1 hr.	1717.6
Lima bean flour	300 mg flour; pH 8.5: 65°C; 1 hr.	27.1
Lima bean flour	300 mg flour + 34 mg NAC; pH 8.5; 65°C; 1 hr.	130.5
Soy flour	dialyzed in H$_2$O	67.0
Soy flour	700 g + 21 g NAC; pH 8.5; 45°C; 1 hr.	139.4

[1] Determined on an amino acid analyzer as cysteic acid after performic acid oxidation and acid hydrolysis (60).

Conclusions

Disulfide bonds in trypsin inhibitors from soybean (STI), lima bean (LBI), and hen egg white (OMI) were reductively alkylated with tri-n-butylphosphine and 2-vinylpyridine. The extent of S-pyridylethylation was determined by both amino acid analysis and ultraviolet absorption spectroscopy. Modification resulted in loss of inhibitory activity against trypsin or chymotrypsin. Observed losses in resistance to proteolysis and in stability to heat denaturation, measured by differential scanning calorimetry, showed that structural order was decreased by modification of the inhibitors.

Sensitivity of measurements of the extent of disulfide bond modification can be increased by a factor of about 50 if a fluorescent adduct is produced, as was shown in similar studies with an S-quinolylethylated analog of STI. These reactions should find application in studies of structural and functional consequences of partial or complete alteration of protein disulfide bonds.

The cooperative effect of thiols and heat in inactivating STI suggests that the principle of inactivating plant protease inhibitors via sulfhydryl-disulfide interchange merits nutritional and toxicological evaluation, especially since autoclaving of soybean flour may result in destruction of cystine, arginine, and lysine (55). In fact, the amount of heat required to destroy growth inhibitors in raw soybeans may destroy sufficient cystine to make this amino acid first-limiting. The use of thiols to facilitate inactivation at lower temperatures may, therefore, be especially valuable for heat-sensitive commodities because it may: 1) decrease protein damage; 2) improve digestibility and increase the sulfur (half-cystine) content of sulfur-poor legume proteins; and, 3) save energy.

Acknowledgement

This chapter is paper 3 in a series on "Inactivation of Enzyme Inhibitors". For papers 1 and 2, see 17 and 18. Presented at the American Chemical Society Meeting, New York, New York, August 23-27, 1981 (Abstract AGFD 23) and at the Second Chemical Congress of the North American Continent, Las Vegas, Nevada, August 25-29, 1980 (Abstract AGFD 28).

Literature Cited

1. Richardson, M. Phytochem., 1977, 16, 159. Food Chemistry, 1981, 6, 235.
2. Liener, I. E.; Kakade, M. L. In "Toxic Constituents in Plant Foodstuffs"; Liener, I. E., Ed., Academic Press: New York, 1980; p. 7.
3. Liener, I. E., In "Protein Nutritional Quality of Foods and Feeds", Friedman, M., Ed.; Marcel Dekker: New York, Part 2, 1975; p. 523.

4. Bozzini, A.; Silano, V. In "Nutritional Improvement of Food and Feed Proteins"; Friedman, M., Ed.; Plenum Press: New York, 1978; p. 249.
5. Swartz, M. J.; Mitchell, H. L.; Cox, D. J.; Reeck, G. R. J. Biol. Chem., 1977, 252, 8105.
6. Silano, V. Cereal Chem., 1978, 55, 722.
7. Rouleau, M.; Lamy, F. Can. J. Biochem., 1974, 53, 958.
8. Ryan, C. A. Ann. Rev. Plant Physiol., 1973, 24, 173.
9. Booth, A. N.; Robbins, D. J.; Ribelin, W. E.; DeEds, F. Proc. Soc. Exp. Biol. Med., 1960, 104, 681.
10. Rackis, J. J.; McGhee, J. E.; Booth, A. N. Cereal Chem., 1975, 52, 85.
11. Anderson, R. L.; Rackis, J. J.; Tallent, W. H. In: "Soy Protein and Human Nutrition", Wilcke, H. L.; Hopkins, D. T.; Waggle, D. H., Eds.; Academic Press: New York; 1979, p. 209.
12. Tschesche, H. Angew. Chem. Internat. Ed., 1974. 13, 10.
13. Whitaker, J. R.; Feeney, R. E. In "Toxicants Occurring Naturally in Foods", National Academy of Sciences, Washington, D.C., 1973; p. 276.
14. Laskowski, M., Jr.; Sealock, R. W. In "The Enzymes", Boyer, P. D., Ed.; Academic Press: New York, 1971; Vol. III, p. 375.
15. Hogle, J. M.; Liener, I. E. Can. J. Biochem., 51, 1973, 1014.
16. Means, G. E.; Ryan, D. S.; Feeney, R. E. Accounts Chem. Res., 1974; 7, 315.
17. Friedman, M.; Zahnley, J. C.; Wagner, J. R. Anal. Biochem., 1980; 106, 27
18. Friedman, M.; Grosjean, O. K.; Zahnley, J. C. J. Sci. Food Agric., 1982; 33, 165; Nutrition Repts. Int., 1982, 25, 743.
19. Weder, J. K. P. In "Advances in Legume Systematics", R. M. Polhill, Ed.; Raven Press, New York, 1981; p. 533.
20. Friedman, M. "The Chemistry and Biochemistry of the Sulfhydryl Group in Amino Acids, Peptides, and Proteins", Pergamon Press: Oxford, England and Elmsford, New York. 1973; p. 199.
21. Friedman, M.; Noma, A. T. Text. Res. J., 1970, 40, 1073.
22. Krull, L. H.; Gibbs, D. E.; Friedman, M. Anal. Biochem. 1971, 40, 80-85.
23. Donovan, J. W.; Beardslee, R. A. J. Biol. Chem., 1975, 250, 1966.
24. Donovan, J. W.; Ross, K. D. Biochemistry, 1973, 12, 512.
25. Beardslee, R. A., Zahnley, J. C. Arch. Biochem. Biophys., 1973, 158, 806.
26. Zahnley, J. C.; Friedman, M. submitted for publication.
27. Willard, H. H.; Merritt, L. L., Jr.; Dean, J. A. "Intrumental Methods of Analysis", 4th edition, Van Nostrand: New York, 1965; p. 94.

28. Smith, C.; Van Megen, W.; Twaalfhoven, L.; Hitchcock, C. J. Sci. Food Agric. 1980, 31, 341.
29. Mikola, J; Suolinna, E. M. European J. Biochem., 1969, 9, 555.
30. Chase, T., Jr.; Shaw, E. Methods Enzymol., 1970, 19, 20.
31. Zahnley, J. C.; Davis, J. G. Biochemistry., 1970, 9, 1428.
32. Erlanger, B. F.; Kokowsky, N.; Cohen, W. Arch. Biochem. Biophys., 1961, 95, 271.
33. Wilcox, P. E. Methods Enzymol., 1970, 19, 64.
34. Rupley, J. A. Methods Enzymol., 1967, 11, 905.
35. Tanford, C. Adv. Protein Chemistry. 1962, 17, 69.
36. Wolf, W. J. In "Food Proteins" Whitaker, J. R.; Tannenbaum, S. R.; Eds., Avi: Westport, Connecticut, 1977; p. 291.
37. Ferdinand, W. H.; Moore, S.; Stein, W. H.; Biochim. Biophys. Acta., 1965, 96, 524.
38. Haynes, R.; Feeney, R. E. J. Biol. Chem., 1967, 7, 2879.
39. Zahnley, J. C. Biochem. Biophys, Acta, 1980, 613, 178.
40. Donovan, J. W. Biochemistry, 1967, 6, 3918.
41. Stevens, F. C.; Doskoch, E. Can. J. Biochem., 1973, 51, 1021.
42. Kress, L.; Laskowski, M., Sr. J. Biol. Chem., 1967, 242, 4925.
43. Kress, L. F.; Wilson, K. A.; Laskowski, M., Sr. J. Biol. Chem., 1968, 243, 1758.
44. Liu, W. K.; Meienhofer, J. Biochem. Biophys. Res. Commun., 1968, 31, 467.
45. DiBella, F. P.; Liener, I. E. J. Biol. Chem., 1969, 244, 2824.
46. Plunkett, G.; Ryan, C. A. J. Biol. Chem., 1980, 255, 2752.
47. Jones, G.; Moore, S.; Stein, W. H. Biochemistry, 1963, 2, 66.
48. Sjöberg, L. B.; Feeney, R. E. Biochim. Biophys. Acta., 1968, 168, 79.
49. Wu, Y. V.; Cluskey, J. E.; Krull, L. H.; Friedman, M. Can. J. Biochem., 1971, 49, 1042.
50. Kato, I.; Kohr, W. J.; Laskowski, M., Jr. Proc. 11th FEBS Meeting, "Regulatory Proteolytic Enzymes and their Inhibitors", Magnusson, S.; Ottesen, M.; Foltmann, B.; Dano, K., Eds.) 1978, p. 197.
51. Rupley, J. A. Methods Enzymol., 1967, 11, 905.
52. Asquith, R. S.; Otterburn, M. S. In "Protein Crosslinking: Nutritional and Medical Consequences", Friedman, M., Ed., Plenum Press: New York, 1977, p. 93.
53. Friedman, M. In "Nutritional Improvement in Food and Feed Proteins", Friedman, M., Ed., Plenum Press: New York, 1978, p. 613.
54. Friedman, M.; Zahnley, J. C.; Masters, P. M. J. Food Sci., 1981, 46, 127.
55. Rios Iriarte, B. J.; Barnes, R. H. Food Technol., 1966, 20, 835.

56. Mihalyi, E. "Application of Proteolytic Enzymes of Protein
     Structure Studies", 2nd edition, CRC Press: Boca Raton,
     Florida, 1978; p. 129.
57. Kwong, E.; Barnes, R. H. J. Nutr., 1963, 81, 392.
58. Frost, A. B.; Mann, G. V. J. Nutr., 1966, 89 49.
59. Kakade, M. L.; Arnold, R. L.; Liener, I. E.; Weibel, P. E.
     J. Nutr., 99, 34.
60. Moore, S. J. Biol. Chem., 1963, 238, 235.

RECEIVED June 16, 1982.

Nutritional Quality of Deteriorated Proteins

LOWELL D. SATTERLEE

University of Nebraska, Department of Food Science and Technology, Lincoln, NE 68583

KOW-CHING CHANG

Chinese Culture University, Department of Food and Nutrition,
Yam-Ming-Shan, Taiwan, People's Republic of China 113

During the processing and/or storage of proteins, any alteration of the amino acid residues within the proteins almost
always affects the protein's nutritional quality. High temperatures or prolonged exposure to heat or alkali cause losses in the
essential amino acids (EAA), which are due to: 1) Maillard reactions, 2) formation of isopeptide bonds, and 3) racemization of
the L-amino acid residues. Secondly, the presence of certain
modified amino acids in the diet of the rat have been implicated
in both toxic side reactions and growth inhibition. The presence
of pro-oxidants during protein processing results in the partial
and/or complete oxidation and loss of availability of the EAA
methionine and cysteine/cystine. Protein hydrolysis under the
controlled conditions of the plastein reaction can be beneficial
to protein quality. Plastein gels made from damaged or crude
protein sources have improved sensory, functional and nutritional
properties.

Presence of Oxidized Sulfur Amino Acids in Food Proteins

Incidence. It is known that several key essential amino
acids (EAA), methionine, cysteine/cystine (Cys) and lysine, can
exist either partially or completely in biologically unavailable
forms within proteins in their native state, as well as in proteins having undergone processing. Oxidized Cys (cysteic acid)
has been shown to be inherent in two isolectins found in unprocessed Haricot dry beans (1). Methionine sulfoxide, S-methyl
cysteine and S-methyl cysteine sulfoxide have been found to be
inherent in kidney beans and several other cultivars of Phaseolus
vulgaris (2, 3). The presence of γ-glutamyl methionine sulfoxide
in seeds other than dry beans has been reported by Kasai et al.
(4). Evans and Bauer (5) noted that the methionine and Cys in
autoclaved Sanilac beans were only 40-50% available, while
Sgarbieri and coworkers (6) found that only 30-40% of the methionine was available in four other cultivars of Phaseolus vulgaris
tested.

0097-6156/82/0206-0409$06.25/0

Kunachowicz et al. (7) and Strange et al. (8) have shown
that the processing of soy isolate, casein and meat (frank-
furters) transforms some of the methionine and Cys in these protein sources into unavailable forms. Heat, hydrogen peroxide
and lipid produced hydroperoxides can oxidize both methionine and
Cys to partially or completely oxidized forms (9, 10, 11).

The bioavailability of the methionine and Cys residues in a
protein is determined by two factors: 1) the degree to which the
protein can be digested by proteases and peptidases, a measure of
the total quantity of essential amino acids released via digestion, and 2) the chemical structure of the amino acids (i.e.,
oxidized or reduced, and complexed with sugars) released, a determinant of their bioavailability for use in synthesizing new
tissue protein.

The bioavailability of the various oxidized forms of methionine and Cys differ, depending upon their degree of oxidation.
Cysteic acid and methionine sulfone, as well as S-methyl cysteine,
are completely unavailable to the rat, whereas L-methionine sulfoxide is partially utilized by the rat (60% as effective as
L-methionine), (12, 13). Protein-bound methionine sulfoxide has
been suggested to be as available as protein-bound methionine
(9, 14). Methionine sulfoxide may, in fact, be partially or
completely reduced to methionine in the digestive tract, which
could account for its high degree of bioavailability. The ability of the rat to adapt to and efficiently use methionine sulfoxide was described by Miller et al. (15). Miller and coworkers
demonstrated that data from short term rat bioassays (17 days)
showed methionine sulfoxide to be only 66% as available as
methionine. But when fed over a 132-day period, there was no
significant difference between methionine and methionine sulfoxide in promoting rat growth.

Recent data discussed by Scrimshaw and Young (16) indicate
that soy isolates free of anti-nutrients (lectins and trypsin
inhibitors) had a higher protein nutritional quality when
measured in humans, than when measured in the rat. The variation
in protein quality assessment between man and the rat could be
the result of their differing abilities to use altered (oxidized
in this case) sulfur amino acids. Since methionine is the
limiting EAA in soy products, any oxidation of methionine and Cys
(Cys can partially make up a methionine deficiency) would lower
the nutritional quality of soy products. This is especially true
if the species being fed (the rat) could only partially utilize
some of the intermediate methionine oxidation products (i.e.,
methionine sulfoxide).

Detrimental effects of processing. Proteins derived from
legumes (i.e., soy, peanut, dry beans and select leaf proteins)
are first limiting in the essential amino acid methionine and
often undergo extensive processing before incorporation into
human foods. The effects processing has on the sulfur amino

acids in these four legume protein sources was studied by
Marshall et al. (17). Their results obtained on the products
obtained from the processing of soy and dry beans are discussed
below.

Preliminary work by Marshall et al. (18) as illustrated in
Figure 1, indicated that the extent of Cys loss in model proteins
(casein, egg white and soy isolate) varies between proteins, but
the loss of Cys always increased with increasing exposure to the
oxidizing agent (H_2O_2) and temperature. Figure 2 illustrates
that the extent of methionine loss was not as dependent on protein type, as it was for Cys. Figures 3A and 3B do note that
methionine sulfoxide is rapidly formed under mild oxidizing conditions, whereas strong oxidizing conditions are needed for the
production of the biologically unavailable methionine sulfone.

Table I contains the results obtained on samples taken
during the processing of dry beans. Reactive Cys (available
cysteine and cystine) is slightly lower than the total Cys in the
original bean flour and appears to be substantially lower in the
laboratory and pilot plant produced bean protein concentrate
(BPC). Only minor amounts of methionine sulfoxide (MetSO) were
found in the BPC products.

Table I.   The various forms of methionine and Cys found in dry
bean products prepared in the laboratory and pilot
plant (17).

Sample	Cysteine/Cystine g/16gN		Methionine g/16gN	
	Total	Reactive	Total	MetSO[1]
Bean Flour	1.07	0.92	1.57	0
Laboratory				
BPC 25	0.91	0.45	1.74	0.05
BPC 90	0.91	0.49	1.63	0.06
Pilot Plant				
BPC 25	0.86	0.60	1.54	0.10
BPC 80	0.81	0.57	1.47	0.06

[1] Methionine sulfoxide was the only form of oxidized methionine
present.

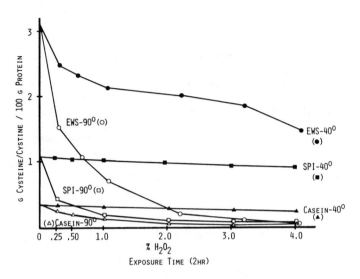

Figure 1.   The loss of both cysteine and cystine during the exposure of egg white solids (EWS), soy protein isolate (SPI), and casein to H_2O_2 at 30 and 90°C. (Reproduced, with permission, from Ref. 18. Copyright by Institute of Food Technologists.)

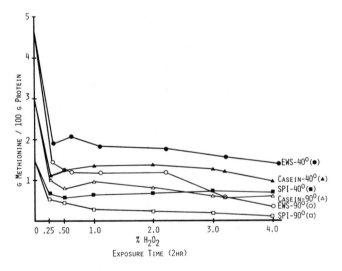

Figure 2.   The loss of methionine during the exposure of egg white solids (EWS), soy protein isolate (SPI), and casein to H_2O_2 at 40 and 90°C. (Reproduced, with permission, from Ref. 18. Copyright by Institute of Food Technologists.)

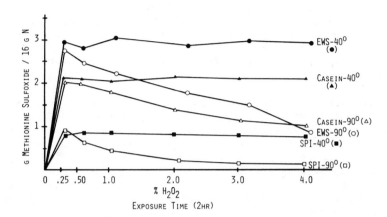

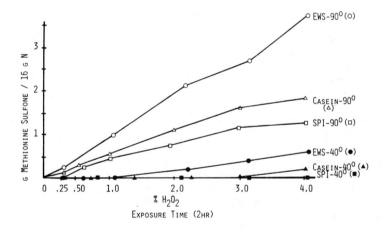

Figure 3. The production of methionine sulfoxide (top) and methionine sulfone (bottom) in egg white solids (EWS), soy protein isolate (SPI), and casein during exposure to H_2O_2 at 40 and 90°C. (Reproduced, with permission, from Ref. 18. Copyright by Institute of Food Technologists.)

The lower concentrations of reactive Cys in the BPC products are consistently found. As one views the simple NaCl extraction, acid precipitation and spray drying process to obtain PBC, it is not apparent that there are any specific steps that should cause losses of any of the amino acids. This is also indicated by the methionine data, which show little to no loss. Yet, we do see appreciable losses of reactive Cys in all BPC products, for which we have no good explanation at this time. It should be noted that for this, as well as other processes studied, the only oxidized form of methionine present was MetSO. No MetSO$_2$ was detected in any of the processed legume samples or starting materials.

The unprocessed soybeans contain small amounts of cysteic acid and MetSO, which end up in: 1) white flakes and flour following oil removal, 2) acid wash (whey) following soy protein concentrate production and 3) acid whey following soy isolate production. Soybean proteins rich in sulfur amino acids (enzymes, protease inhibitors) and peptides containing oxidized sulfur amino acids are present in the whey fractions (19, 20). Green gram seeds and kidney beans have been shown to contain peptides possessing oxidized sulfur amino acid residues (2, 4). There is no evidence at present to indicate that any measurable losses of methionine and Cys occur at any one step in soybean processing, only that the MetSO and cysteic acid present in various soy products is generally derived from the raw soy bean.

The above statement is true for all products and by-products listed in Table II, except the commercial soy isolate 1. This product has about 90% of its total Cys content present as reactive Cys and yet over 50% of the total methionine is present as MetSO. The model protein studies described at the beginning of this section show a similar effect for the oxidation of the three model proteins. At low concentrations of H$_2$O$_2$ (0.25%), methionine residues are readily oxidized to MetSO, while Cys residues remain relatively unchanged. It is entirely possible that the procedures employed to produce this isolate involve a mild oxidation step that results in the preferential oxidation of methionine residues in contrast to Cys. Perhaps this may be the addition of a mild oxidizing agent to remove off-colors from the final product.

Significance of findings. A review of the literature and the recent findings of Marshall et al. (17) indicate that partially or completely unavailable forms of Cys and methionine can be found to be significant, they either are present as the unavailable cysteic acid or the partially available MetSO. Methionine sulfone, which is not available biologically, is very difficult to produce and therefore is rarely, if ever, seen in typical foods or food ingredients.

Table II. The various forms of methionine and Cys found in
soybeans and its protein products and by-products
(17).

Sample	Cysteine/Cystine (g/16gN)		Methionine (g/16gN)[1]	
	Total	Reactive	Total	MetSO
Raw Soybeans	1.49	1.43	1.59	0.04
Hexane Extracted White Flakes	1.52	1.32	1.48	0.06
Commercial				
Soy Flour 1	1.50	1.28	1.44	0
Soy Flour 2	1.40	1.32	1.46	0
Soy Flour 3	1.41	1.26	1.48	0
Laboratory				
Soy Concentrate	1.15	1.14	1.44	0
Whey from Soy Concentrate	3.26	2.35	1.55	0.35
Commercial				
Soy Concentrate 1	1.36	1.16	1.43	0
Laboratory				
Na Soy Isolate pH 10	1.23	1.09	1.47	0
Whey from Na Soy Isolate (pH 10)	3.36	2.51	1.59	0.35
Commercial				
Soy Isolate 1	1.13	1.00	1.37	0.70
Soy Isolate 2	1.09	1.04	1.32	0
Soy Isolate 3	1.16	1.10	1.34	0
Soy Isolate 4	1.07	1.18	1.28	0

[1] The only form of oxidized methionine found was MetSO.

The Maillard Reaction -- New Concerns

     The Maillard non-enzymatic browning reaction which occurs
between free amine and carbonyl groups, is ubiquitous in food
processing. This reaction is known to create both positive as
well as negative characteristics in food colors and flavors.
However, the nutritional quality of foods undergoing the Maillard
reaction is almost always reduced. The Maillard reaction has
been found to occur over a broad range of temperatures, ranging
from high temperatures obtained during processing down to ambient
storage temperatures. The nutritional changes induced by the
Maillard reaction include: 1) decrease in the bioavailability
of lysine as well as of several other essential amino acids,
2) decrease in protein digestibility and 3) possible formation of
growth inhibitory and/or toxic compounds. The loss of lysine,
arginine, serine, threonine, isoleucine and leucine has been reported during storage of egg albumin in the presence of glucose
at 37°C for 10 days (21), with lysine having the greatest loss.
The availability of all essential amino acids in the egg albumin
were reduced during the 30 days of storage with glucose at 37°C
(22). The digestibility of the browned egg albumin was substantially reduced, and it retained less than 10% of its original
nutritive value, as was determined by Relative Nutritive Value
(RNV) and Protein Efficiency Ratio (PER) bioassays. The decrease
in protein digestibility of the Maillard browning products is due
to the formation of cross-linkages between amino acid residues.
These cross-links decrease the number of protease specific sites
for peptide bond cleavage and also serve as a steric hindrance
for those enzymes which attack non-specific peptide bonds.
Reductions in amino acid availability, protein digestibility, and
nutritive value as a result of the Maillard reaction have been
reported by other researchers (23, 24, 25). The N-substituted
aldosylamines and the Schiff bases formed early in the Amadori
rearrangement have been shown to be 100% biologically available
(26). Essential amino acids used to produce the Amadori compounds, i.e., N-substituted 1-amino, 1-deoxy, 2-Ketose-amino
acids have been shown to be poorly absorbed and unavailable for
the growth of experimental animals. Several free Amadori compounds, i.e., fructose-lysine, fructose-tryptophan, fructose-
leucine, fructose-methionine and fructose-phenylalanine have been
observed to be resistant to digestion and absorption in the small
intestine, but easily assimilated by the microorganisms in the
large intestine. A portion of these free Amadori compounds are
absorbed by passive diffusion through the wall of the large
intestine, but are rapidly excreted as intact molecules in the
urine (22, 27-30).
     The decrease in the nutritional quality of proteins having
undergone Maillard browning is not only due to the reduction in
the essential amino acid content of the protein and the decrease
in protein digestibility, but is also due to the suspected

formation of toxic and/or growth inhibitory compounds by the
Maillard reaction (31). Egg albumin stored with added glucose
under mild conditions (37°C for 10 days) readily browned, and
was fed to rats for a one-year period (31). The results from
the one-year feeding study showed the rats fed the browned protein gained less weight (Figure 4) and possessed enlarged internal organs, particularly the liver and kidneys. These rats also
exhibited increases in serum glutamate-oxalate transaminase,
serum alkaline phosphatase, blood urea nitrogen and urine specific gravity, and decreases in blood hemoglobin and hematocrit
levels. Abnormal kidney function also resulted because of the
high specific gravity of urine excreted by the rats fed the brown
egg albumin diet. The accumulation of a brown-black pigmentation
in the rat's liver was also observed. Further isolation and
identification of the inhibitory and toxic compounds produced by
the Maillard reaction are needed in order to substantiate the
results given above and to determine if such compounds are
present in man's normal diet. Secondly, the toxic levels of such
compounds should be determined.

Mutagenic compounds can be produced when Maillard reaction
products react with nitrites. The Amadori compound, fructose-
tryptophan, when incubated with excess sodium nitrite for 24
hours at 37°C, was shown to be mutagenic. The mutagenicity of
the Maillard products in the absence of nitrite was also observed
for a mixture of glucose and lysine after being heated at 100°C
(32). A carcinogenic nitrosamine has been reported to be formed
by the nitrosation of the Maillard product thiazolidine (a condensation product of glucose and cysteine) in an acidic sodium
nitrite solution. Nitrosamines could be formed in heated foods
containing nitrites or by the in vivo nitrosation of Maillard
reaction products in the gastrointestinal tract, since nitrite
is present in saliva (33). Amadori compounds which are secondary
amines can react with nitrite to form nitrosamines which are of
significance in the etiology of cancer; and deserve further
investigation.

Contrary to the above report, egg albumin having undergone
Maillard browning under mild conditions (37°C, 40 days) displayed
no mutagenic properties (34). The conflicting reports on the
mutagenic properties of the Maillard products imply that further
research is needed in order to determine the factors which influence the mutagenic response. Since the Maillard reaction can use
many different reactants (free amino acids, peptides, proteins,
reducing sugars and aldehydes) and reaction conditions (degree of
heat, pH, A_w, and time), extensive research is needed to determine which of these variables leads to the possible production of
mutagenic compounds.

Heat Induced Protein-Protein Cross-Bridging

Protein deterioration during heating in the absence of

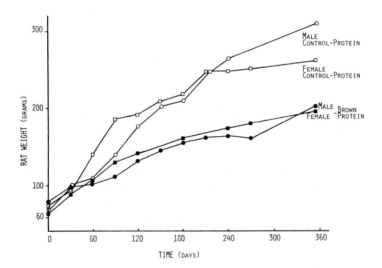

Figure 4. The growth of white rats fed both control and Maillard browned egg white protein over a one year period. (Reproduced from Ref. 31. Copyright 1980, American Chemical Society.)

reducing sugars and carbonyls, proceeds at a slower rate than
when such compounds are present (35). Yet the nutritional
quality of proteins can be lowered when they are exposed to
severe and/or prolonged heat treatment without the Maillard
reaction operative. The reduction in protein nutritional quality
in these cases is primarily due to a drop in protein digestibility, caused by the formation of protease resistant cross-links
between amino acid residues within or between proteins. The
decreased protein digestibility reduces the bioavailability of
almost all of the amino acid residues in the protein. Isopeptide
bonds (peptide bonds between amino acid side chains) readily
formed by the condensation of the ξ-amino group of lysine with
the amide groups of asparagine or glutamine, have been found in
heated (131°C, 27 hours) chicken muscle proteins (35). Lysinoalanine, a well known cross-linkage formed by lysine and degraded
cystine, can also be induced by heating both in the absence and
presence of alkaline, will be discussed in the following section.

Data obtained from running the Net Protein Ratio (NPR)
bioassay on heated chicken muscle (121°C, 27 hours) noted that
it had only 35% of the nutritive value of the unheated muscle
(35). The drop in protein nutritional quality could only be
partially explained by a drop in digestibility, since the isopeptide bond can delay release of amino acids, but the isopeptides
once released from the protein chain are easily digested and the
amino acid once again becomes available. Unlike the Amadori
compounds (i.e., fructose-lysine) the glutamyl-lysine type compounds have been shown to be available to the rat as a source of
lysine. Other factors that may have contributed to the low NPR
of the protein are: 1) destruction of Cys, 2) racemization of
the amino acid residues, and/or 3) formation of growth inhibitory
substances during heat treatment.

Lysinoalanine Residues Within Proteins

The alkaline treatment of food ingredients has been used in
food processing for several desirable purposes. Yet is is known
that alkaline treatments can undesirably alter the structure and
chemical composition of food proteins. Alkali-induced reactions
of nutritional significance include: 1) racemization of amino
acid residues, 2) destruction of several essential amino acids,
and 3) formation of unusual amino acids, such as lysinoalanine
(LAL), lanthionine and ornithinoalanine, etc. Factors influencing the degree to which the above mentioned reactions occur
include protein source, concentration of alkaline salt used,
temperature, and time of exposure (36). The overall nutritional
quality of proteins treated with alkali is lowered because of the
destruction of essential amino acids, formation of unavailable
D-amino acids and the development of cross-links between amino
acid residues. Cross-links decrease the number of enzyme specific peptide bonds that can be cleaved and also serve as a

steric hindrance to the proteolytic enzymes trying to cleave at amino acid residues further down the chain, both of which results in a decrease in protein digestibility.

Research in the past has primarily focused on the mode of formation of lysinoalanine and its possible biological effects, since it was found to induce nephrocytomegaly in the rats. Nephrocytomegaly was found in all rats fed semi-purified diets containing either alkali-treated alpha protein from soy or synthetic LAL, at LAL levels ranging from 250-3000 ppm (37). Interperitoneal injection of free LAL (30 mg/day) for seven days resulted in renal lesions similar to those obtained by oral administration. DeGroot et al. (38) found that protein-bound LAL had much less of an ability to induce cytomegalia than did either free LAL or protein hydrolyzates containing LAL. Diets containing 100 ppm free LAL were toxic to rats, but diets containing 2400 ppm protein-bound LAL had no cytomegalia effects. This difference could be attributed to differences in the bioavailability of free and protein-bound LAL. Struthers et al. (39) also observed that cytomegaly cannot be induced at protein-bound LAL levels less than 3000 ppm in the diet.

Different strains of rats also varied in their susceptibility to nephrocytomegal (40). LAL has two assymetric carbon centers, therefore it has two pairs of enantiomeric diastereoisomers: LL, LD, DL, and DD-LAL (41). The LD-LAL enantiomer has been shown to be the most toxic to the rats.

It appears that the rat is the specific animal susceptible to nephromegaly induced by LAL. It was reported that 1000 ppm free LAL was unable to induce nephromegaly in mice, hamsters, dogs, rabbits, monkeys, and quail (38, 42). Sternberg and Kim (42) found that LAL existed in a number of home-cooked and commercial foods which had not been alkali-treated. Finally, renal lesions have not been observed in humans, although many foods containing LAL have been consumed by humans for a long time. Struthers et al. (40) stated that the level of LAL in commercial food products is well below health hazard level. Therefore, the greatest problem associated with LAL would be the lowered nutritional quality of proteins which contain high levels of LAL, especially if these proteins were the major source of protein in the diet. It has been reported that the presence of LAL in several food proteins is associated with decreased contents of cystine and lysine and decreased net protein utilization (NPU) values (42). The LAL content of soy protein isolates treated under varying conditions of alkalinity showed an inverse relationship ($r = -0.96$) to NPU values obtained on the same isolates (Table III). Most of the essential amino acid residues in the alkali treated soy protein isolates were released at relatively slow rates by a pepsin-pancreatin digestion. More severe alkali treatment (pH 12.2, 60° or 80°C) caused a decrease in the serine content of the isolates. Feeding rats high levels of dietary protein treated at pH 12.2, 40°C, for four hours, revealed only an increased degree of nephrocalcinosis in females which could be prevented by addition of calcium in the diet.

Table III. The lysinoalanine (LAL) content and nutritive value (NPU, digestibility) of isolated soy protein after heat and neutral-to-alkaline pH treatment (43).

Sample Number	Treatment			LAL (g/16gN)	NPU	True Digestibility (%)
	pH	Temp (°C)	Time (Hrs)			
1	----	--	-	0	35	97
2	7	23	0	0	35	97
3	12.2	40	1	0.42	19	95
4	12.2	40	2	0.68	12	92
5	12.2	40	4	0.83	12	93
6	12.2	40	8	1.09	- 2	86
7	12.2	23	4	0.4	19	94
8	12.2	60	4	1.71	- 3	76
9	12.2	80	4	2.08	-14	70
10	7	40	4	0	34	98
11	8	40	4	0	38	95
12	10	40	4	0.1	36	96

Diets containing low contents of protein-bound LAL did not show any adverse effect on rat growth, feed efficiency, or kidney function (44). A slight destruction of arginine, Cys, lysine, threonine, and serine was observed when sunflower protein was treated with low concentrations of alkali at low temperatures. Extensive destruction of these amino acids occurred under severe alkali treatment conditions (1 N NaOH, 80°C, 16 hours).

In vitro digestibility of the alkali-treated proteins decreased as a function of the severity of the alkali treatment (45, 46). Struthers et al. (40) reported that the PER for diets supplemented with 3000 ppm LAL (30% alkali-treated soy protein) was 1.8, compared to 2.8 for the untreated soy protein control diet.

Both diets were supplemented with amino acids to meet the NRC amino acid requirement. The lower PER value of the alkali-treated soy protein diet was due to poor digestibility of the cross-linked proteins. Since racemization of amino acids occurred along with the formation of LAL during alkaline treatment, the digestibility was also affected by the formation of D-amino acid residues in the soy protein.

Racemization of Amino Acid Residues

The treatment of food proteins with heat and alkali, alone or in a combination, leads to the racemization of a portion of the amino acid residues in these proteins. Kossel and Weiss (47) and Dakin (48) over 60 years ago described the formation of D-amino acid enantiomers by the alkaline treatment of proteins. Masters and Friedman (49) showed that five common foods/food ingredients, texturized soy protein (TSP), soy based baby formula, soy bacon analogue, corn chips and caseinate based dairy creamer, contained significant quantities (from 9 to 17% of the amino acid form present) of D-amino acids (Table IV).

Table IV.   D-Aspartic acid (D-Asp) content of commercial foods and food ingredients.

Product	D-Asp/L-Asp	D-Asp[1]/D-Asp + L-Asp
Texturized Soy Protein	0.095	0.09
Baby Formula (soy based)	0.108	0.10
Simulated Bacon (soy based)	0.143	0.13
Corn Chips	0.164	0.14
Dairy Creamer (casein based)	0.208	0.17

[1]Relative ratios of total aspartic acid to its D- and L-enantiomers (D-Asp, L-Asp).

Provansal et al. (45) noted that the treatment of isolated sunflower protein with heat and alkali caused a conversion of from five to over 50% of the total lysine to the D-form. Since lysine is the first limiting essential amino acid in sunflower, the racemization of lysine would cause a significant loss in protein nutritional quality for that protein source.

Friedman et al. (46) noted in a study on amino acid racemization in four isolated proteins (casein, soy isolate, wheat gluten

and lactalbumin), that the rate of D-Asp, D-alanine, D-glutamic
acid and D-phenylalanine formation differed between the proteins.
The soy isolate was generally the most susceptible to racemization while lactalbumin was the most resistant. The differences
noted in the ability of alkaline pH to racemize the L-amino acid
residues in different proteins can be largely attributed to the
structure of the proteins. The R-groups on the amino acid residues affect their ability to enhance (δ-amide of glutamine and
asparagine) or slow (alanine, leucine, valine and proline) the
overall rate of racemization.

The nutritional significance of the presence of the D-amino
acid residues in food proteins is definitely two and possibly
three fold:

1) Provansal et al. (45), Hayashi and Kameda (50) and
   Friedman et al. (46) described the ability of racemic
   amino acid residues within a protein to slow the rate of
   in vitro protease digestion of that protein. D-amino
   acid residues that are the site of protease attack prevent cleavage of the peptide bond because of the stereo-
   specificity of proteases such as trypsin, chymotrypsin,
   pepsin, etc. The presence of D-amino acid residues
   adjacent to protease susceptible residues are also able
   to hinder proteolysis.

2) If D-amino acid residues are released by digestion into
   the gut, both their rate of absorption by the intestine
   and their conversion to the L-enantiomer by the body are
   slow (52). Ohara et al. (52) noted that L-amino acids
   are actively transported across cell membranes while the
   D-enantiomers rely only on diffusion for movement across
   cell membranes. Ohara et al. (52) noted that D-trypto-
   phan was only 21% as effective as L-tryptophan when fed
   to the chick, whereas it was 99% as effective as L-
   tryptophan in the rat. Studies conducted on humans by
   Kies et al. (53) and Zeyulka and Calloway (54) indicated
   that D-methionine is poorly utilized when used to supplement oat and soy based diets. Rose (54) noted that
   of the eight essential amino acids, only two D-enantiomers (D-phenylalanine and D-methionine) could maintain
   a human in nitrogen balance.

3) Masters and Friedman (49) and Friedman et al. (46) postulate that some or most of the histo-pathological
   changes seen in rat kidney cells (nephrocytomegaly, and
   cytomegaly) and the predisposition of these cells to become preoplastic, can be attributed to D-amino residues
   in the protein fed and not solely due to the presence of
   lysinoalanine as noted by Woodard (37) or Erbersdobler
   et al. (56, 57).

After reviewing the literature, it is very evident that D-
amino acid residues, if present in food proteins, may have significant nutritional implications. Yet, a recent study conducted

by Bunjapamai and coworkers (58) examined six foods, both
processed and unprocessed for D-amino acids. The six foods and
their controls were peanut butter (raw peanuts), taco shells
(cornmeal), fried hamburger (raw hamburger), dark toast (white
bread), textured soy protein (soy flour) and irradiated chicken
(raw chicken). A small amount of D-amino acids were found both
in the processed foods and their controls, with the exception of
dark toast which had twice the content of D-aspartic acid (10.5%
of total aspartic acid) than did the control (white bread). The
authors demonstrated that most of the D-amino acids seen in both
the processed foods and their controls were formed from L-amino
acids during protein hydrolysis in 6 N HCl. Additional studies
are needed to see if the problem of amino acid racemization in
processed food proteins exists, and if it does exist for certain
foods, then measures to correct these problems could be investigated. Most likely the methods used to process those protein
sources most affected could be altered in order to minimize
amino acid racemization.

Plasteins From Protein Hydrolyzates

Background. As a result of the world's population growth
there is an ever increasing deficiency in the world food supply,
with protein deficiencies being serious in several countries.
Because of this deficiency, attention is now being drawn to the
use of non-conventional food proteins. These proteins, often
high in nutritional quality, are not always acceptable in food
because of their undesirable odor, taste, and color. Proteins
may carry impurities that cause these unacceptable characteristics. Therefore, methods for producing proteins free from impurities are needed. The plastein synthesis procedure has been
found to be just such a method.
Plastein synthesis involves two basic steps. First, a protein is enzymatically hydrolyzed to a low molecular weight mixture of peptides. After hydrolysis, the impurities which cause
undesirable odors, colors, and flavors can be easily removed
(59). Hydrolysis, however, results in a characteristic bitterness which has been attributed to peptides with more than three
amino acids (60). This material serves as the substrate for the
third step, plastein synthesis. During plastein synthesis, a
high concentration of the hydrolysate (30 to 50%; w/v) is incubated with an enzyme or heated to form a viscous gel-like
material. Fujimaki et al. (61) and Yamashita et al. (62) found
that the plastein reaction eliminated the bitterness present in
the hydrolyzate. Plastein synthesis can also be used to incorporate limiting essential amino acids into the plastein to form
a protein-like material with increased nutritional quality (63).
The plastein reaction can improve the acceptability and
nutritional quality of a protein as well as alter its physical
properties. However, the plastein reaction is not thorougly

understood, and there is controversy concerning the exact mechanism of the plastein reaction. Because of the heterogeneous
peptide substrate, conflicting results, and difficulties encountered in examining an insoluble material such as a plastein, the
formulation of a specific theoretical model for the plastein
reaction has been difficult and speculative at best. Yet, a few
theories explaining the mechanism have been proposed. It was
originally thought that an increase in molecular weight due to
peptide bond exchange and reformation occurred during plastein
synthesis (65, 68). This molecular weight increase helped to
explain the viscosity increase and solubility decrease characteristics of plastein material. However, conflicting data and a
different interpretation of these results led others to believe
that the increased viscosity and decreased solubility were a
result of enzyme or heat induced hydrophobic interactions (69,
72). This controversy has not been resolved, and researchers
still have various opinions concerning the mechanism of plastein
synthesis.

Amino acid fortification of plasteins. Amino acids can be
added to foods in the free form. This type of supplementation
is not highly efficient because free amino acids can degrade or
react with other food components (63). It has also been found
that the solubility of free amino acids facilitates their loss
during processing and subsequent cooking. The plastein reaction
decreases these negative effects because the amino acids are incorporated into the plastein structure.

Yamashita et al. (63) studied the incorporation of methionine into soy protein via the plastein reaction. Soy protein is
first limiting in the sulfur amino acids and, therefore, the
incorporation of methionine can greatly increase its quality.
This group tested L-methionine and several derivatives (N-acetyl-
L-methionine, L-methionamide, L-methionine ethyl ester and L-
methionyl-methionine) to determine the most effective form of the
amino acid for its incorporation into soy plastein. Free L-
methionine was not reactive. But, L-methionine ethyl ester and
L-methionyl-methionine were very effective and increased the percent methionine in soy protein from 1.2% to 7.2% and 5.1%,
respectively. The other derivatives, N-acetyl-L-methionine and
L-methionamide, were only slightly reactive. They attributed
the effectiveness of incorporation to the hydrophobicity of the
side chain of the amino acid and its derivative.

Aso et al. (73) produced corn zein plasteins and incorporated either tryptophan, threonine or lysine into the plasteins.
The amino acids were effectively incorporated as ethyl esters.
Arai et al. (74) improved two photosynthetic protein sources,
Spirulina maxima and Rhodopseudomonas capsulatus, with the
plastein reaction. The impurities were removed from each hydrolyzed protein, as previously described. The ethyl esters of
lysine, methionine and tryptophan were then incorporated into

each of the two hydrolyzates during plastein formation. The result was two plasteins having essential amino acid profiles
similar to the 1973 FAO/WHO (75) suggested profile. Yamashita
et al. (68) worked with several proteins: ovalbumin, gluten,
blue-green algae (S. maxima), nonsulfur purple bacteria (R.
capsulatus) and a variety of white clover (Trifoleum repens L.).
Their work with ovalbumin was intended to determine which amino
acid ethyl esters were effectively incorporated during synthesis.
The results indicated that the reactivity of the amino acid depends primarily on the hydrophobicity of its side chains. Those
with hydrophobic side chains were more reactive. This group also
prepared a gluten plastein with a high lysine content. They, as
well as Fujimaki (64), simultaneously incorporated three essential amino acids (methionine, lysine and tryptophan) into the
plasteins of the three previously mentioned unconventional protein sources to produce plasteins with an essential amino acid
profile similar to the 1973 FAO/WHO (75) pattern.

The ability of the plastein reaction to incorporate amino
acids into various proteins has also been used to prepare protein
sources for special diets. Yamashita et al. (76) proposed a
method to prepare a low-phenylalanine, high tyrosine plastein for
incorporation into food for phenylketonuria (PKU) patients. A
fish protein concentrate and a soy protein isolate were used as
protein sources. Each protein was slightly hydrolyzed with pepsin, then further hydrolyzed with pronase under special conditions to facilitate the removal of aromatic amino acids. The
aromatics, including phenylalanine, were removed by absorption
to Sephadex G-15, resulting in an aromatic amino acid free protein hydrolyzate. Both of these treated hydrolyzates were substrates for the plastein reaction where tyrosine ethyl ester and
tryptophan ethyl ester were incorporated. This work resulted
in two protein sources (plasteins) which contained no free amino
acids and were almost tasteless. The plasteins were low in
phenylalanine (<.25%) and high in tyrosine (>7.6%). These characteristics make these plasteins acceptable as food materials for
patients with PKU.

Plasteins from combined protein hydrolyzates. Improving the
amino acid profile of proteins via the plastein reaction has been
accomplished by incorporating essential amino acids as described
above. But, the amino acid profile can also be improved by combining one or more proteins with complementary essential amino
acid profiles. The production of combination plasteins was first
studied by Yamashita and coworkers (63). This group made two
combination plasteins, one with soy protein hydrolysate and
hydrolyzed wool keratin, and the other with hydrolyzed soy protein and hydrolyzed ovalbumin. Hydrolyzed wool keratin and
hydrolyzed ovalbumin contain high concentrations of the sulfur
amino acids, for which soy protein is first limiting. Arai et al.
(77) made several types of plasteins by combining: 1) casein

hydrolyzate with milk whey protein hydrolyzate, 2) casein hydrol-
yzate with soy globulin hydrolyzate, 3) milk whey protein hydrol-
yzate with zein hydrolyzate, 4) soy hydrolyzate with zein hydrol-
yzate, 5) casein hydrolyzate, and 6) casein hydrolyzate with
ethyl esterified soy globulin hydrolyzate.

The nutritional quality of plasteins. Yamashita et al. (78)
tested the in vitro protein digestibility of isolated soy protein and a soy plastein with the proteases, pepsin and trypsin.
Data from enzyme assays indicated that the protein and plastein
were comparable in digestibility. The soy plastein and soy
protein were also approximately 80% soluble in 10% trichloro-
acetic acid after in vitro protease digestion. Rat studies
showed an average in vivo digestibility of 90.3% for the soy
plastein, a value comparable to that for soy protein isolates.
These measurements suggest that the soy plasteins are highly
digestible.
     Studies were designed and run to compare the ability of
plasteins to support growth in weanling rats (78). The diets fed
to three groups were: 1) soy plastein, 2) essential amino acid
supplemented soy plastein, and 3) casein. Group 1 had an average
weight gain of 27.4 g in 20 days. The average weight gain for
Groups 2 and 3 was 68.8 g and 71.0 g in 20 days, respectively.
The latter two values were not statistically different. The high
nutritional quality of amino acid supplemented plastein demon-
strates the potential for using plasteins in foods. During this
study it was also noted that the rats fed the three different
proteins/plasteins had almost equal values for diet consumed.
The rats did not hesitate to eat the soy plastein diet, indicating the acceptable flavor present in the plastein.
     A PER bioassay was run to determine the protein nutritional
quality of: 1) casein control, 2) soy protein, 3) soy protein
hydrolyzate and 4) a combination plastein made from one part
(50:50; w:w) hydrolyzed soy protein and hydrolyzed wool keratin
(50:50; w:w) and two parts soy protein. The combination plastein
had added soy in order to adjust the sulfur amino acid level down
to approximately 5%. The PER values obtained were: 2.40 ± 0.05,
1.20 ± 0.20, 1.28 ± 0.18 and 2.86 ± 0.10 for the casein, soy
protein, soy hydrolyzate and plastein-protein, respectively.
The plastein-protein diet had a PER greater than that of casein.
     In summary, the plastein reaction can be used to improve
several characteristics of proteins. It can produce bland,
tasteless, nonbitter protein products with improved amino acid
profiles.

Conclusions

     The nutritional quality of a protein will almost always
change as that protein changes its configuration and/or when one
or more of its amino acid residues undergo chemical change. The

change can be from a small increase or decrease in protein digestibility as a protein unfolds and forms aggregates during
heating, to a large decrease in digestibility and quality when
essential amino acid (EAA) residues react with non-protein components (i.e., sugars, phenols and peroxides). The processing
of raw plant materials into foods and food ingredients significantly alters both the characteristics and the nutritional
quality of the proteins in those materials. Alkaline and heat
alone or in combination can create reaction products from amino
acid residues which may not only possess a lower nutritional
quality, but may be toxic when released from the protein during
proteolysis. Enzymatic modification of plant and animal proteins
can be used to develop desired functional properties in a food,
and at the same time be useful in enhancing protein digestibility
and serve as a means of incorporating needed EAA into the protein.

Understanding the mechanisms by which proteins are altered
during isolation, processing and storage, will allow for the
modification of these processes in order to protect the inherent
nutritional quality present in the proteins found in the raw
material.

Literature Cited

1. Pusztai, A.; Watt, W. B. Biochim. Biophys. Acta, 1974,
    365, 57.
2. Zacharius, R. M. Phytochem., 1970, 9, 2047.
3. Baldi, G.; Salamini, F. Theoret. Appl. Genetics, 1973,
    43, 77.
4. Kasai, T.; Sakamura, S.; Sabamoto, R. Agr. Biol. Chem.,
    1972, 36, 967.
5. Evans, R. J.; Bauer, D. H. J. Agric. Food Chem., 1978,
    26, 779.
6. Sgarbieri, V. D.; Antunes, P. L.; Almeida, L. D. J. Food
    Sci., 1974, 44, 1306.
7. Kunachowicz, H.; Pieniazek, D.; Rakowska, M. Nutr. Metab.,
    1976, 20, 415.
8. Strange, E. D.; Benedict, R. C.; Miller, A. J. J. Food
    Sci., 1980, 45, 632.
9. Slump, P.; Schreuder, H. A. W. J. Sci. Fd. Agric., 1973,
    24, 657.
10. Tannenbaum, S. R.; Barth, H.; LeRoux, J. P. J. Agric.
    Food Chem., 1969, 17, 1353.
11. Schnack, U.; Klostermeyer, H. Milchwissenschaft, 1980, 35,
    206.
12. Block, R. J.; Jackson, R. W. J. Biol. Chem., 1932, 97,
    CVI.
13. Anderson, G. H.; Ashley, D. V. M.; Jones, J. D. J. Nutr.,
    1976, 106, 1108.
14. Sjoberg, L. B.; Bostrom, S. L. Br. J. Nutr., 1977, 38, 189.

15. Miller, S. A.; Tannenbaum, S. R.; Seitz, A. W. J. Nutr.,
     1970, 100, 909.
16. Scrimshaw, N. S.; Young, V. J. Am. Oil Chem. Soc., 1979,
     56, 110.
17. Marshall, H. F.; Chang, K. C.; Miller, K. S.; Satterlee,
     L. D. J. Food Sci., 1982, (in press).
18. Chang, K. C.; Marshall, H. F.; Satterlee, L. D. J. Food
     Sci., 1982, (in press).
19. Catsimpoolas, N. Arch. Biochem. Biophys., 1969, 131, 185.
20. Weil, J.; Pinsky, A.; Grossman, S. Cereal Chem., 1966,
     43, 392.
21. Tanaka, M.; Kimiagar, M.; Lee, T. C.; Chichester, C. O.
     Adv. Exp. Med. Biol., 1977, 86B, 321.
22. Sgarbieri, V. C.; Amaya, J.; Tanaka, M.; Chichester, C. O.
     J. Nutr., 1973, 103, 1731.
23. Hansen, L. P.; Millington, R. J. J. Food Sci., 1979,
     44, 1173.
24. Hurrell, R. F.; Carpenter, K. J. Adv. Exp. Med. Biol.,
     1977, 86B, 225.
25. Dworschak, G. CRC Critical Reviews in Food and Nutrition,
     1980, 13, 1.
26. Finot, P. A.; Bujard, E.; Mottie; F.; Mauron, J. Adv. Exp.
     Med. Biol., 1977, 86B, 343.
27. Erbersdobler, H. F. Adv. Exp. Med. Biol., 1977, 86B, 367.
28. Lee, C. M.; Lee, T. C.; Chichester, C. O. Comp. Biochem.
     Physiol., 1975, 56A, 473.
29. Horn, M. J.; Lichtenstein, H.; Womack, M. J. Agric. Food
     Chem., 1968, 16, 741.
30. Johnson, G. H.; Baker, D. H.; Perkins, E. G. J. Nutr.,
     1979, 109, 590.
31. Kimiagar, M.; Lee, T. C.; Chichester, C. O. J. Agric. Food
     Chem., 1980, 28, 150.
32. Shinohara, K.; Wu, R. T.; Jahan, N.; Tanaka, M.; Moringa,
     N.; Murakami, H.; Omura, H. Agric. Biol. Chem., 1980,
     44, 671.
33. Coughlin, J. R. Diss. Abstr. Int. B., 1979, 40, 719.
34. Pintauro, S. J.; Page, S. V.; Solberg, M.; Lee, T. C.;
     Chichester, C. O. J. Food Sci., 1980, 45, 1442.
35. Hurrell, R. F.; Carpenter, K. J.; Sinclair, W. J.; Otter-
     burn, M. S.; Asquith, R. S. Br. J. Nutr., 1976, 35 383.
36. Struthers, B. J. J. Am. Oil Chem Soc., 1981, 58, 501.
37. Woodard, J. C.; Short, D. D.; Alvarez, M. R.; Reynier, S. J.
     "Protein Nutritional Quality of Foods and Feeds, Part II";
     Marcel Dekker: New York, 1975; p. 595.
38. DeGroot, A. P.; Slump, P.; Feron, V. J.; Van Beek, L. J.
     Nutr., 1976, 106, 1527.
39. Struthers, B. J.; Dahlgren, R. R.; Hopkins, D. T. J. Nutr.,
     1977, 107, 1190.

40. Struthers, B. J.; Dahlgren, R. R.; Hopkins, D. T.; Raymond, M. L. "Soy Protein and Human Nutrition"; Academic Press: New York, 1979; p. 235.

41. Tao, A. C.; Kleipool, R. J. Liebenm. Wiss. U. Technol., 1976, 9, 360.

42. Sternberg, M.; Kim, C. Y. "Protein Crosslinking: Nutritional and Medical Consequences"; Plenum Press: New York, 1977; p. 73.

43. DeGroot, A. P.; Slump, P. J. Nutr., 1969, 98, 45.

44. Van Beek, L.; Feron, V. J.; DeGroot, A. P. J. Nutr., 1974, 104, 1630.

45. Provansal, M. M. P.; Cuq, J. L. P.; Cheftel, J. C. J. Agric. Food Chem., 1975, 23, 938.

46. Friedman, M.; Zahnley, J. C.; Masters, P. M. J. Food Sci., 1981, 46, 127.

47. Kossel, A.; Weiss, F. Z. Physiol. Chem., 1909, 59, 492.

48. Dakin, H. D. J. Biol. Chem., 1912-13, 13, 357.

49. Masters, P. M.; Friedman, M. J. Agric. Food Chem., 1979, 27, 507.

50. Hayashi, R.; Kameda, I. Agric. Biol. Chem., 1980, 44, 891.

51. Gibson, Q. H.; Wiseman, G. Biochem. J., 1951, 48, 426.

52. Ohara, I.; Otsuka, C. I.; Ygari, Y.; Ariyoshi, S. J. Nutr., 1980, 110, 634.

53. Kies, C.; Fox, H.; Aprahamian, S. J. Nutr., 1975, 105, 809.

54. Zezulka, A. Y.; Calloway, D. H. J. Nutr., 1976, 106, 1286.

55. Rose, W. C. Fed. Proc., 1949, 8, 546.

56. Erbersdobler, H. F. "Protein Crosslinking: Nutritional and Medical Consequences"; Plenum Press: New York, 1977; p. 367.

57. Erbersdobler, H. F.; vonHangenheim, B.; Hanichen, T. "Adverse Effects of Maillard Products -- Especially of Fructoselysine in the Organism"; XI Internat. Congress of Nutr.; Rio de Janeiro, Brazil, 1978.

58. Bunjapamai, S.; Mahoney, R. R.; Fagerson, I. S. J. Food Sci., 1981, (in press).

59. Fujimaki, M.; Utaka, K.; Yamashita, M.; Arai, S. Agric. Biol. Chem., 1973, 37, 2303.

60. Lalasidis, G. Ann. Nutr. Alim., 1978, 32, 709.

61. Fujimaki, M.; Yamashita, J.; Arai, S.; Kato, H. Agric. Biol. Chem., 1970, 34, 1325.

62. Yamashita, M.; Arai, S.; Matsuyama, J.; Gondo, M.; Kato, H.; Fujimaki, M. Agric. Biol. Chem., 1970, 34, 1484.

63. Yamashita, M.; Arai, S.; Tsai, J.; Fujimaki, M. Agric. Biol. Chem., 1970, 34, 1593.

64. Fujimaki, M. Ann. Nutr. Alim., 1978, 32, 233.

65. Weiland, T.; Determan, H.; Albrecht, E. Ann. Chem., 1960, 633, 185.

66. Determan, H.; Bonhard, K.; Kohler, R.; Weiland, T. Helv. Chim. Acta, 1963, 46, 2498.

67. Determan, H.; Heuer, J.; Jaworek, D. Ann. Chem., 1965, 690, 182.
68. Yamashita, M.; Arai, S.; Fujimaki, M. J. Agric. Food Chem., 1976, 24, 1100.
69. Holfstein, B.; Lalasidis, G. J. Agric. Food Chem., 1976, 24, 460.
70. Edwards, J.; Shipe, W. J. Food Chem., 1979, 27, 1281.
71. Monti, J. C.; Jost, R. J. Agric. Food Chem., 1979, 27, 1281.
72. Lynam, E. M.Sc. Thesis, 1980, University of Nebraska-Lincoln.
73. Aso, K.; Yamashita, M.; Arai, S.; Fujimaki, M. Agric. Biol. Chem., 1974, 38, 679.
74. Arai, S.; Yamashita, M.; Fujimaki, M. J. Food Sci. Vitaminol, 1976, 22, 447.
75. FAO/WHO. "Report of Ad Hoc Expert Committee on energy and protein requirements"; Tech. Report, Series 522, 1973; World Health Org., Rome.
76. Yamashita, M.; Arai, S.; Fujimaki, M. J. Food Sci., 1976, 41, 1029.
77. Arai, S.; Yamashita, M.; Aso, K.; Fujimaki, M. J. Food Sci., 1975, 40, 342.
78. Yamashita, M.; Arai, S.; Gondan, M.; Kato, H.; Fujimaki, M. Agric. Biol. Chem., 1970, 34, 1333.

RECEIVED June 1, 1982.

INDEX

INDEX

Jacket design by Kathleen Schaner
Indexing and production by Deborah Corson and Paula Bérard

Elements typeset by Service Composition Co., Baltimore, MD
Printed and bound by Maple Press Co., York, PA